THE WORLD REVOLUTION
OF WESTERNIZATION

The World Revolution of Westernization

The Twentieth Century in Global Perspective

THEODORE H. VON LAUE

New York   Oxford
OXFORD UNIVERSITY PRESS
1987

Oxford University Press

Oxford   New York   Toronto
Delhi   Bombay   Calcutta   Madras   Karachi
Petaling Jaya   Singapore   Hong Kong   Tokyo
Nairobi   Dar es Salaam   Cape Town
Melbourne   Auckland

and associated companies in
Beirut   Berlin   Ibadan   Nicosia

Library of Congress Cataloging-in-Publication Data
Von Laue, Theodore H. (Theodore Hermann)
The world revolution of Westernization.
Includes index.
1. History, Modern—20th century.   I. Title.
D445.V73   1987   909.82   86-33246
ISBN 0-19-504906-3

1 3 5 7 9 8 6 4 2

Printed in the United States of America
on acid-free paper

TO ANGELA

Acknowledgments

For this book, the product of many years of reading, writing, and teaching, I owe a debt of gratitude to more people than can be listed here. Among those who have made specific contributions, I would like to mention the following institutions: the John Simon Guggenheim Foundation, for awarding me a Fellowship (my second) which made possible, in 1974/75, a year's immersion in West African (especially Ghanaian) society and culture; and Clark University, for letting me follow the humble pursuits of a seemingly nonproductive generalist—I could teach whatever I needed to learn for my own benefit, for understanding the world into which I was born. At Clark, more specifically, I have benefited from the scholarly advice of two colleagues in American history, George Billias and Douglas Little. From Karen Gottschang at Clark and the neighboring College of the Holy Cross I have learned much about China. I am also indebted, at the final stages of preparing this book, to the advice given by the two readers chosen by the Oxford University Press, as well as by its senior editor, Nancy Lane, and its editorial staff. All along there has been no greater source of encouragement than my wife, Angela, an experienced and ever supportive reader, editor, and critic.

More generally, I would like to express my gratitude to all of those— teachers, colleagues, students, friends and Friends (Quakers)—who over the years have treated me, a culturally self-conscious German who came to the United States in 1937, to the best qualities in American life. Pursuing with their help an academic career in the United States has been an immense privilege. No other country in the world could have inspired the elevated perspectives guiding this work. In some ways this book represents an effort to be worthy of that privilege.

Contents

Introduction

The nature of the inquiry pursued in this book and the problems encountered in that inquiry can best be introduced by taking an imaginary journey from the ground floor to the moon, that is, from a local to a global perspective.

Suppose we first place ourselves at a busy intersection in our neighborhood and look about: what do we see? We see people, including friends and acquaintances; even strangers we can look in the eye and assess as individuals. We also notice the material settings of life, the places where people dwell and do business—houses, shops, offices; we observe the traffic, people coming and going, on foot, by car, by public conveyances. In a park nearby we see trees, gardens, flowers; we smell the fragrance of the earth. This is life on the ground floor with its infinite dense detail, the subject of the psychologists' analyses, the novelists' art, and the ethnographers' "thick description." Here we celebrate our existence, the intimate human environment in which personality and psyche are shaped and asserted. In essentials we never leave that ground floor. And yet . . .

Suppose we next take the elevator to the top of the highest skyscraper in a big city. Below lies a teeming city set into a wide landscape often bordering on a river, a lake, or the ocean. Here ground-floor perspectives no longer suffice. Street-corner and household details have been absorbed into larger shapes, the grid of streets, city blocks, the downtown section, the suburbs. Human beings now seem insignificant objects crawling down below. But taken as a whole, in the enlarged vision, the city and its structured environment represent an impressive collective human achievement. Overlooking the city from its highest building adds to our civic pride. It is a thrill to catch a glimpse of the larger contexts of our lives.

Ground-floor persons also fly in airplanes. What do they see? Below, on a clear day, lies a wide expanse of rivers, fields, hills or mountains, with towns and cities scattered among them, linked by the thin threads of interstate highways. Lower-level perspectives are now superseded. Human beings have entirely disappeared, yet not their collective works. Air travellers look down on man-made landscapes as on a map, the physical space for millions of people kept in peace and order by invisible authority.

A select few ground-floor people rise even higher, travelling in a space capsule and taking pictures of the sights below, say, of the whole of California

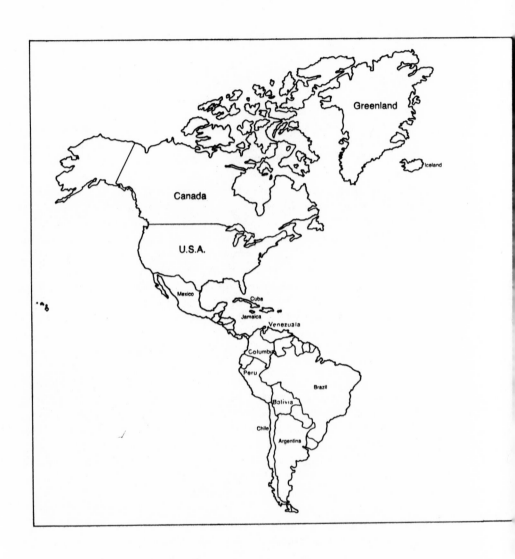

or the southern United States, or of central an
from an altitude that renders all previous per
human has now vanished amidst the bare co
minds the space travellers are keenly aware o.
of states and nations that let untold millions o.
incidentally, assure their own safe return.

And, finally, the first view from the moon as reco
in 1969. There, bright in the black sky, hangs Planet Eart
a whole, utterly inhuman in its stellar distance from the kno.
tat, the most terrifying of all oversimplifications of life as we kn
the minds of the astronauts it is still a human universe responsible i.
the most spectacular human exploits—ground-floor people elevated
highest point of vantage achieved thus far.

To catch our breath, let us repeat this mental journey in rapid elevation
from a slightly different angle. The ground floor, we said, represents our core
community; here we are formed in all essentials of our being. From the skyscraper we observe a larger community, hardly less essential. It shelters us,
provides order, and furnishes basic services. We are members of a civic organization, identified when we travel by the location we call home. Presumably
we know what happens in local government; its problems are our problems
too—and charges on our purse as well. In that setting we wear a public face;
the perspectives of home and neighborhood are enlarged.

From the airplane or space capsule we look down on even larger communities, on regional or national government in charge of tens or even
hundreds of millions of people, each person invisible like individuals seen
from the airplane. As individuals in this multitude we are tiny atoms, civic
atoms, lost among the mass of fellow citizens. And yet, however impersonal
our civic status, the whole country is part of ourselves, through the news services, party affiliation, government and the many obligations it imposes
(including dying for our country in war), and, more subtly, through myth and
symbol and a collective memory reaching deep into the past; our ego contains
a collective dimension. When our country is humiliated in international
crises, we feel personally humbled; when our astronauts reach the moon, we
cheer. Our American awareness and identity is stretched, somehow, to cover
over two hundred million other human beings. In other Western countries,
too, we find a similarly expanded awareness linking, through national culture,
the individual with the huge community of the nation.

On then, finally, to the top story of contemporary existence, up to the
world now seen as a single unified whole, the patternless and blurred mosaic
of a thousand "culture-scapes." In its shaky interdependence it supports over
ve billion human lives, some of them in exceptional splendor, many others
n hunger and misery. Yet the relative orderliness prevailing at lower levels
as disappeared. Despite all interdependence—and in part because of it—
e world suffers from a rising propensity for ever larger human sacrifice
rough war and civil war, seemingly headed for the nuclear holocaust. The
ilable perspectives—indeed, the skills of peaceful cooperation evolved for

the lower stories—are patently inadequate for that all-encompassing top story. The nationalism flourishing on the floor below, although a training ground for large-scale human cooperation, is insufficient for coordinating human wills around the globe; it is now the driving force behind the nuclear arms race and many other forms of inhumanity. Obviously, an all-inclusive global awareness transforming the patternless and blurred mosaic of the world's many cultures into a meaningful image inspiring worldwide cooperation does not exist, though the survival of humanity depends on it.

II

Yet, however crucial the need for an all-inclusive global awareness, what lunatic ground-floor person should want to venture into that transcendent global unknown? The risks of such super-Toynbeean flight of the imagination are indeed enormous. As the German sociologist Ralf Dahrendorf, concerned with the intermediary reaches of contemporary society, has observed: "Whoever makes his own society the subject of an all-encompassing analysis must be capable of distancing himself from the everyday world in which he lives. That effort is not for everybody; distancing oneself from the obvious realities brings not only better chances for insight but also threats to one's existence."[1]

The distancing, of course, is far more perilous when one ventures beyond the nation. The perspectives of the ground floor are no guide to living in larger contexts; the rules creating order in the nation-state bring havoc to global politics. What, then, are the paths to understanding, control, and peace in the worldwide contests of ideologies, conflicting national creeds, and competing religions? Rising to a comprehensive overview means a drastic distancing from family, neighbors, colleagues, compatriots, from the social universe in which one lives. The moon-bound spectator becomes an isolated human being and perforce a super-abstract intellectual, out of breath in the global stratosphere, bereft of familiar landmarks and groping in utter uncertainty among explosive issues of human existence.

In the upward thrust the details crucial to lower-level analysis with its impressive sophistication become blurred and indistinguishable, or even irrelevant, just as in modern skyscrapers the ornamental clutter of small-scale architecture has been eliminated. The search is for comprehensive patterns, for larger frameworks—in the elevator dialectics between detailed facts and enlarged frameworks the latter always have the upper hand, creating new facts, new truths. If, as Goethe observed, "everything factual is already a theory,"[2] every detail, every fact, has its generative framework. Conversely, every framework, every perspective, creates its own facts; in changing perspectives we generate new landmarks. Yet what recognizable pattern or framework do we find as we search for global perspectives? Physical elevation automatically reduces lower-level details and reveals new facts in the larger contours. By contrast, the intellectual effort of rising to a higher overview entails an endless inconclusive process of abstraction and conceptualization in search of meaningful patterns (or proper theory). It disestablishes existing convictions, but leads to no universally recognized (or objective) facts or truths. There exists, alas, no objectivity in the assessment of human landscapes, only agreement regarding reality based on consensus. Where do we find consensus when we rise above the roofs of the nation-state?

Adjusting perspectives upward indeed provokes bitter protest, because it runs counter to the need for lower-level certitude; it even sets off a contrary search for the ground-floor verities, for roots, particularly in times of rapid change. Yet nothing could be more dangerous in an age when we must grope away from lower-level landmarks and forget the memories, the roots, that divide humanity. The overview, the mainland of survival, lies ahead; returning to our islands of origin leads to conflict and inhumanity. In the age of global interdependence any preoccupation with former times as seen from the ground floor is an act of irresponsibility. We need drastic reinterpretation based on all-inclusive perspectives turned forward, a view of the past seasoned with prophesy; "where there is no vision, the people perish" (Proverbs 29:18).

Obviously, a fundamental reorientation of outlook with its progressive simplifications opens a treacherous frontier of insight and speculation. As the art historian E. H. Gombrich has observed: "to see all we must select and isolate." But select and isolate what? Our choices are bound to be marred by a thousand limitations. We cannot escape, for instance, from a Western perspective that, for all its cosmopolitan inclusiveness, can claim no universal validity; it cannot provide ready access to the inwardness of non-Western cultures. Refined in these pages by a penchant for a transcendence-oriented cultural relativism, that Western cosmopolitanism defines the limits of understanding this age; it is the best we can achieve (aware that it reaches over the heads of ground-floor people everywhere who insist on *their* version of global reality).

Even within our own culture an expanded overview advances the pitch of abstraction in language and political analysis far beyond current usage or comprehension. The "nation" and the "nation-state" are capacious abstractions which, in the course of a long evolution, have been endowed with experiential concreteness. But "global interdependence" and "globalism" still remain elusive and unreal terms, even though, by an open-eyed view of the daily news, worldwide interdependence shapes countries and peoples down to individual destiny.

Likewise, the overall concept of "culture" (used here synonymously with "ways of life" or "traditional ways") remains vague and abstract, like a distant landscape seen through myopic eyes. Yet in the inclusive "culturalist" approach attempted in these pages it has its uses. It helps to bring into a single focus the entire range of human action, including the invisible realms of cultural conditioning in the human psyche, the conscious regions of human existence on the ground floor of life, the advanced collective discipline of urban-industrial nation-states, and the distant, outermost framework of global interdependence affecting men, women, and children alike.

Who but a lunatic would try to bring the endless information about contemporary life into a single focus?

III

There exists no compass, obviously, for the *terra incognita* of global perspectives but subjective personal conviction and experience. Yet some basic human concerns, evolved on lower levels, afford broad guidance for exploring the top story of human association as well. These basics are a concern for power, for morality, and for a sense of mastery over human destiny, all interrelated. These concerns are viewed most clearly from a high altitude, where indeed they operate on their own, jet-stream fashion, influencing human behavior below.

First of all in that all-inclusive perspective we are guided by a concern for power, political power and cultural power, and for *all* the factors that contribute to power. Social life is *not* shaped by the ways in which, by Marxist analysis, human beings engage nature through production; it is conducted in communities that organize and control *all* aspects of life (including production) and compete with other communities for security, prosperity, and survival. While appreciating the Marxist alertness to the importance of economic factors, we find no benefit in the concept of "class" or "class struggle." The agents of power that count in the global perspective are states and, more impersonally, established ways of life politically and culturally organized for self-protection. In the wide perspectives here attempted, Marxism (like liberal thought) is both too culture blind and too power blind to be of analytic value. I will therefore always put the term "capitalism," which yields no useful insights, into quotation marks.

In any case, power at its core is not a material but a psychological force. It represents the vital desire of individuals or communities to prevail, to impose change on others while remaining unchanged, to dominate rather than be dominated, whether by violence or by the subtler arts of peace, and always in the compelling contests of invidious comparison. Viewed in this manner, power is foremost a concern of human consciousness: "I am what I feel to be in comparison with other human beings." All analyses of power—or all power-oriented studies of history—must therefore begin with human consciousness, whether individual or collective. Power is a matter of human perception (culturally conditioned perception, to be sure). The material aspects of power originate in the mind, in human ingenuity, in the desire to be less vulnerable.

Seen in this light, power is inseparable from morality. Morality, most centrally expressed in religion, regulates that libidinous assertion of selfhood in the interest of peaceful social cooperation on which individual and collective survival depends. Responsible analysis of human relations in all their complexity must therefore reflect also a live moral awareness. For that reason the issue of morality—a transcendent morality suited to an age of globalism—will constantly crop up in the pages that follow, guided by the maxim that the foremost moral obligation lies in understanding people, even the most repulsive, before judging them. Morality in sociopolitical analysis is not an unscientific luxury but a necessity of survival. Moral indignation, all too fashionable these days, tends toward violence; it prevents clarity of vision. How can we control our destiny when ignorant moralizing keeps us from grasping the forces that shape our age?

Considering the craving for moral judgment in our society, let there be no doubt about the morality of the transcendence-oriented cultural relativism here pursued. Our need is for the utmost moral sensitivity as well as moral fortitude. In the 20th century we deal with unprecedented inhumanity; it surpasses our inherited capacity for comprehension, tempting us callously to argue away millions of slaughtered lives by facile explanations. Yet let the outrage not drive us into irrationality either. Salvation lies in the *proper* remembering. We are morally obliged to prevent a recurrence by endowing our memories with a rational understanding of the causes no matter how painful the investigation, how repugnant the conditions investigated. Resisting an uncomprehended reality by an unpremeditated and outraged response abstracted from limited experience of the world surely breeds further inhumanity. We need experiential knowledge.

Yet does insistence on such knowledge not rule out any understanding of realities lying beyond our horizons? It is indeed one of the contentions of this essay that when we have no experiential access to alien realities, we should be extremely careful and compassionate in our judgments, aware that all around us lie cognitive barriers of the utmost magnitude; we must examine these barriers before proceeding. Carrying the experiential knowledge of one culture thoughtlessly across these barriers into other cultures constitutes an act of "cognitive imperialism" of which even sophisticated anthropologists are guilty. In politics the resulting misjudgments may well produce catastrophic conflict.

But, then, will the fullness of rational understanding (or even the effort to attain such mastery) not lead to an all-forgiving moral surrender? Admittedly, within the family and even within well-constituted states, the moral standards necessary for collective existence must be preserved. Understanding must lead to a more effective affirmation of the moral premises of community (though even here forgiving helps to soften human wills toward greater charity all around, which improves the quality of any human relationship). When, however, we deal with the anarchic global community above and outside all constituted authority, we find no common standards of social and political morality. In that setting the affirmation of moral judgment is inescapably partisan. It means taking sides; it tends to increase hostility, which counteracts the emergence of more universal rules of peaceful cooperation. Under these conditions moral judgment must be raised to a higher potency, to an all-inclusive forgiveness which alone can overcome the global anarchy. In the subsequent pages, therefore, we will cross over many times from understanding to forgiveness, in line with Goethe's maxim: "one cannot understand anything which one does not love"—with compassion (hopefully) even for those who refuse to love their enemies.

War and Peace observed that the greater the distance from which we observe human affairs and the greater our knowledge, the more the area of freedom shrinks and the area of necessity expands. Real freedom can emerge only from a better knowledge of *all* the human necessities under which human beings operate. As Francis Bacon observed,[3] we master nature by obeying her. How could human beings fly to the moon without having mastered the necessities of gravity and a thousand other physical laws? In this spirit we must masterfully obey the necessities of globalism in human affairs. Drifting into worldwide interdependence we have not, unfortunately, even reached the equivalent stage of Newtonian physics, neither in our historical studies, nor in our social sciences, nor in our statecraft. Viewed in this light, the determinist transcendence-oriented cultural relativism practiced in these pages represents a search for mastery over the yet uncomprehended necessities of the new globalism through greater obedience to them. It is one of the assumptions guiding the analysis set forth in these pages that mastery requires submission to necessity, and liberation, consciously or unconsciously, imposes a yet tighter discipline.

Fortunately, the focus of this search for obedience-oriented liberation can reasonably be limited to essentials. We need not feel overwhelmed by the unmanageable profusion of human knowledge accumulated in the Western tradition and around the world; we need not be specialists in science and technology. What matters is our sense of control over the social order that maintains all human knowledge. Our thinking is social thinking; our knowledge is social knowledge. If our society collapses, all knowledge, all science, all art—the universe as we know it—also collapses. For that reason we must give top priority to the skills and sensibilities on which social cooperation depends. If that base of our existence is sound, the infinitely larger universe of human knowledge is safe too. These pages are therefore devoted to the sociopolitical-cultural framework of human cooperation on whose continued function all else depends. That framework of peaceful interaction within global interdependence is still highly fragile; it cannot be taken for granted. Its preservation and improvement call for far larger resources of insight and perspectives than are currently available. This essay offers an opening in that direction.

V

I will conclude this introduction on a double note of humility and defiance. As this volume deals with big subjects barely penetrated by scholarship or moral sensibility, its coverage is bound to be impressionistic as well as spotty. This book attempts to interpret the present century; it intends to provide deeper insights into its dynamics rather than offer a survey covering all parts of the world. It concentrates on areas and issues that have caused the most trouble, paying scant attention, unfairly, to many parts of the world, including the lands of Islam and Latin America; it disregards many worthy peoples. And, despite its efforts to escape from ethnocentrism, it is still tied to a West-

As for the enemies, these pages follow the maxim of a German pher living under the memory of the fierce religious wars of the 17th G. W. Leibniz, who wrote that "the place of the other is the better view in both politics and morals." Putting ourselves into our opponen and trying to see, for once, the world from their angle endows us wi ble vision, ours plus theirs, which certainly enhances our grasp of re our command over it. Is this humanly possible? How, it will be asked love the mass murderers of Auschwitz, the perpetrators of the Ho But how can we prevent a repetition if we cannot fathom the causes to Auschwitz? How is it possible in the age of nuclear weapons to pre worse outrage if we do not have adequate control over our destiny? T searches for that deeper understanding, aware that understanding co moral component.

IV

Next, the search for control through morally guided understanding r this perspective-raising experiment a pessimistic deterministic bent. *miste, pessimiste*, in the words of a perceptive Frenchman. Yet the f insistence that at every fatal turn in modern history a happier end been possible appears more a product of shallow ideology than of analysis. We learn more about the complex dynamics determining the of history by assuming that events represent the balance of forces : than by toying with might-have-beens that merely foist our limited insi uncomprehended realities.

The balance of forces at work becomes more visible in global persp Although within our own culture we feel free, we are blind to the factors of cultural conditioning that shape our actions. Only by lool ourselves from the outside, in the inter-cultural contexts, can we se deeply enmeshed we are in the network of hidden factors that constitu cultural identity. In that enlarged perspective we become aware th actions are determined by an almost infinite number of forces beyo range of our consciousness. Any historical analysis which attempts to in all the world's cultures must therefore include a strong dose of cu determinism.

In addition, we need to become aware of the subtle psychological anisms at work in the pervasive and ever more intense inter-cultural co ison. Inter-cultural comparison is a contest of invidious comparison; the testants intend to win. Inevitably they impose their own ground rule comparison, applying the cognitive equipment of their own culture to cultures structured altogether differently. The result is incomprehensio hostility; the comparers cannot let go of their own cultural assumpt whether conscious or unconscious. In short, whether we know it or no are suspended in a dense network of causative factors, almost all of invisible and hence outside our analyses and efforts to control our des We need to probe into those invisible depths.

Searching for global perspectives we follow Tolstoy, who at the en ern, even an American, experience (though an unrepresentative, immigrant one). So be it. In the midst of a profound transition in human fortunes, traditional ways of thinking, no matter how buttressed by traditional knowledge and hardened by political power, no longer suffice; in the largest dimensions of contemporary life, certitude escapes us. What is offered here, then, is, viewed minimally, merely a search for a global overview stitched together from individual insights (or opinions, if you will) acquired on the run—on the run through a world and a vastness of information that has grown over everybody's head.

Yet, in a world so out of control, are we not free to take risks in exploring the unexplored with a minimum of scholarly baggage and a maximum of reflective speculation? We certainly cannot afford to let ignorance, produced by an overload of information, become an excuse for not thinking creatively and inclusively at all. May therefore all learned authors, living and dead, without whose labors this book could not have been written, forgive their being left unnamed. May they also condone this author's ignorance (as well as accept their own). And more: may they be persuaded to take a cautionary look at traditional scholarly method based on the flimsy grounds of documentary evidence. Documents frequently omit the better part of reality: the feeling of the times, the unspoken assumptions of the author, and, above all, the proper sense of contexts (especially important in the study of alien cultures). Inescapably, historians have to assess the documentary evidence with the intuitive insights gained with time and a compassionate openness to human events. Meaningful history resembles the fine arts rather than the exact sciences.

And yet, if attempting so much in plain language and with so little scholarly certitude be considered madness or arrogance, the excuse is obvious: the incentive for rashness lies not in the author but in the times in which we live (governments conduct their business in even greater ignorance). When humanity plunges into an unprecedentedly perilous future, intellectuals who don't think big—and don't think big in simple terms—don't earn their keep.

So let there be heretics. The risks of heresy seem minor as compared with the penalties—the guilt, the treason—of not searching for ways of promoting a peaceful collective transition to a global consciousness. Surveying the 20th century and frightened by the inhumanities revealed ever more clearly from a high altitude, we are entitled to an existential curiosity: Why Hitler? Why Stalin? Why the rush toward nuclear extinction? In search of overall perspectives, without which no adequate answers are possible, all intellectual and personal risks are justified.

How much, anyway, does the individual—heretic or conformist—count these days? At the end of the 20th century it is our historic fate to live with the possibility of man-made human extinction. Under that grim gallows the best we can do is to act with the utmost responsibility. As Samuel Johnson observed: "Depend upon it, Sir, when a man knows he is to be hanged in a fortnight, it concentrates his mind wonderfully."[4] In this liberating defiance of fate, convention, and scholarly timidity, the present volume concentrates on thinking globally.

A Note on Organization

A book-length essay covering world history in the 20th century to the present inevitably runs into problems in organizing its huge subject matter. As the world is constituted in separate polities, is a world history to be a collection of parallel national histories? Or is it a single story with its own dynamics into which the evolution of the separate countries has to be fitted piecemeal, in mini-essay segments? Here the latter prescription has been followed. For the sake of a commanding overview and the inter-cultural comparisons for which it calls, the best procedure seemed keeping the focus on the unity, on the common dynamics, of global developments. The treatment of the selected key countries for that reason is discontinuous up to the end of World War II, on the assumption that the internal continuities are less significant than the innovations imposed from without by the logic of global interdependence. Moreover, key agents, whether individuals or countries, are treated in full only when they have reached their prime influence. Readers will have to cope with the complexities of shifting time frames as a minor—a literary—inconvenience compared with the troubles they face in the real world.

The treatment of the post-1945 years follows a simpler order, with a substantial analysis, up to the 1970s, devoted first to the United States, next to the Soviet Union, and finally to the Third World. The book concludes with an assessment of the present and of the challenge of the future. The appendix spells out the book's underlying theoretical assumptions, which are merely hinted at in the course of the historical analysis.

In view of the fact that this book offers not a monograph but an essay of interpretation, footnoting has been kept to a minimum. Sources are given for all quotations, except for those within the public domain, that is, those commonly quoted without reference.

THE WORLD REVOLUTION
OF WESTERNIZATION

"Each age, it is found, must write its own books: or rather, each generation for the next succeeding. The books of an older period will not fit this."—Ralph Waldo Emerson

"To ask larger questions is to risk getting things wrong. Not to ask them at all is to constrain the life of understanding."—George Steiner

"All history which is not contemporary is suspect."—Blaise Pascal

"The first category of historical consciousness is not memory, but what has been promised, what is expected, and what is hoped for."—G. W. F. Hegel

"No theory, no history."—Werner Sombart

"The American (and world) public badly need new visions, new generalizations, new myths, global in scope, to help us navigate in our tightly interactive world. If historians fail to advance suitably bold hypotheses and interpretations, then politicians, journalists, and other public figures will continue as now, to use unexamined cliches to simplify the choices that must be made."—William H. McNeill

"There is no knowledge except this which can make human nature truly benevolent and kind to the whole of the species, and, with the certainty of a mathematical demonstration, render all men charitable, in the most enlarged and best sense of the term. It will force on the human mind the conviction that to blame and to be angry with our fellow-men for the evils which exist, is the very essence of folly and irrationality, and that the notions which can give rise to such feelings never could enter into the composition of any human being that had been made once rational."—Robert Owen

The Thesis

This book offers a novel look at the conditions of the anarchic world community in which we live, a look reasonably free of the illusions buttressing liberal complacency. It aims at the detachment that allows true impartiality as well as a better sense of control over our destiny. It explains the 20th century, the most momentous and still largely uncomprehended age in all history, in terms of the central force that shaped it: the world revolution of Westernization. That gigantic, all-inclusive, and still incomplete historic process may be briefly outlined as follows.

The Expansion of the West

For the first time in all human experience the world revolution of Westernization brought together, in inescapably intimate and virtually instant interaction, all the peoples of the world, regardless of their prior cultural evolution or their capacity—or incapacity—for peaceful coexistence. Within a brief time, essentially within half a century, they were thrust into a common narness, against their will, by a small minority commonly called "The West"—the peoples of Western Europe and their descendants in North America. As a result, the human condition in the present and the future can only be understood within the framework of the Westernized world.

This massive confluence of the world's peoples, infinitely exceeding in intensity all previous interdependence and transforming the world's ecosystem on which human life depends, was started by irresistible force, by guns, supported by a vast and complex array of cultural skills, adding up to an overwhelming political presence that excelled also in the arts of peace. In creating an interdependent world through conquest, colonization, and expanded opportunities for all, that Western minority imposed its own accomplish-

3

ments as a universal standard to which all others, however reluctantly, had to submit. Robbed of their past freedom to go their own ways politically and culturally, non-Western peoples were subjected to a world order that perpetuated or even deepened their helplessness. Henceforth equality could be attained only on the terms imposed by the West.

Western ascendancy was so complete that it left only one rational response: abject imitation as a condition of survival and self-affirmation. Decolonization and the formation of Western-inspired nation-states among the former colonial and semi-colonial peoples merely escalated the imitation and hardened the grip of Western institutions and values over the entire world. Even the most heated protests against Western power—and they were never lacking—were expressed in Western concepts and propagated by Western technology in Western languages. Yet inequality continued.

Submission and cultural imitation, however, were not without elemental advantage, individually and collectively. Overall material conditions improved, increasing individual welfare and tripling the volume of human life on earth within less than a century. Yet the survival of the ever growing and more demanding human multitudes led to yet greater dependence, calling for a further copying of essential aspects of Western culture. For the sake of feeding, housing, transporting, educating, and employing the world's population, "Westernization" is now pressed forward by non-Westerners themselves. Culturally neutralized, it has become "modernization" or simply "development," the common goal of all peoples and governments no matter how handicapped in achieving it.

Seen in a superficial light, the unification of the world was the work of all humanity In expanding their power around the world the Europeans freely drew on the ingenuity, riches, and labor of non-Europeans; all humanity gained as the achievements of its separate peoples became accessible to all. Yet, examined more closely, the Europeans merely copied on their own terms whatever strengthened their might; they exploited the world's resources, hitherto mostly dormant, for their own gain; they enlisted the prowess and resilience of people around the world to make themselves masters. The will to power and the capacity for taking advantage of all opportunities for their own aggrandizement—the initiative for the world revolution of Westernization—sprang from Europe, from the hothouse competition among the Europeans themselves. In expanding around the world and enlarging their base from Europe into the "the West," they foisted their singular qualities on the unwilling and unprepared majority of humanity, dynamically transforming the entire world in their own image and establishing a hierarchy of prestige defined by the success of imitation. In the world revolution of Westernization, Western political ambition and competitiveness became universal—and fiercer because of the fury born of persistent inequality.

The Consequences of Western Expansion

The major effect of the world revolution of Westernization—generally downplayed in the West—has been to undermine and discredit all non-Western cultures. The victorious Westerners, their own ways and self-confidence boosted by their worldwide sway, left the rest of the world humiliated and in cultural limbo. Under the Western impact traditional authorities and local customs had no future; they crumbled away. Meanwhile the imported ways of the West remained superficial or even incomprehensible; they did not fit societies whose cultural sovereignty had been crushed.

The subversion of traditional cultures admittedly took different forms in different parts of the world. It was perhaps least painful in Japan, a unique case where native tradition proved miraculously compatible with Western imports. Elsewhere, among the majority of peoples around the world, the results were cultural chaos characterized by a loss of purpose, moral insensibility, and a penchant for violence; by social and political fragmentation, and by the psychological misery of knowingly belonging to a "backward" society. Never before had the extremes of inequality been so great. Now the Great Confluence carried the wealth and glory of the richest instantly into the presence and conscience of the poorest. Expectations were raised and crushed, opportunities advanced and denied in the same instant. No wonder that violence spread and turned more vicious.

Traditional culture and society in non-Western parts of the world were subverted just as the pace of political competition was accelerated to worldwide intensity. After World War I, and even more after World War II, Western democracy, with the United States in the lead, stood out as the universal model of power in the world, of good government, and of economic prosperity; however imperfect, unfinished, and vilified by its critics, it proved elementally persuasive by comparison. Who among the onlookers did not feel envious? Who among the envious did not feel morally entitled—and tempted—to make the world even safer by their own ideals? Equality in political power, though deftly omitted from the Western-inspired inventory of human rights, certainly was claimed as an entitlement by proud non-Westerners. Few states possessed the wherewithal to take up the challenge, yet the aspiration became embedded in the new competition for global power, a source of heightened instability and conflict. But how could ambitious leaders of polities caught in cultural disorientation and social fragmentation mobilize their peoples to match this mighty model on their own terms?

Futile Counterrevolutions

The search for answers among culturally subverted countries with political potential, so this book argues, led to the totalitarian experiments of communism and fascism in the wake of World War I; from them, after World War II, communist Russia emerged as the most powerful challenger to the Western model. Simultaneously, the totalitarian experiments inspired state-builders among the new states emerging from decolonization. What was wanted, it turned out, was not self-affirmation but further Westernization through "reculturation" (to coin an ugly term for an ugly process), through an unnatural revamping of unsuitable indigenous institutions and human values under external pressure to meet alien goals. Leaving people to their own devices under these conditions, as tried by short-lived democratic regimes, merely enhanced the common disorientation and political weakness. Command and indoctrination had to be substituted for the lacking appropriate voluntary motivation, while a furious anti-Westernism covered up the blatant imitation.

In the West the experiments of totalitarianism have been castigated as outrages of inhumanity and terror; yet as counterrevolutions they merely carried the world revolution of Westernization one step further. Western-oriented indigenous leaders impressed by the fullness of Western power applied to their own peoples the violence that had characterized the elemental expansion of the West. They tried to convert their subjects by force into organization-minded citizens as disciplined, loyal, and cooperative as their counterparts in the Western democracies. They had to accomplish, in a hurry and by conscious design, what in the West had been achieved over centuries of largely invisible cultural conditioning (accompanied too, we should remember, by ferocious violence in war, civil war, and revolution). Seen in this light, communism and fascism were no more than idealized versions of Western (or "capitalist") society dressed up to inspire the humiliated and disadvantaged. Their statecraft was merely a disguised form of cultural colonialism creating, out of helpless and resentful people, a "new man" and a "new society" capable of competing with the West on more equal terms. Yet the results of all totalitarian experiments have refuted the high hopes for matching, let alone outpacing, the Western model. Inasmuch as they relied on compulsion, they remained tragically non-Western and inferior because compulsion can never match the cultural creativity of spontaneous civic cooperation.

Not all past experiments of Westernization (or modernization), fortunately, were as extreme in ambition or execution as Hitler's or Stalin's. Some were favored by newfound wealth, others by long association with Westerners or, as in the case of Japan (always the exception), by an element of cultural compatibility. Yet all of them suffered from intense partisanship, violence, and mismanagement; indigenous ways did not harmonize with the imported modern ways. Alienation and anger brooded under the surface (even in Japan, certainly in the twenty years before 1945), directed against Westernized natives or the ultimate source of all misery, the West itself. Tragic indeed is the record of state-building and development through the non-Western world in the past and in the present; ominous are the prospects for the future. The political competition and cultural disorientation of the Great Confluence continue to accelerate unabated, promoted in large part by Western ignorance of the consequences of Westernization.

Cultural Incomprehension and Confusion

The world's population is now organized in states patterned after the European nation-state. The relations between states are both cooperative and competitive within a common framework of an anarchic community governed by the rules of power; preparation for war and war itself characterize the new world system as much as it did the old European system of states. For better or worse, the whole world has now become a furnace of human creativity in intense competitive political and cultural interaction. The worldwide rivalry for wealth and power has stimulated an unprecedented rise in material prosperity; it has quickened the pace of scientific and technological discovery; it has advanced human mobility and built, on the surface, a global metropolis remarkably uniform in appearance and standards, glittering with the splendor of human ingenuity, forever eager to advance its vision of human rights in all lands.

Yet that global city is also crammed full with deadly fears and explosive anger. The larger the volume of life on earth, the greater also the common willingness to sacrifice human beings for the sake of power, as one can see from the rising casualty rates of two world wars and the projected carnage of nuclear conflict. The tensions, playing havoc with human rights, are constantly escalating—for obvious reasons.

To this day all-too-few people in the immensely privileged West realize the depths of despair, frustration, and fury to which the world revolution of Westernization has reduced its victims; public opinion, denying all responsibility, still prefers to look only at its positive aspects. All-too-few observers admit the legitimacy of the counterrevolutions which, in bloody experiments, have tried to recreate the fullness of Western power from recalcitrant non-Western peoples—peoples gladly taking the latest fruits of modernity (including the pride of leadership in the world) yet unwilling or unable to submit to the demanding work routines and civic obligations which constitute the invisible aspects of effective power. In the old Europe cultural diversity was a source of tension and war. In a shrunken world the tensions are even greater. Incompatible cultures and historical traditions are compressed into even closer association among people utterly incapable of dealing constructively with cultural incompatibility. Their ignorance has raised hostility to hitherto unmatched ferocity.

The world revolution of Westernization, in short, has not created a peaceful world order guided by the ascetic and all-inclusive humane rationalism, the best quality in Western civilization. Universalizing the tensions inherent in its own dynamic evolution, it has rather produced a worldwide association of peoples compressed against their will into an inescapable but highly unstable interdependence laced with explosive tensions. Underneath the global universals of power and its most visible supporting skills—literacy, science and technology, large-scale organization—the former diversities persist. The traditional cultures, though in mortal peril, linger under the ground floors of life. Rival political ideologies and ambitions clash head-on. The world's major religions vie with each other as keenly as ever. Attitudes, values, lifestyles from all continents mingle freely in the global marketplace, reducing in the intensified invidious comparison all former absolute truths to questionable hypotheses.

Viewed in this manner, the global confluence thus far has produced not a shiny global city but a global Tower of Babel in which the superficial and ignorant comparison of everything with everything else is undermining all subtle distinctions between right and wrong, good and evil, worth and worthlessness. Even the centers of Western culture, whose traditions were until recently affirmed by their political power, are now inundated by alien ideas and practices, with subversive effects on the very convictions responsible for Western ascendance. An aimless cultural relativism threatens all moral energies; it encourages withdrawal into an illusory shelter of tradition while glorifying brute force perfected by sophisticated technology as the ultimate authority and source of security. Where in this global Babylon do we find the transcendent moral absolutes that can restrain the rising penchant for violence?

Prospects for the Future

Clearly, the world revolution of Westernization has saddled all humanity, if it is to survive, with an elemental necessity to readjust its most fundamental assumptions about collective human existence. Escape from the nuclear holocaust now requires an expansion of human awareness and social responsibility to cover the entire global framework. In the past, civic obligations were limited to family and lineage. Even now most people around the world, though forced into the more inclusive citizenship of nation-states—often at a high cost in human lives—hardly transcend these parochial bounds. Nation-states at their best represent a majestic advance in socialization, tying millions of people together in a hugely expanded sense of self-interest and in correspondingly refined personal discipline. For this discipline they are indebted to the Judeo-Christian religion, which assisted in the process of transcendence, expanding the individual sense of selfhood to the spiritual unity of all believers regardless of family or ethnicity. In the tight harness of contemporary global interdependence, however, exclusive loyalty limited to the state or to religious tradition patently promotes hostility and violence that easily rises to the extremes of genocide; in the rivalry of the two superpowers it may end all civilization

Thus we live in perilous suspension between loyalty to limited associations and dependence on transcendent global cooperation. Governments everywhere are caught between the need for international cooperation in the long-run national interest and contrary short-run pressures from ignorant and fickle electorates. Because there can be no effective internationalism without firm support from reasonably integrated nation-states acting as intermediaries between the individual and global humanity, states are indispensable. Yet their governments remain tied to their subjects' limited vision; for better or worse, nationalism takes precedence over internationalism, at least until major catastrophes demonstrate the necessity of more inclusive civic awareness. A considerable advance in perspectives and inclusive organization took place as a result of the chastisements of World War II, embodied in the European Economic Community and, more globally, in the United Nations. Support for that enlarging vision, however, has shrunk again in the fat years that followed; it was further reduced, as economic growth lagged, by cultural iso-lationalism and economic protectionism.

Admittedly, any advance from limited to more inclusive association rep-resents a major collective achievement, commonly spread over several gen-erations and accompanied by massive crises. The cruel difficulties of such adjustment are currently illustrated by the endless troubles of reculturation haunting the developing countries. The transition from nationhood to glob-alism foreshadows similar tensions for all humanity. Wherever we look, the adjustment of individual and collective wills to an enlarged perspective is made only as a last resort, and always at an appalling price in human suffer-ing. In the age of nuclear weapons, do we have sufficient control over the factors shaping our destiny to contain the suffering that softens selfish wills and advances human consciousness and collective discipline to a yet higher pitch? Can we accomplish by reason and in peace what formerly was accom-plished by force and inhumanity?

There is more to the burden of globalism: the task of integrating the explosively growing new technologies into community and culture, into the contexts of human survival. Individual awareness and social organization must match the proliferation of technologies whose consequences cannot be assessed except over several generations. The "global factory," with all its technological complexities and implications for the structure of individual and collective identity or for population control, has put humanity under an overload hardly comprehended at present.

And, finally, there looms ahead the practical task of bringing the world's population into reasonable balance with the earth's limited resources. In the 20th century the inexorable advance in population has wrought havoc with the available supply of forests, arable land, water, clean air; the busy human locusts and their machines have eaten dangerously into the human biosphere, in a manner by no means fully understood even now. The task of balancing people and knowledge against the world's resources can be solved only when the institutions of international cooperation have been sufficiently strength-ened. And that will happen only when the run of people have become willing to order their private lives and their consciences according to the dictates of their worldwide common welfare; when they will be ready to fuse their per-sonal egos with the egos of billions of other human beings, even in intimate matters like procreation and family size.

Huge, indeed repulsive and barely graspable by present ways of thinking, is the ultimate burden imposed upon humanity by the world revolution of Westernization. Thus—to use extreme terms for an extreme condition—the present generation lives in unprecedented utter precariousness, as on a knife's edge. On one side flashes the threat of nuclear doom as the culmina-tion of all the havoc wrought by Westernization. On the other side looms the unwelcome, distasteful, and infinitely demanding challenge of a massive advance in the West's ascetic and self-enlarging rationality.

I

Beginnings

1

How the Concept Originated

Sometimes an individual in high position addressing matters of state unwittingly rises to historical perspectives far larger than warranted by the immediate occasion. One occasion for such a fortuitous and subsequently forgotten grasp of historic essentials unexpectedly bridging disparate cultural universes was the speech which Lord Lytton, viceroy of India (1876–80), gave before his Legislative Council in Calcutta on March 14, 1878.

The speaker as well as the circumstances were significant. The first viceroy appointed after the elevation of Queen Victoria to empress of India, Lord Lytton was a high-minded Tory who, despite uncertain health in the heat of Bengal, attended vigorously and honestly to his duties, familiar with the history of the British *raj* in India. The British, he knew, had come to India when the power of the Mughal emperors, the previous conquering outsiders, was in decline. They meant to be masters, absorbing the territories of native princes and expanding their sway into Burma and up to the Khyber Pass; simultaneously, British ways of life penetrated into Indian life. By mid-century railways, telegraph lines, and major irrigation works were stimulating the economy and raising British revenues, but also displacing or undermining established convictions, authorities, and vested interests.

Hostility to the unwanted changes erupted in the Sepoy Mutiny of 1857 and continued to smolder in the depths of Indian society, as Lord Lytton learned to his dismay. The business he laid before the Council on that day in March 1878 was the passage of a bill permitting the British authorities to suppress seditious newspapers published in native languages (perhaps imperfectly understood by the censors). The challenge before him personally was the conflict between the suppression of free speech and the affirmation of the British imperial mission with its ideals of freedom and human dignity.[1]

The viceroy did not flinch in the face of his dilemma. Speaking in the measured cadences of mid-Victorian political rhetoric he admitted:

> By association, by temperament, by conviction, I should naturally find my place on the side of those to whom the free utterance of thought and opinion is an inherited instinct and a national birthright. I should have rejoiced had it fallen to my lot to be able to enlarge, rather than restrict, the liberty of the press in India; for neither the existence nor the freedom of the press in this country is of native origin or growth. . . . It is one of the many peculiarly British institutions which British rule has bestowed upon a population to whom it was previously unknown in the belief that it will eventually prove beneficial to the people of India, by gradually developing in their character those qualities which have rendered it beneficial to our own countrymen.

Yet moments later, with remarkable insight, he faced up to the cultural incompatibility in bringing the British version of freedom to India:

> It must be . . . remembered that the problem undertaken by the British rulers of India . . . is the application of the most refined principles of European society, to a vast Oriental population, in whose history, habits, and traditions they have had no previous existence. Such phrases as "Religious toleration," "Liberty of the press," "Personal freedom of the subject," which in England have long been the mere catchwords of ideas common to the whole race, and deeply impressed upon its character by all the events of its history, and all the more cherished recollections of its earlier life, are here in India to the vast mass of our native subjects, the mysterious formulas of a foreign, and more or less uncongenial, system of administration, which is scarcely, if at all, intelligible to the greater number of those for whose benefit it is maintained.

In that setting, Lord Lytton continued, full freedon of speech was hardly warranted, considering what abuses it entailed in the semi-literate popular press just emerging:

> Written for the most part by persons very imperfectly educated and altogether inexperienced; written, moreover, down to the level of the lowest intelligence, and with an undisguised appeal to the most disloyal sentiments and mischievous passions—these journals . . . have begun to inculcate combination on the part of the native subjects of the Empress of India for the avowed purpose of putting an end to the British *raj*.

In the face of that threat the viceroy saw no alternative but to invoke "that supreme law—the safety of the State." Rebellion in India, combined with the expansion of Russia through Afghanistan (a prospect even then causing intense anxiety), touched the jugular of British imperial rule. Lord Lytton felt no qualms therefore about suppressing the disloyal newspapers (the English-language press, considered more responsible, was not touched). Such minor and presumably temporary abridgement of free speech seemed justified in view of the immense boons of civilization introduced, admittedly as an uncongenial alien force, into India.

But what a massive impact, according to the viceroy, those British boons had on the country! Never did British imperialism speak with a more prophetic insight than in the following reflection:

> It is a fact which there is no disguising . . . and also one which cannot be too constantly or to anxiously recognized that . . . we have placed, and must permanently maintain ourselves at the head of *a gradual but gigantic revolution— the greatest and most momentous social, moral, and religious, as well as political revolution which, perhaps, the world has ever witnessed.* (italics added)

In these words (disregarding the cautionary "perhaps") Lord Lytton caught hold of the most important single force in modern history running at high tide from the late 19th century to the very present: the world revolution of Westernization.

That gigantic revolution of which Lord Lytton spoke meant imposing the best of British institutions (as the British saw it) as a "foreign and more or less uncongenial" and at best a "scarcely intelligible" system of administration upon potentially seditious and disloyal subject peoples for their ultimate benefit. It was done by an act of force, "by a conquering race," as Lytton had said, justifying in the name of "the safety of the State" the continued use of brute force. And most significantly, it implied a vast recasting of all aspects of indigenous life, a total cultural revolution. The fact that, as a landed aristocrat, Lord Lytton did not mention the mundane subject of commerce hardly mattered. The revolution of which he talked was an all-encompassing process of reculturation, gradual but irresistible. Unlike revolutions in the classical sense of that term, which originated in the internal dynamics of society and promised liberation, this permanent cultural revolution was instigated from without and, under the guise of liberation, led to further repression.

We further note that to Lord Lytton this gigantic process of reculturing an unresponsive people was a high-powered competitive enterprise in which his country was determined to stay in the lead. As he was aware, by 1878 other European countries were pressing the same revolution, above all France, and soon Germany and even Italy. Indeed, through several centuries all of western Europe had competitively expanded its power into the non-European parts of the world. Spain, Portugal, the Netherlands, England, and France—the Atlantic states—had gathered major colonial empires, while at the eastern end of Europe the tsars of Russia had pushed their version of European culture through Siberia, to the Pacific, and into central Asia, threateningly close to the borders of India. In the Western Hemisphere the European settlers ever since Columbus had practiced on the Amerindians a cultural revolution far more drastic than that of the British in India. From their conquered continent the English-speaking Americans reached out even farther. Twenty years after Lord Lytton's speech and spurred by the British example, the United States joined the European empires in the Westernization of the world, expanding across the Pacific to the Far East.

The Westernization of the world, though accomplished by a tiny minority relying largely on the labors and fighting capacities of non-Westerners, was, when seen in broad perspectives, a common enterprise of all Europeans and their descendents settled elsewhere; it summed up centuries of European culture. Yet in their own eyes and those of the world, by the end of the 19th century the British stood at the head of the outward thrust, setting a model for all the world.

And with what zest! While Lord Lytton, in India, outlined the essence of the revolution, Cecil Rhodes, a twenty-four-year-old dreamer drawing on Aristotle and the social Darwinists of his day, envisioned "the extension of British rule throughout the world, . . . the colonization by British subjects of all lands wherein the means of livelihood are attainable by energy, labour, and enterprise. . . ." He was convinced of God's purpose to make the Anglo-Saxon race predominant for a universal reign of liberty, justice, and peace. In this spirit he dedicated himself to assist in "the foundation of so great a power as to hereafter render wars impossible and promote the best interests of humanity."[2] Although not advocating a world state, a mature liberal statesman, Lord Rosebery, echoed similar sentiments in 1893, observing: "We have to remember that it is part of our responsibility and heritage to take care that the world, so far as it can be moulded by us, shall receive the Anglo-Saxon, and not another character."[3] These were heady words, bound to leave their mark also on non-Anglo-Saxons.

Yet even more was at stake than a vision of world domination. Already in 1872 Benjamin Disraeli, the Conservative leader and subsequently Lord Lytton's sponsor, had outlined a specter of an ultimate either-or. In a famous speech at the Crystal Palace in London he told his listeners: the issue is "whether you will be content to be a comfortable England, modelled and moulded upon Continental principles and meeting in due course an inevitable fate, or whether you will be a great country—an imperial country—a country where your sons, when they rise, rise to paramount positions, and obtain not merely the esteem of their countrymen, but command the respect of the world."[4] It was to be either a world power or a nonentity in the annals of world history. That message, too, was bound to be heard around the world, associated sometimes with crude assertions about the superiority of the Anglo-Saxon race.

At the beginning of the present century there was no question which way the British were headed. They basked in the glories of their peaceful and equitable domestic institutions, developed in the unrepresentative security of a small island, their self-confidence boosted by the fact that their empire claimed roughly one-quarter of the worlds' land surface and that their ships ruled the oceans. Yet in their smug conviction of superiority did they realize that they also disseminated a passion for world domination among other peoples?

The world revolution of Westernization, in short, carried a double thrust. It was freedom, justice, and peace—the best of the European tradition—on the one hand; on the other hand (and rather unconsciously), raw power to reshape the world in one's own image. How were the two aspects related?

What effects did these messages have upon the non-Western world—or even upon the West itself?

In search of answers and defining "the West" for the moment merely as the states of western and northwestern Europe (with the United States still at the periphery), we will first outline, from a high-altitude perspective, the basic facts which, as Lord Lytton had said, "cannot be too constantly or too anxiously recognized."

2

The Political Revolution

To begin the familiar story with a comment on the obvious: expansionism was not a peculiarity of the Europeans. Admittedly, they carried it to an extraordinary pitch; they have been correspondingly hated for it. But a self-affirming expansionism is a central ingredient in all individual and collective existence; it is the universal ingredient in an otherwise highly differentiated human nature.

In order to make the most of their lives individuals assert themselves, radiating their personalities into their environment. Heads of families want more progeny, more land, more trade, more influence. The ambition to prevail and enlarge one's identity becomes collectivized, growing in scale with the size of the community. It becomes part of its religion, its essence an omnipresent, all-powerful God mightier than all other gods. What, in short, holds together all societies, religions, and cultures down to the core of individual wills is a pervasive faith in the practical and metaphysical superiority of the common bonds: in pursuit of community people need to universalize their ways to the limit or else community falls apart for lack of common conviction. The forms of such aggrandizement have varied throughout history; in the past nature's adversity perhaps more often than the resistance of neighbors set narrow limits. The capacity for expansion, moreover, varies greatly among different cultures. But the expansionist urge, whether expressed in the arts of war or of peace, through command or service, is a constant factor in human existence. Inevitably the expansionist urges collide. They have done so since the beginning of history, with Europe in the lead since the 16th century.

The outpouring of European power began in response to Turkish conquests, which in turn followed massive Mongol conquests; in northern Africa and Asia it collided with a virulent Islamic expansionism. In the Western

Hemisphere the conquering Spaniards subdued empire-building Aztecs and Incas. By their conquest of Africa, the Europeans brought a measure of peace to people harassed by war, civil strife, or slave raids, all fomented by indigenous ambition for domination. And in the Far East they encountered the Middle Kingdom of China, the biggest empire of all; it claimed dominion over the entire world. Natives and foreigners alike kowtowed before the emperor in submission to his role as the universal mediator between Heaven and Earth. What counts, then, in our context are not the basic motives of expansionism but the forms of power that make them effective. In this respect indeed the Europeans rose above all others in the world. They possessed forever unmatched advantages.

These advantages, briefly outlined, resulted from a unique combination of cultural unity and diversity in a relatively small geographic area favored by climate and natural resources (including easy lines of communication). Unity was provided by the Judeo-Christian tradition with its heritage of Graeco-Roman culture, both of them products of that competitively interactive cradle of civilization called "the fertile crescent," both of them a source of further cultural creativity in the even more interactive competition emerging, through Italy and Spain, in western Europe. Geography and the common cultural inheritance allowed speedy communication of all essential cultural accomplishments. The rivalry of craftsmen, artists, savants; of cities, regions, and eventually of nation-states set off an upward spiral of challenge and response spreading with an ever more rapid pace of emulation and imitation throughout society, the results of which were quickly circulated from country to country thanks to the overall cultural unity. With the help of an ascetic— or otherworldly—religion, a major flywheel of cultural creativity, the discipline of individual conduct and social cooperation was advanced to an intensity unique in the world, in the name of human dignity and freedom and for the purpose of building up power in all its aspects

Viewed in global perspective, Western Europe, and eventually the United States on the other shore of the North Atlantic, resembled a singularly prolific cultural hothouse in which all human accomplishments were advanced at a forced clip. The arms race originated in the 15th century with the rise of the nation-state; escalating through wars without number, it has not ceased since. But the race was to the swift also in technological ingenuity, industrial productivity, artistic creativity, intellectual innovation, political organization, and even humanitarianism—all contributing to power. Access to the world's riches certainly enhanced the fertility of that cultural hothouse. But the capacity, in the competition with the non-Europeans, to channel these riches into Europe originated in the ceaseless and often bloody interaction among the Europeans themselves. The rest of the world contributed to the West's ascendancy on Western terms.

From the start European territorial expansion was worldwide because it was seaborne. For seafaring people skilled also in warfare, the oceans set no limits; on the high seas the European gun-carrying ships had no equals. In the late 15th century Portuguese explorers, traders, and missionaries moved south; they touched a corner of South America (now Brazil), rounded Africa, and entered the Indian Ocean with its active coast-wise traffic linking east Africa, Arabia, India, and the Far East. They gained many footholds along their far-flung commercial empire, growing rich in the spice trade. Their neighbors and competitors, the Spaniards, crossed the Atlantic into the virtually defenseless Western hemisphere. By design (through their superior military skills) and by accident (through their diseases, which ravaged the indigenous peoples who possessed no immunity) they made themselves masters; they settled thinly populated lands with their own kind and with slaves traded in Africa. From New Castile (as Mexico then was called) they extended their rule to the Philippines, recently discovered by Magellan. When they were repulsed by their Portuguese rivals in the Spice Islands on the other side of the globe, they had reached the limits of their expansion.

In due time, other seafaring Europeans with stronger domestic bases joined the outward thrust. The English, Dutch, and French carved out their own colonial dependencies, often displacing the Iberian pioneers. By the end of the 17th century the Western Hemisphere was completely controlled by Europeans: English in the north, with a French presence at the mouth of the St. Lawrence River; Spanish in the south, except for the Portuguese in the Amazon region; a motley of European nationalities in the Caribbean Islands; plus the ubiquitous African slaves to do the hard work. Elsewhere in the world the European impact was less drastic, because indigenous peoples showed more capacity for resistance or adverse climates discouraged European settlers.

But the Europeans never ceased probing for opportunities. With time they grew more resourceful. They increasingly drew on the wealth of the lands under their control. More important, their wrangle for ascendancy on their own continent constantly perfected the arts of domination in war and peace; no other parts of the world could match Europe's hothouse pace of cultural evolution. From the 18th century onward the Europeans gradually gained the upper hand in all encounters, empowered by the mechanical arts of the industrial revolution and the sociopolitical arts of representative government and the nation-state. The democratic revolutions begun in the late 18th century helped to give independence to hitherto closely controlled overseas settlers; they ended the original structures of colonial empire. But they also boosted the forces prompting the expansion of European culture, providing it with more broadly based political support and furnishing more popular energies for the outreach from Europe itself and from its scattered overseas outposts. There was never a lack of opportunity.

Take India, for example. In the 18th century the decline of the Mughal emperors had led to both French and English intervention in Indian affairs. The English—meaning the East India Company and increasingly the government too—prevailed in part because they helped to defeat the French on their home ground in Europe. Thereafter they gradually expanded their rule, trying, as we have seen, to transform the teeming multitudes of that large and varied subcontinent after their own principles. As rulers they satisfied the desire for gain among the merchants of the East India Company and, later, of a greater variety of English entrepreneurs. But their ambition also reached further. Under the influence of liberal idealism and the evangelical revival of the early 19th century they assumed a moral responsibility for the people now at their mercy. In this spirit well-established indigenous customs repulsive to English sensibilities, like *suttee* (widow-burning) or *thugee* (caste-based murder and robbery), were suppressed and the penal code accordingly reformed.

For the purpose of enlightened administration, which increasingly enrolled Indians in its lower ranks, they made English the official language. What else could they do? "We have to educate a people who cannot at present be educated by means of their mother tongue," wrote Thomas Babington Macaulay, the famous historian, when consulted on the conditions of India's progress in 1835. He concluded with a plea "to form a class who may be interpreters between us and the millions whom we govern; a class of persons, Indian in blood and color, but English in taste, in opinion, in morals, and in intellect."[1] The well-intentioned onslaught on Indian traditions came to a climax in mid-century under the administration of James Ramsey, marquis of Dalhousie, but was soon brought to an abrupt end by the Sepoy Mutiny, an elemental rebellion against British rule. Lacking effective weapons and organization, the rebels failed, but the British, though persisting in their efforts to reculture the Indians, became more cautious. In his flash of illumination Lord Lytton had begun to realize the vastness of the cultural revolution under way.

Aware of their helplessness and attracted by the novel opportunities, more Indians sought education in the British fashion, still tied to the past (whether Hindu or Muslim) yet eager to make a career under the new masters. They swallowed their resentment against the persistent discrimination, flattered by the interest shown by Queen Victoria and cheering her when, by the design of the Tory prime minister, Disraeli, she became empress of India in the very years that Lord Lytton assumed his post. Were they now less Indian or more so? Within ten years (in 1885) Western-educated Indians, assisted by a retired Englishman in the Indian service, established the Indian National Congress, the focus of a new nationalism increasingly at odds with the British *raj*. In less than a century the British had taught the Indians, who in the past had evolved neither a common government nor a common political identity, how to clamor for political unity and nationhood. The revolution of Westernization, however, had to grow much stronger before self-determination on the Indian subcontinent could be achieved.

Take China next. The Chinese empire, in decline after the peace and prosperity under the great emperors of the Qing dynasty, had traditionally spurned contact with European traders. In 1785 the emperor Qian Long had haughtily declared foreign wares unworthy of Chinese attention. English traders buying Chinese tea were thus forced to pay with scarce bullion rather than merchandise. The most profitable item they could sell was opium, its import wisely forbidden by law but smuggled into the country all the same. In 1839 a conflict between Chinese authorities and British merchants over the illegal opium trade broke into war, the First Opium War. The Chinese, who had at first laughed at the ridiculous appearance of British sailors, suddenly found themselves at the mercy of their guns. Thereafter China was opened up to foreign goods, people, and ideas (including Christianity), hesitantly at first, but after further opium wars in chastened helplessness.

In the upshot the Chinese government lost control over its seaboard centers and foreign trade. The Chinese Customs Service was placed under British administration; run with exemplary efficiency, it became the empire's steadiest source of revenue. At the same time a British general helped government forces to defeat the Taiping rebels. In 1860, while that bloody rebellion raged, Beijing, the imperial capital, was occupied by British and French troops and the emperor's summer palace burned. In the same year, taking advantage of China's weakness, the tsar of Russia acquired the Maritime Province located north of Manchuria. And in the following year, in a major break with hallowed tradition, the emperor established a government agency for handling foreign affairs, forced at last to recognize the existence of strangers who would not kowtow to the Son of Heaven. Once admired even in Europe, the Middle Kingdom found itself shrinking in size, reputation, and even self-esteem, a tempting prey to Western penetration.

After many more humiliations by Westerners and Westernized Japanese the Qing dynasty, together with the two-thousand-year-old imperial tradition, collapsed in 1911. It was replaced, on the surface, by a republic governed under a Western constitution and, in reality, by a variety of warlords in a simmering civil war that left the country even more at the mercy of foreign entrepreneurs and their governments. Patriots feared that their once proud country might dissolve into colonial dependencies. In that crisis some mandarins of the old school committed suicide. But a new brand of patriots copied Western nationalism, rebelled against their families and the Confucian order, dressed in Western clothes, and left for study abroad—destined for further agonies. Deploring their country's utter helplessness, they became helplessly suspended in the confusion between their defeated cultural heritage and the victorious Western influences. Given the depth of traditions and the immutability of a huge population kept insulated from the outside by the size and geography of the country, the travail caused by the revolution of Westernization was bound to be unending and exceedingly costly in human lives.

By contrast, in Japan that revolution seemed miraculously successful, and in record time. While the Chinese empire, with its huge landlocked, loosely-governed territory and population as well as its millennial cultural continuities, was cemented into cultural rigidity, Japan was prepared for the encounter. A rather small chain of islands with a shore-based smallish population vulnerable to foreign warships and accustomed to absorbing cultural stimuli from the Chinese mainland, Japan also had enjoyed—or suffered—the benefits of a competitive feudal order that had advanced both the mechanical and martial arts and built up a tradition of tight collective discipline never achieved in China. Like China, it had defensively closed itself to commercial and cultural exchanges with Europe; like China, it was pried open against its will in a show of force by American and European warships firing occasional cannonades.

Yet, unlike their neighbor, the Japanese, between 1853 and 1868, by a remarkable consensus marred only by a brief civil war, recognized the inevitable: they had to learn all the skills of the Westerners if they wanted to preserve their political independence. While reaffirming the continuity of tradition by restoring the power of their emperor, they threw open their gates to Western influences. For a time the Westerners limited Japanese sovereignty; their residents in Japan were governed by Western not Japanese laws. But as for cultural self-determination, the Japanese had seemingly surrendered for good. Taking the best from England, France, Germany, and the United States, they copied Western constitutional government and the rule of law complete with individual rights, plus a string of other essentials, including science, technology, universal education, industry, modern organizations for their armed forces—everything, in short, that contributed to power, including the self-confidence of being accepted by the West as equals. Wrote Fukuzawa Yukichi, the foremost Westernizer of the time, "the final purpose of all my work was to create in Japan a civilized nation, as well equipped in both the arts of war and peace as those of the Western world."[2] In 1911 Western extraterritorial rights were repealed because Japan was now judged a civilized nation.

In its rapid progress of adaptation to Western ways the country was aided by the fact that it possessed two essential prerequisites: mechanical and intellectual sophistication not too different from that of the West and a habit of collective cooperation immune to subversion by foreign influences. On these essential grounds Japan and the West enjoyed a unique compatibility. Yet in other, more central aspects of culture, in religion, and in the "sacred inviolable" authority of the emperor, Japanese traditions continued; even more than in China Christianity failed to make many converts. At the deepest levels Western and Japanese culture remained at odds, though not in open conflict.

What reconciled Western ways to Japanese traditions was their patent capacity for leading Japan to the greater glories of political expansion, the ultimate sanction and healer of domestic tensions. A Westernized Japan defeated China in the mid-1890s; it dominated Korea. In 1904–05 it even trounced one great European country, the Russian empire, on land and sea, to the cheers of all non-Westerners under Western domination who hoped one day to match that feat. Even before the Russo-Japanese War Japan had been accepted as a Far Eastern ally by Great Britain, gaining world-wide respectability. In the unique case of Japan, in short, Westernization had quickly and relatively painlessly accomplished its double purpose. It had made the country more civilized in its own and Western eyes and also more powerful, without apparent loss to national identity. Yet the revolution continued under ever new challenges issuing from the West.

Bypassing the other regions of East and Southeast Asia—by 1900 they were firmly under Western control: English Burma and Malaya, French Indo-

China, Dutch Indonesia, and the American Philippines, with small German footholds on the China coast and in small Pacific islands—let us now shift to Africa.

A large continent, located for the most part outside the troubled yet culturally stimulating contacts linking India (and even China) with Europe, Africa was the last part of the world to fall under Western domination. Except for its southern tip, colonized by the Dutch in the 17th century, its climate proved fatal to Europeans until the development of tropical medicine toward the end of the 19th century. The early Europeans, traders accompanied by missionaries, obtained what they wanted—gold, slaves, ivory—through commerce, maintaining from the late 15th century onward fragile outposts on the West African coast, first the Portuguese, then the English, Dutch, French, and Spanish. Through their trade Europeans from the start influenced economic and political life in the African hinterland, matching Islamic influences penetrating through the Sahara, up the Nile or along the east coast.

In their contact with Westerners trained in the highly refined arts of statecraft and nationhood, the African peoples in their small-scale oral cultures based on family and lineage were at a decided disadvantage. Survival under African conditions was precarious; people therefore put a high premium on bodily vitality. Bodily vitality, however, impeded the rise of ascetic practices that promoted the cumulative evolution of literacy and cerebral knowledge in Western culture. The fragility of life also hampered civic continuity. What political tradition or power structure was left of the fabled empires of Ghana, Mali, Songhai, or Zimbabwe when the Europeans arrived? The famous Benin bronzes were buried and forgotten until discovered by Europeans, their craftsmanship not passed on or perfected over time like that of Chinese, Mediterranean, or European metallurgists. In mechanics and, above all, in political organization the Africans were no match for the conquering outsiders, whether Islamic or Christian. When in the 19th century the Europeans came with the organizational and technological resources of large nation-states, what chances did the Africans have? Under these conditions the European takeover of Africa—and even of its Islamic parts—required little effort. It was accomplished by minute numbers of Europeans armed with superior weapons and employing ever willing African manpower against other Africans determined to resist the destruction of their customary ways.

What gave the European partition of Africa its special character was the heightened competition among the major European governments. The dynamos of the world revolution of Westernization were always located within Europe itself (at least until World War II); at the end of the 19th century they were revved up to full speed. The competition for power within Europe now spilled over massively into the world at large; European politics became world politics. In 1871 the creation of a powerful united Germany in the heart of the Continent changed the European balance of power and raised political competitiveness to the alternative between world power and insignificance, as indicated in Disraeli's speech of 1872. Henceforth not only governments were aroused, but their subjects as well; power politics went global and democratic at the same time. Meanwhile, new sciences and technologies had speeded transportation and communication around the world. As a result, Europeans and Americans found themselves at overwhelming advantage wherever they went, even in tropical Africa.

As the other non-Western parts of the world had already been apportioned (except for minor openings on the China coast or in the Pacific), Africa offered the last major opportunity for colonial expansion. A scramble began for occupation of its hitherto unclaimed coasts, which led to penetration of the hinterland as well. By the Treaty of Berlin in 1885 Africa was formally allocated to European states, most of them with a long record of dealing with Africans: the bulk to Britain, a lesser share to France, small parcels to Portugal and Spain (the earliest imperialists). Two determined newcomers also profited; imperial Germany acquired a number of colonies, and an ambitious king of Belgium laid claim to the Congo. By the end of the century all of Africa found itself under European control, except for the ancient kingdom of Abyssinia (which had just defeated an attempted Italian takeover) and Liberia, an independent state formed by American blacks prouder of their American than of their African heritage and enjoying American protection. In North Africa European control of Morocco was still under dispute, but Algiers was a *département* of France; Tripoli officially belonged to the Ottoman Empire but was destined to become an Italian colony; Egypt was firmly in English hands. While all colonial powers had agreed in 1885 to suppress slavery and the slave trade in their African colonies, they now began, each in their own manner, to enslave their African subjects to their alien ways. In sub-Saharan Africa especially, the world revolution of Westernization was destined to plow exceedingly deep, with profound tragic results inconclusive to the present.

By the end of the 19th century—to look at a final case—the revolutionary expansion of Europe had also subverted the major Islamic power, the Ottoman Empire. Once master of the Eastern Mediterranean, covering North Africa, Arabia to the Persian Gulf, southern Russia, and the Balkans to the gates of Vienna, it had since shrunk and decayed, sinking into heavy indebtedness to European banks and saddled with stringent foreign controls. Like China, it escaped outright subjection, but it merited being labelled as semi-colonial. It was hollowed out from within by rebellious subjects—the Young Turks—infected with revolutionary ideals of freedom, democracy, and nationalism, and headed for collapse.

Wherever we look, then, by the beginning of the 20th century the entire world was dominated by the political controls and the ways of life radiating outward from western Europe and North America. Some 85% of the world's land surface had been absorbed into Western colonial empires; the remaining parts were irradiated with Western influences (the high seas, in any case, had always been a Western preserve).

Thus for the first time in all human experience the whole world was united under the auspices of a single culture which, though unfinished, imperfect, and politically divided, constituted a reasonably coherent single entity. Arrogant in its superiority, it contrasted with other cultures evolved under different conditions in different parts of the world. The others were now deprived of their past freedom of cultural evolution and emasculated by the unending agonies of reculturation. In danger of being reduced to "people without history," they were forced through a catastrophic collective trauma utterly alien to the West.

The new unity, marred by greater and more visible inequality then the world had yet witnessed, was still loosely wrought and superficial. Yet it also imposed some crucial restraints on all. It had created a single worldwide framework for essential human activities. Wars and rumors of wars in one part of the world rattled the nerve centers of power in all other parts; governments had to assess power relations in the entire world. Trade in essential commodities had also become worldwide; a world economy emerged dependent more than ever on Europe. The needs and wants of the Europeans began to reshape even the world's ecology; tropical forests were cut down, soils eroded, traditional food staples replaced by export crops headed for the tables of spoiled Europeans. And underneath the raw power relationships, and to some extent independent of them, a new globalism arose of diseases and pests, of plants and animals, promoted by the accelerating circulation of human beings.

Even more important, the new unity had joined the peoples of the world, most of them under alien compulsion and without preparation, into inescapable interaction with each other. All non-Westerners were now forced to cope uncomprehendingly with the ubiquitous yet equally uncomprehending Western culture. At the same time they could not escape closer—and often hostile—contact with each other. In the long view, the new interdependence even overburdened the Westerners. Against their preference or awareness they had saddled themselves with a political and cultural, as well as moral, responsibility for the unsettled alien multitudes whose miseries their own privileged circumstances and patent superiority prevented them from understanding.

3

The Cultural Revolution

While the world revolution of Westernization created a political world order radically above the horizons of all past human experience, it also unhinged, in the revolutionary manner sensed by Lord Lytton, the depths of non-Western societies constituting the bulk of humanity. As he had said, "the application of the most refined principles of European government and some of the most artificial institutions of European society to a . . . vast population in whose history, habits, and traditions they have had no previous existence" was a risky enterprise, perhaps more than he had anticipated.

Examining the history of colonial expansion, one can discern a rough but generally applicable pattern for the revolutionary subversion of non-Western societies. Subversion began at the apex, with the defeat, humiliation, or even overthrow of traditional rulers. The key guarantee of law, order, and security from external interference was thus removed. With it went the continuity of tradition, whether of governance or of all other social institutions down to the subtle customs regulating the individual psyche. Thus ended not only political but also cultural self-determination. Henceforth, the initiatives shaping collective existence came from without, "mysterious formulas of a foreign and more or less uncongenial system" not only of administration but also of every aspect of life.

Once the authority of the ruler (who often was the semi-divine intermediary between Heaven and Earth) was subverted, the Western attack on the other props of society intensified. Missionaries, their security guaranteed by Western arms, discredited the local gods and their guardians, weakening the spiritual foundations of society. At the same time, colonial administrators interfered directly in indigenous affairs by suppressing hallowed practices repulsive to them, including human sacrifice, slavery, and physical cruelty in its many forms. Meanwhile, Western businessmen and their local agents redirected the channels of trade and economic life, making local producers and consumers dependent on a world market beyond their comprehension and control. In a thousand ways the colonial administration and its allies, though not necessarily in agreement with each other, introduced a new set of rewards and punishments, of prestige and authority. The changeover was obvious even in the externals of dress. Africans became ashamed of their nudity, women covered their breasts; Chinese men cut off their queues and adopted Western clothes. The boldest even tried to become like Westerners "in taste, in opinion, in morals, and intellect."

The pathways of subversion here outlined indicate the general pattern and the directions which it followed over time. Its speed depended on Western policy and the resilience of local society. Things seemingly fell apart quickly in the case of the most vulnerable small-scale societies of Africa and much more slowly in India or China, if at all in Japan. Often the colonial administration itself, under a policy of "indirect rule," slowed the Western impact for fear of causing cultural chaos and making trouble for itself. In all cases, tradition (however subverted) persisted in a thousand forms, merely retreating from the external world into the subliminally conditioned responses of the human psyche, its last refuge. It is still lurking in the promptings of "soul" today.

And did things really fall apart? The world revolution of Westernization prevailed by the arts of both war and peace. Certain aspects of Western power possessed an intrinsic appeal which, even by indigenous judgment, enhanced life. New crops often brought ampler food; European rule often secured peace. Through their command of the seas and of worldwide trade Europeans and Americans opened access to survival and opportunity in foreign lands to countless millions of people in China and India. Or take even the persuasion of raw power: once convinced of the superiority of European weapons, who would not crave possession of them too? And more generally, being associated with European power also carried weight; it patently held the keys to the future. More directly perhaps, doing business with Westerners promised profit. If they played it right, compradors would get rich.

More subtly, certain categories of the local population eagerly took to foreign ways. Missionaries sheltered outcasts: slaves held for sacrifice, girls to be sold into prostitution or abandoned, or married women feeling abused and oppressed. The struggle for sexual equality is still raging in our midst, yet by comparison even Victorian England offered hope to women in Africa or East Asia. Regarding Japan, Fukuzawa related the story of a highborn dowager lady who "had had some unhappy trials in earlier days." She was told of "the most remarkable of all the Western customs . . . the relations between men and women," where "men and women had equal rights, and monogamy was the strict rule in any class of people. . . ." It was, Fukuzawa reported, "as if her eyes were suddenly opened to something new. . . ."[1] As a messenger of women's rights he certainly had Japanese women, "especially the ladies of the higher society," on his side. In China liberated women rushed to unbind their feet.

In addition, the Westerners introduced hospitals and medicines that relieved pain and saved lives, a fact not unappreciated. Besides, whose greed was not aroused by the plethora of Western goods, all fancier than local products: stronger liquor, gaudier textiles, faster transport? Simple minds soon preferred Western goods merely because they were Western. Given the comparative helplessness of local society, was it surprising that everything Western tended to be judged superior?

The Westerners with their sense of mission also introduced their education. It was perhaps not enough, according to anti-Western nationalist suspicious of European desires, to keep the natives down, yet it offered access to Western skills at some sacrifice on the part of teachers willing to forgo the easier life in their own culture. Privileged non-Westerners even attended schools and universities in the West. Thus, as part of the general pattern of Westernization, a new category of cultural half-breeds was created, the Westernized non-Western intelligentsia. It differed somewhat according to cultural origins, but shared a common predicament. Product of one culture, educated in another, it was caught in invidious comparison. As Thomas Hobbes observed "Man, whose Joy consisteth in comparing himselfe with other men, can relish nothing but what is eminent."[2] Riveted to Western preeminence, this intelligentsia struggled for purpose, identity, and recognition in the treacherous no-man's-land inbetween—and most furiously in lands where skin color added to its disabilities. Talented and industrious, these intellectuals threw themselves heroically into the study of Western society and thought so alien to their own.

Along the way they soon acquired a taste for the dominant ideals of the West, foremost the liberal plea for equality, freedom, and self-determination and the socialists' cry of social justice for all exploited and oppressed peoples and classes. They were delighted by the bitter self-criticism they discovered among Westerners—Western society produced many doubters, especially among its fringes in central and eastern Europe. At the same time, non-Western intellectuals quickly perceived the pride that lurked behind Western humanitarianism. They might be treated as equals in London or Paris, but "east of Aden" on the Indian circuit or anywhere in the colonies, they were "natives"—natives hypersensitive to the hypocrisy behind the Western mission of exporting high ideals without the congenital ingredient of equality. Thus they learned the lessons of power not formally taught by their masters. They needed power—state power—not only to carry the Western vision into practice on their own but also to make equality real.

Inevitably, the non-Western intellectuals turned their lessons to their own use. The ideals of freedom and self-determination justified giving free rein not only to the promptings of their own minds and souls, but also to protests over the humiliation of their countries and cultures. As a result of their Westernization they became anti-Western nationalists, outwardly curtailing, in themselves and their compatriots, the abject imitation of the West. Yet, as an 18th-century German wag had said, "to do just the opposite is also a form of imitation." Anti-Western self-assertion was a form of Westernization copying the cultural self-assertion of the West. Moreover, limiting western influence in fact undercut any chance of matching Western power (and the issue of power was never far from their minds). Thus anti-Western intellectuals were caught in a love-hate attitude toward the West, anti-Western purveyors of further Westernization.

Take Mohandas Gandhi, perhaps the greatest among the Westernized non-Western intellectuals. Born into a prominent tradition-oriented Hindu family and of a lively, ambitious mind, he broke with Hindu taboo and studied English law in London, fashionably dressed and accepted in the best society, though by preference consorting with vegetarians and students of Eastern religion. After his return he confessed that "next to India, [he] would rather live in London than in any other place in the world."[3] From 1892 to 1914, however, he lived in South Africa, using his legal training for defending the local Indian community against white discrimination. There he put together from Indian and Western sources a philosophy as well as a practice of non-violent resistance, strengthening the self-confidence and civil status of his clients.

On a trip to India in 1909 he spelled out his political program, called Indian Home Rule *(Hind Swaraj),* in which he curtly announced: "If India copies England, it is my firm conviction that she will be ruined," not because of any special flaw in the English (he rather liked them), but because of the nature of modern civilization in general.[4] It was materialist, hedonist, and mechanical, and therefore diseased and in decline, as many European authors familiar to Gandhi had said for a long time. By contrast, Indian civilization was superior (as he documented from a long list of British and continental European works). His patriotic plea was for a mighty Indian soul-force *(satyagraha)* that was to overthrow the British *raj* in order to create—what? Home rule for an Indian nation in a subcontinent whose intellectuals—admittedly half-anglicized Indians—had copied the ideal of nationalism and statehood from their colonial master.

One of Gandhi's precursors, Narendranath Datta, better known as Swami Vivekananda, had gone even further. At a lecture in Madras he exhorted his audience: "This is the great ideal before us, and everyone must be ready for it—the conquest of the whole world by India—nothing less than that. . . . Up India and conquer the world with your spirituality."[5] Western globalized nationalism, obviously, was working its way around the world, escalating political ambition and cultural messianism to novel intensity.

Contradiction, perhaps even more sharply accentuated, spoke from the writings of Westernized Africans or descendents of African slaves. By 1900 a West African intellectual tradition was well established. Its first spokesman, James Africanus Horton, was an Edinburgh-trained physician from Sierra Leone, a British colony established for freed slaves; he served in the British West African Army Medical Service. In his book *West African Countries and Peoples* (1868) he set out to dispel European prejudice and "to prove the capability of the African for possessing a real political government and national independence, and that a more stable and efficient government might yet be formed in Western Africa, under the supervision of a civilized nation." Admitting that currently African institutions were "rude and barbarous," he foresaw a time when self-government would be established and "Africa . . . be left for the Africans." He added that the language of self-government would be English, "because it is now the universal language of the colony." He also noted with satisfaction the progress of missionary education and the resulting rise of a national spirit.[6] As "Africa, in ages past, was the nursery of science and literature," so he assured "the nations of Western Africa" of a proud future: "Their turn will come when they will occupy a prominent position in the world's history, and when they will command a voice in the council of nations."[7]

A few years later (1881), a lively descendent of African slaves born at St. Thomas in the Danish West Indies, named Edward W. Blyden, sounded a more strident note in his inaugural address as president of Liberia College in Monrovia. He had been raised by literate parents in a Jewish neighborhood, a member of the Dutch Reformed Church, whose pastor had encouraged his education. Refused admission to theological school in the United States on grounds of race, he eventually moved to Liberia, the first militant representative of a new pan-Africanism based in an independent African state.

Mapping out a program for Liberia College (a Western institution), he deplored the estrangement from African tradition and the self-deprecation caused by European education. Instead he urged the students to go to the interior in order to recover there "the laws of growth for the race." Yet, quoting English authors, he also urged that "the instruments of culture which we shall employ in the college will be chiefly the Classics [Greek and Latin languages and their literature] and Mathematics"; in addition he recommended the study of Arabic. After pleading, in a revolutionary break with Islamic and African tradition, "that arrangements . . . be made by which girls of our country may be admitted to share in the advantages of this college," he concluded with a Westernizing flourish: "The suspicions disparaging to us will be dissipated only by [our] . . . successful efforts to build up a nation, to wrest from Nature her secrets, to lead the van of progress in this country, and to regenerate a continent."[8]

As these glaring and inescapable contradictions indicate, the run of Westernized non-Western intellectuals led awkward lives—"in a free state," as V. S. Naipaul has put it—forever in search of roots, and certitude; inwardly split, part backward, part Western, camouflaging their imitation of the West by gestures of rejection; forever aspiring to build lofty halfway houses that bridged the disparate cultural universes, often in all-embracing designs, never admitting the fissures and cracks in their lives and opinions; and always covering up their unease with a compensating presumption of moral superiority based on the recognition that the promptings of heart and sould are superior to the dictates of reason. Knowing their own traditions and at least some of the essentials of the West, they sensed that they had a more elevated grasp of human reality; the future belonged to them rather than to the "decadent" West. Out of that existential misery of "heightened consciousness" (as Dostoyevsky called it) have come some of the most seminal contributions to the intellectual and political developments of the 20th century, including the anti-Western counterrevolutions.

However divided in their cultural origins and contradictory in their convictions, the Westernized non-Western intellectuals tended to agree on one point: the necessity of combining forces against Western domination. A straw in the winds of change was the first Pan-African Conference held in London (where else?) in 1900. A small group of Africans and Afro-Americans met for the common defense of Negroes everywhere. One of its prominent members was W. E. B. Du Bois, an unusually fascinating case of a highly gifted intellectual, part-white part-black, who shared all the contradictions and dilemmas as well as the glories of that type. Having become familiar with pan-Germanism and pan-Slavism while studying in Berlin, he subsequently coined the term "pan-Negroism." Rephrasing that term, the conference set up the first Pan-African Association. Eleven years later, again in London, a "Universal Races Congress" met, attended by delegates from many non-Western lands. The gathering provided an opportunity for pooling resources plus global visibility for the impassioned protests against the human havoc wrought by Western imperialism. An obscure beginning, it foreshadowed a new solidarity of the West's victims and also offered testimony to the relentless logic of Westernization.

There can be no question of the validity of the anger expressed at that congress or on subsequent occasions. Indigenous authorities and traditions had been overthrown, peoples and cultures hitherto mapping their own course in reasonable self-sufficiency were now reduced to dependence and helplessness, their cultural creativity downgraded or minimized by comparison with the victorious West. What Westerners can fathom the agony that goes with the unending disorientation and cultural chaos imposed by reculturation under alien auspices? Moral sensibility anywhere revolts against such massive destruction of custom and cultural continuity; it is an act of profound inhumanity and the source of endless calamities. Yet can we afford to give free rein to our indignation?

A historic outrage of such magnitude—to slide from issues of power into issues of morality—deserves a moment's reflection so that we do not add the furies of the past to the furies of the present. Let it be said first that the relations between the colonized and the colonizer are exceedingly subtle and complex, subject to keen controversy among all observers, all of them partisans, all of them now judging not by indigenous but by Westernized standards. Western ideals and practices have shaped and intensified the protests of Westernized non-Western intellectuals taking full advantage of the opportunities offered by Western society. Their protests, incidentally, were hardly ever turned against past inhumanities committed by their own kind (because traditionally they were not considered as such).

Next, having already surveyed the not inconsiderable side benefits of Western domination, let us ask: did the Westerners in their expansion behave toward the non-Westerners worse than they behaved toward themselves?

While they never treated their colonial subjects as equals, they never killed as many people in all their colonial campaigns as they did in their own wars at home (the brutality of Europe's cultural evolution has been carefully rinsed out of all current historical accounts). And in their peaceful intercourse with non-Westerners we find the whole range of emotions common in Western society. It was darkness at heart on one extreme and saintliness on the other, and every mix in between, with the balance perhaps tending toward darkness. As one colonial officer in East Africa confided to his diary: "It is but a small percentage of white men whose characters do not in one way or another undergo a subtle process of deteriorization when they are compelled to live for any length of time among savage races and under conditions as exist in tropical climates."[9] The colonial district commissioner, isolated among people whose ways sharply contradicted his own upbringing, often suffering from tropical sickness, and scared at heart, found himself perhaps in a worse dilemma than the Westernized non-Western intellectuals. Some of them, no doubt, were unscrupulous opportunists seeking escape from the trammels of civic conformity at home; they turned domineering sadists in the colonies. On the other hand, missionaries often sacrificed their lives, generally among uncomprehending local folk. It was perhaps a credit to the Westerners that the victims of imperialism found considerable sympathy in their own midst. The evils stood out while the good intentions were taken for granted.

Yet—to take a longer view—even compassionate Western observers generally overlook the fact that among all the gifts of the West the two most crucial boons were missing: cultural equality as the basis for political equality and reasonable harmony in the body politic. The world revolution of Westernization perpetuated inequality and ruinous cultural subversion while at the same time improving the material conditions of life. More people survived, forever subject to the agonies of inequality and disorientation resulting from enforced change originating beyond their ken. Collectively and individually, they straddled the border between West and non-West, on the one side enjoying the benefits of Western culture, on the other feeling exploited as victims of imperialism. Indigenous populations always remained backward and dependent, unable to match the resources and skills of a fast-advancing West.

What we should weigh, then, in any assessment of Western colonial expansion before World War I is perhaps not only the actions, good or evil, of the colonial powers, but also the long-run consequences thereafter. The victims of Western colonialism do not include only the casualties of colonial wars but also the far greater multitudes killed or brutalized in the civil commotions in the emerging modern nation-states. Whatever the mitigating circumstances, the anti-Western fury has its justifications indeed.

And yet, in the all-inclusive global perspective, is it morally justified? Was the outreach with all its outrages planned by the Westerners? Was it based on a deliberate design of conquest? Or was it the accidental result of stark imbalances in the resources of power for both war and peace which had come about through circumstances beyond human control? Why were the Westerners so powerful? Their stock answer has been: because of their ideals embedded in their religion, culture, and political institutions, adding up to their overwhelming material superiority. That answer, however, will not suffice for the overview appropriate to this age. In the enlarged contexts of global interaction human beings appear far more helpless than in their smaller settings, where they may claim a measure of control. As argued above, it was merely by historical and geographic accident that the Europeans were enabled to create the cultural hothouse that made them uniquely powerful in the world.

If we grant the vital premise of human dependence on forces beyond human control, the moral issue in the confrontation of the colonizers and the colonized, of the Westerners and non-Westerners, is elevated to the level of tragedy. In the fateful encounter both sides shared a common ambition: to enlarge their dominion. But the crucial difference, the flagrant inequality in their pursuit of self-affirmation, stemmed not from deliberate human intention but from nonhuman external factors, from the natural variety of geographic and historical circumstances in different parts of the earth. Inequality was a fact predetermined long before the expansion of Europe; its roots extend deep into history.

Thus in the tragic perspective—the only one that can claim a transcendent global validity—all parties in the Great Confluence were drawn innocently—or with the common sin of self-affirmation—into the tensions of the contemporary age. Viewed in this light, the moral obligation—the liberating catharsis—requires diminishing rather than raising the volume of moral indignation. It demands compassionate understanding of the long-range factors responsible for the diversity of human resources as well as recognition of how limited the current insights on all sides of the argument are. Above all, it calls for reduced expectations and infinitely greater patience. The accidental precolonial heritage of inequality cannot be set right within a century, especially as many divisive geographic factors persist.

Yet the tragic perspective still adds up to a moral challenge. As we now see the grand connections more clearly, we also understand that the burden of responsibility for bringing about cultural equality falls more heavily on those who have been so privileged, so spoiled, by circumstances beyond their control. They have furnished the energies behind the world revolution of Westernization; they carry the obligation to complete it according to their ideals of freedom, equality, and human dignity and in a manner beneficial to all humanity.

Back then to earth, to the realities of that revolution; back to Europe, where at the eve of World War I the locomotives of world history were gathering speed. A close-up view of Europe provides further insight into the dynamics of that revolution.

4

The Revolution at Work
in Europe

I

Thus far the capacious term "the West" has been used rather loosely, though not without justification. By the experience of peoples around the world, it indeed possessed a concrete meaning. Seen from Africa or Asia, or even from Russia (where the term first assumed its present connotation), the difference between the Westerners themselves hardly counted as compared with the differences between them and the indigenous populations. Yet at this point a more precise focus becomes necessary, revealing both that "the West" was hardly a monocultural entity and that its internal divisions played a significant role in its revolutionary outreach.

More precisely, then, in the days of Lord Lytton up to the outbreak of World War I, "the West" meant Europe; despite its expansiveness, the United States remained on the sidelines into the 20th century. And Europe meant a collection of nation-states coexisting in a keenly competitive balance of power in which the biggest, called Great Powers, stood out. Nations and nation-states in the European state system possessed a concrete reality, most obviously so in case of war. National loyalty in practice as well as an ideal surpassed all rival loyalties, as, for instance, membership in a social class or allegiance to a common European heritage shaped by classical antiquity and Christianity. What mattered most was state power as exercised by national governments in their relations with other governments.

Among the Great Powers two towered over the others as models of power and prestige: France and Great Britain—France from the late 17th to the early 19th century, Britain for a century thereafter; both of them located on the Atlantic, both west of their chief continental rivals. France supplied the model of the centralized powerful nation-state as well as of *civilisation*, the source of the Enlightenment, the intellectual foundation of modern Euro-

35

pean philosophy and political theory. In the course of its revolution of 1789, it pioneered in peace and war the essentials of liberalism and democracy, with spin-offs subsequently into socialist theory. As an externally more secure and therefore internally more peaceable polity Britain added to the accomplishments of parlimentary government the material wealth created by the industrial revolution, by sea power, and by world-spanning empire.

Whatever the differences between these two countries, whatever their internal flaws, both shared the unique privilege of never submitting to the dictates of an alien culture. In their exceptional freedom of self-determination they set the ideal of power and good taste for rulers and their aspiring subjects throughout the continent. French and English became the preferred languages of the social elites, reaching into the depths of Eurasia.

Invidious comparison of wealth and power had been part of the competitiveness built into European history and culture. The winners in this competition—and Britain foremost in the 19th century—were responsible for a continuous escalation of political ambition. As shown above, by 1871 that competitiveness had risen to a new height. The struggle for supremacy within Europe was now deliberately expanded around the world and, perhaps more ominously, given extreme forms: it was either world-spanning empire or historic insignificance. The British vision of domination left its mark around Europe and the world: the late 19th century witnessed the culmination of the expansion of Europe.

Seen thus in context, by 1900 the world revolution of Westernization was more than ever the product of the competition among the Great Powers of Europe; the central contest was not between the Europeans and the other peoples of the world, but among the Europeans themselves. Compared with the arms race and preparations of war raging within Europe, imperialist expansion was but a sideshow, often conducted, despite all nationalist verbiage, by private interests at the edge of government. Thus the world was unified in a crescendo of European rivalry; competition for power, preparation for war, and war itself were the dynamos behind the revolutionary outreach into the world at large.

In that fierce competition the two Atlantic states France and England were pacesetters, the others ambiguous imitators, showing both admiration and resentment in the enforced cultural dependence. That difference struck deep, far deeper than appears in the specialized and insufficiently comparative accounts of European history. The difference set up a challenge-and-response relationship within Europe which affected state, society, the sense of national identity, and even the individual ego always in search of opportunities for enlargement.

The winners in the competition of invidious comparison suffered no discomfort. They proudly glorified their traditions and systems of government which had carried them to the top, assigning to their own privileged experience a universal significance by which to measure the entire world. Thanks to their external security and political preeminence they had been able, through the centuries, to follow their own paths, developing, at their own pace and in reasonable congruence, their institutions and basic values—and even their revolutions—in the free interplay of domestic forces. Secure in their ways, they had borrowed from abroad what suited them, easily absorbing alien accomplishments. Their acknowledged preeminence filled their citizens with pride; it thereby cemented loyalty to their polity, resolving or mitigating their not inconsiderable domestic tensions; it built goodwill and consensus, which despite all social discord made the emerging liberal-democratic institutions prosper—yet another proof of their superiority.

The Western preeminence of power was built invisibly into the liberal creed which, especially in its genial British variety, set the tone for all of Europe (if not the world). Taking the good fortune for granted, liberals forever downplayed or even disregarded the facts of raw power, preferring to think of human relations in terms of peaceful cooperation and compromise. They believed in the essential goodness of man, projecting their humanitarian values around the world: the Western record of "progress" heralded the evolution of all humanity; the Western sense of human nature defined human nature everywhere.

The key ideal of liberalism was freedom naively abstracted from its underpinnings and universalized. When encountering contrary practices in societies which had not shared their uniquely privileged cultural breeding, Western liberals were nonplussed, impotent, or forced to stray from their convictions in the name of national security. Outside their sanctuary they encountered doubts, disagreements, or even failure, without, however, losing their faith. Never forced to question the underlying assumptions of their society even in their sharpest self-criticism, they possessed no insight into the invisible substructures of individual and collective discipline which promoted their achievements and prevented easy assimilation abroad.

II

Consider, by contrast, the losers in the comparison, who were compelled to copy the winning ways of England and France in order to hold their own. Of copying for fear of falling behind there was no end: the nation-state, industrialism, big business, constitutional government, middle-class culture, freedom of the press, a high standard of living, empire, and more. Yet what of the traditions displaced by these innovations? What of the humiliations of taking rather than giving directions of collective development? The foreign imports raised doubts about collective identity, fragmented opinion, and undermined confidence in the existing system of government. The nagging sense of inferiority magnified all ills in the body politic, stimulating hypercritical or subversive attitudes. In addition, the visible structures of traditional society and its invisible substructures in the depths of the human psyche did not fit the new order drifting in from western Europe; the discrepancy caused endless disorientation.

Seen in this light, the revolution of Westernization began in Europe itself, as the advanced ways of western Europe descended, irresistibly and at a fast clip, down the cultural slope into central, southern, southeastern, and eastern Europe, into the fringe lands of the continent, as it also spilled overseas into the non-European world.

East of the Rhine River, along the West-East slope—the main axis of receptive resistance (or resisting reception)—the first people to suffer were the inhabitants of the politically weak and divided Germany. A land of barbarians at the height of the Roman Empire, a frontier for centuries thereafter, subsequently drawn into the mainstream of early European culture (though never entirely), it was in the 18th century a part of Europe yet also a cultural and political hinterland under the sway of French power and *civilisation.* Here intellectuals and artists—the guardians of creative spontaneity—began to protest the alien influences that did not express their innermost feelings.

In self-defense, paradoxically inspired by the French *philosophe* Jean-Jacques Rousseau, they searched for the true qualities of the German "soul." From their efforts eventually sprang a "German ideology" defining the essence of Germanness against the foreign intruders. It was said to possess greater spiritual depth—more music—than the materialist or cerebral Western imports, and more insight. Was it not true, as Hegel had said, that the servant, knowing both his role and that of his master, is wiser than the master, who only knows his own? Similarly, Germans under Western influence claimed greater knowledge of human reality than their more powerful neighbors; they were superior cosmopolitans just because they were Germans.

And, as Hegel's example showed, they craved comprehensive philosophical systems to synthesize the discordant worlds and to create stability in their own cultural fracture zone where "community" never harmonized with the fast-moving modern "society" in which one had to live. Hegel's philosophical system, incidentally, revealed yet another German response to Western influence, a yearning for a strong state. As envious outsiders, Germans magnified the significance of an institution which the insiders took more or less for granted; they substituted a potent concept for a reality that thus far had escaped them. England and France indeed were powerful states; the Hegelian concept of the state was even more powerful because it was yet to create such a state.

Another crucial strategy of German cultural defense was an attack on the claim to universality inherent in the presumption of Western superiority. German philosophers preached cultural relativism, pleading the legitimacy of cultural otherness. Each people, so Johann Gottfried Herder argued, possessed not only a unique collective spirit but also an inherent right to it. Ever thereafter the romantic glorification and idealization of indigenous roots has been an integral part of the Westernizing counterrevolutions set off by the revolution of Westernization. All definitions of national or cultural identity among the victims were drawn up to match the Western challenge; they relied on Western cultural stimuli, reflecting, rejecting, and thereby assimilating them; the stereotyped rejections were merely another form of imitation.

In time the relativist prescription became a ready tool for cultural self-defense of Westernized non-Westerners. With its help Western conditions were critically judged by indigenous standards, the criticism bolstered from the ample funds of Western self-criticism easily accessible to outsiders. Yet, while undermining the Western claim to superiority the critics did not resist the temptation, forever insinuated by the Western models, to universalize in turn the uniqueness of their own culture. Again a German philosopher set the example. When the German lands were overrun by Napoleon's armies, Johann Gottlieb Fichte asserted in his *Addresses to the German Nation* (a nation which did not politically exist at the time) that the unique cultural capacities of his compatriots held the key to a complete regeneration of the human race. "If you go under," he thundered, "all humanity goes down too."[1]

At the nearest halfway point between West and East, figuratively as well as geographically, the peoples of Germany were the first to formulate the essential arguments of self-defense. Others learned from them, first the Russian intelligentsia, then Westernized intellectuals in other parts of the world. Each group in turn considered its own cultural inheritance to be profounder in depth and of brighter soul-force, the more so perhaps the greater the cultural distance from Europe. Thus in the dialectics of cultural interaction we observe a West-East progression in which the claims to spiritual superiority steadily increase. Seen from points further downslope, of course, the Germans again became "Westerners," proud models passing their own superior culture into Russia, the Balkans, and points beyond, where German efficiency was both admired and distrusted; here even the Germans were found wanting in "soul." How could it have been otherwise? In all lands only the familiar communicates with the subconscious; the imported innovations are always cerebral and superficial.

Neither the German claim to cultural uniqueness nor their "German ideology" prevented the continued influx of subversive innovations from the West. Although after 1871 united Germany was the most powerful country on the continent, it lacked the cultural integrity—as well as the world-spanning empire—that went with the the British model. Unification, sudden industrialization, urbanization, and a rapid population increase indeed intensified the cultural malaise, sometimes to the point of cultural despair. Modernity did not satisfy the deepest longings of articulate Germans. Outwardly perhaps the new Germany was a model of respectability. Its leadership in the natural sciences was widely acknowledged. In the social sciences, especially sociology and psychology, cultural dissonance proved a decisive asset, broadening insights by extending the range of conscious sensibilities. Yet the body politic remained fragmented, human minds disoriented and restless, the search for panaceas unrewarded.

Perhaps salvation could come, some Germans thought, only when their country had gained its due place in the sun, matching the British Empire. Or would they be saved by the spiritual depths of the East? Influential thinkers probed for a genuinely German pattern of politics, drawing their inspiration from the distant past; Nietzsche called for a transvaluation of all values; expressionist artists yearned for an apocalypse leading forward to an instinct-bound primitive simplicity (and that when the collective order supporting them and their country became ever more complex!).

Throughout we find an undercurrent of frustrated nationalism born from envy of Germany's western neighbors, who had long perfected their unity and power. Characteristically, that nationalism was self-conscious, deliberate, and willful. Whereas in the Western models national coherence maturing over centuries had unconsciously implanted a common outlook, in Germany such unity had to be reproduced artifically and imposed, by an act of conscious will, upon a lively substratum of particularism and rebellious individualism. And more significantly; that nationalism was crassly power-oriented.

The Prussian tradition dominating the newly united Germany perpetu-ated the statecraft of the first down-slope experiment in developing the resources of a small and backward territory by an act of political will. Shaped under adversity by the models of Western superiority (France foremost) and by a Western Calvinist rationality and self-discipline pioneered in Holland, the Prussian experiment glorified the state as the preferred instrument for power-focused imitation. The state took the place of the spontaneous public-spirited cooperative community that provided the bases for Western power (especially in English-speaking countries). The essence of that state, more-over, was military power. How else could a raw newcomer like Prussia estab-lish itself in the competitive state system of Europe?

How else could the united Germany establish itself in the emerging world system? By the end of the 19th century military power, pressed into the mar-row of German society, was projected into the future. As the sociologist Max Weber observed at the start of his academic career (1895): "We move with frightening rapidity towards the point in time when [Germany's share in the world market] will be decided by might, by raw violence. . . . In this fight vic-tory will go to the strongest."[2] That contention, imprinted by the British example upon the most alert contemporaries, left its mark. Germans with sensitive or insecure but ambitious egos were tempted, henceforth, to define their role and that of their country in the largest contexts of *Weltpolitik.* Rab-ble-rousers of the Pan-German League combined that ambition, often expressed after the manner of some Anglo-Saxon social Darwinists in racist language, with anti-Semitism. Among those who listened was a young would-be artist living in Munich, Adolf Hitler.

III

More important in the long run, perhaps, was another German response to British power: the militant revolutionary ideology propounded by the "red Prussian" Karl Marx. An uprooted German intellectual from a family barely one generation out of the ghetto, Marx shifted from Prussia through Paris to London, the hub of the world revolution of Westernization. Remaining true to his central European and preindustrial instincts, he romanticized the French revolutionary tradition and applied it to industrial and imperial England, to the "capitalist system." A permanent outsider amidst Britain's wealth and power, yet benefiting from its freedom and worldwide perspectives, Marx felt that (in the words of Erich Fromm) under "capitalism" "man is made to be a person who *has* much, . . . but who *is* little"[3]—at least by the defensive responses of a romantic from central Europe whose ego was insufficiently socialized for the extensive individual involvement common to Western (or "capitalist") society. Marx wanted to make the unreformed and proud preindustrial—the proletarian—man *big* in the expanded setting of worldwide industrialism.

In this endeavor he prophetically articulated an experience common to all victims of Westernization. For unprepared people any drastic enlargement of their social and political framework causes a feeling of disorientation and loss. In the sudden decompression of communal life human relations and the relations of people with their tools, hitherto close and meaningful, become abstract and impersonal; small-scale comprehension in a large-scale world turns human beings and their belongings into lifeless "commodities"; human relations become "reified." And wage earners who are over-suddenly drawn into uncomprehended work patterns feel "exploited."

The preindustrial values of German intellectuals remained much alive in Marx's mind. They prompted the precarious psychointellectual mechanism of somehow coming to grips with an uncongenial yet obviously inescapable or even admired reality through the concept of "alienation." They also permeated the closed world of Marxist dialectics with its Hegelian ambition for all-inclusive theorizing and ready coinage of capacious but dubious concepts useful for the simplifications—Burckhardt's "terrifying over-simplifications"—needed for modern mass politics. The preindustrial values, moreover, shaped his critique of "capitalism" and his attacks on its callous disregard for the welfare of the working masses. They inspired his vision of a "proletariat" creating the superior social order of "communism." Under "communism," the highest perfection of industrial productivity—paradoxically achieved without a division of labor—allowed the fullest realization of the human potential as envisaged by German romantics (embellished by reports of the untrammelled life on the American frontier).

Yet consider the ambiguities in Marx's concept of the "proletariat." Exploited, impoverished, dehumanized, it nevertheless would rise by its immanent spiritual virtues from the "capitalist system" as a unique historic force destined to save humanity, its self-confidence boosted by the historic inevitability posited by Marxist dialectics. When the internal tensions of "capitalism" reached the explosion point, the "proletariat" would emerge victorious, advancing society and culture to a higher social order. Yet how would the looser norms of small-scale preindustrial society implicit in Marx's thought fit into the necessarily much tighter discipline of "communism"?

Characteristically, Marx's ambiguities also sharpened his vision. He was an exceptionally keen observer of contemporary trends. A generation before Lord Lytton he prophetically outlined essential aspects of the world revolution of Westernization. In the *Communist Manifesto* (1848) he argued that "capitalist" western Europe created "a world after its own image," drawing "even the most barbarian nations into civilization" (a civilization whose validity in the same breath he had denounced as "bourgeois" civilization). Worldwide "capitalism" pitilessly tore asunder all established traditions, which he denigrated as "feudal"; by revolutionizing "the instruments of production" it also revolutionized "the whole relations of society" in Europe and around the world. It produced "uninterrupted disturbance of all social conditions, everlasting uncertainties and agitation. . . . All fixed, fast-frozen relations, with their train of ancient and venerable prejudices and opinions, are swept away, all new-formed ones become antiquated before they can ossify. All that is solid melts into air, all that is holy is profaned. . . ."[4]

Prompted by these insights Marx proved an ambiguous supporter of Western expansion. On the one hand, he affirmed in 1853 that "England has to fulfill a double mission in India, one destructive, the other regenerative— the annihilation of the old Asiatic society, and the laying of the material foundations of Western society in Asia." India, he approvingly predicted, would soon be "annexed to the Western world." On the other hand, he could not hold back his sympathies for the victims of colonialism: "The devastating effects of English industry when contemplated with regard to India . . . are palpable and confounding." He denounced the "profound hypocrisy and inherent barbarism of bourgeois civilization," which exploited "the gentle natives . . . whose submission is counter-balanced by a certain calm nobility . . . whose country has been the source of our language, our religions, and who represent the type of the ancient German in the Jat and the type of the ancient Greek in the Brahmin." In that sentiment a German intellectual's respect for ancient Greeks and ancient Germans was combined with a German romantic's compassion for the oppressed. Never mind the approval with which Marx had just welcomed "the annihilation of old Asiatic society."[5]

Marx's emphasis on the material factors in historical evolution testified to the powerful impression made upon him, the refugee from a "less developed," less socialized part of Europe, by the external aspects of Western prosperity and power. Yet, with his outsider-insider's inclusiveness of observation, he also laid bare the pervasiveness of material and social self-interest in all human activities, a fact overlooked (or covered up) by traditional political theory or idealist philosophy; as a sociologist he introduced vital insights into sociopolitical analysis. Unfortunately, his self-righteous "materialism" blinded him to the far larger range of subliminal factors at work in historical development. What was at issue was not the class-bound "capitalist system" but the infinitely more complex "cultural system" evolved under the most privileged conditions of western Europe, especially in what here has been called "the model" countries; Marx the outsider never understood the dynamics of "capitalism." That much-touted Marxist materialism was but a modish disguise for German idealist intellectualism. Communist society was a utopian goal for revolutionary idealists, especially attractive where sociocultural conditions were less advanced. And Marxist analysis in general remained a high-voltage cerebral exercise in academic social analysis.

Yet, however flawed, Marxism became an influential force in the revolution of Westernization. It offered many gifts: intellectual vigor, sociological sophistication, political activism combined with a Prussian stress on organization, a reassuring historical determinism, an enticing though simplistic globalism, and a powerful moral charge. It carried a message for all the other alienated minds caught in the same predicament: forced to adopt the resources of advanced urban-industrial society yet inwardly rejecting that society and the tight social discipline which, invisible to both insiders and outsiders, supported it. Above all, it supplied a clearly defined enemy, the "capitalist system," and a program for overthrowing it. Because its romanticism was unromantically expressed in revolutionary language derived from the Enlightenment, Marxism was never popular among culturally self-conscious German intellectuals; but it made a profound impression on the Russian intelligentsia and, through it, on the non-Western world.

IV

Russia, of course, stands out as the foremost European victim of the revolution of Westernization. It had long been a cultural hinterland, first of the Byzantine Empire, later of Europe—"the West," as Russians called it by the 19th century. Russians had always been defeated in invidious comparison with the prestigious Western metropolitan centers. Their country, half-European, half-Asian, remained poor, uncivilized, self-consciously weak in power, and defensively expansionist (a point to be elaborated in subsequent chapters).

Invidious comparison with the West always threatened to take the drastic form of military defeat, forcing the government—an autocratic government compelled to combat popular ignorance—to copy forever the resources of European power, "Westernizing" the country in spurts of political and social mobilization considerably more authoritarian than had been the case in Prussia, one of Russia's Western models. Taken as a whole, the tsarist reforms amounted to a veritable permanent revolution from above, conducted by the government in its Westernized capital, Saint Petersburg, against an uncomprehending and hostile population. In the name of defending the political and spiritual identity of Holy Russia, it copied Western methods, desperately trying to create a semblance of the external security and internal continuity responsible for the high culture of western Europe.

The constructive work of the tsars, however, was forever threatened from two sources of trouble. The unceasing revolution from above created endless ill will and rebellion among their subjects forced into an alien discipline of state service. Equally important, the tsars could never prevent invidious comparison between their country and "the West." The West was always more civilized, more humane, more free. In the comparisons of military power the Russian empire managed to hold its own (though with occasional sharp setbacks), but in the invidious comparison over the quality of life it was invariably defeated.

The tsars, however, did not submit passively to such humiliation. Since the

15th century they had claimed superiority by arguing that their country, not the Catholic West, represented true Christianity. In the early 19th century Nicholas I propagated a more elaborate version of that claim called "official nationality," an ideology proving Russia's preeminence because it represented Christian Orthodoxy, Tsarist Autocracy, and Russian Nationality, three solid bulwarks against the political and cultural decadence of the West; that ideology lasted virtually until 1917. Yet it did not provide victory in invidious comparison, even though the tsars tried to prevent comparison by censorship and restrictions on communication backed up by raw repression. Yet they could never avoid contact—they needed it for modernizing their power. Inevitably, therefore, the government of Russia was always culturally at war with the West.

From this cultural battlefield arose the Russian intelligentsia, the classic prototype of Westernized non-Western intellectuals around the world. That intelligentsia—self-willed, immensely gifted, heroic in its efforts to escape from its contradictory entanglements, forever in conflict within itself, with the government, and with the imperious foreign influences—offers a wealth of insight into the creative miseries of life and thought in a backward but ambitious country close to the West.

Suffice it to examine one seminal witness, the hero of Dostoyevsky's *Notes from Underground* (1864).[6] That undergroundling is cast in the role of an unsuccessful retired petty official living in St. Petersburg, "the most abstract and premeditated city on earth." Sometimes likened to a mouse, he lives half below the floorboards of Western civilization and half above; yet in his misery he boasts of a uniquely "heightened consciousness."

From below he looks at the undersides of that alien modernity symbolized by London's famous Crystal Palace Exhibit of 1851. He rails at the age of progress as "our negative century," charging that "civilization has made man, if not always more bloodthirsty, at least more viciously, more horribly, bloodthirsty." Its rationality makes him feel like sending "all this reason to hell" and going back to "our stupid ways," which at least allow the freedom of spontaneity. Hinting that what is irrational in one cultural contest may be rational in another, he contends that "twice-two-makes-four" is an insufferable imposition inferior in quality, at least "now and then," to "twice-two-makes-five." Yet, at the same time, looking at himself from above the floorboards of the cultural divide—as a Russian interpreted through Western eyes—he admits that he is merely a guilt-ridden, detestable nonentity whose opinions need not be taken seriously.

The philosophical self-analysis of his misery at the entrance of the mousehole is followed up, in Dostoyevsky's account, by the tale of his aimless and disoriented life. Is it, then, a sad tale? Not quite: the story ends on an upbeat note of defiant self-justification. The undergroundling as anti-hero is more alive than all the self-righteous and docile philistines in their Crystal Palace. Heightened sensitivity and "raised consciousness" rate above the smugness of Western rationality. They create a more inclusive sense of reality, able, not unlike the Marxist vision, to look both above and below the ground floor of Western superiority.

The *Notes from Underground* are sometimes improperly abstracted from their cultural roots as a statement of a universal human condition. Yet they make proper sense only in the contexts of a backward country at the eastern edge of Europe. Admittedly, they neither discuss political power nor offer practical advice on how to make Russia victorious in the ceaseless contests of invidious comparison. Yet, concerned with the ultimate power of assessing reality, they reveal the hidden doubts and the weaknesses, as well as the superior inclusiveness of perspective, that justify the bold defiance toward the Western model.

And what impassioned defiance! Listen to the protest addressed some years earlier by Alexander Herzen, another quintessential Russian *intelligent,* then in exile, to the French historian Jules Michelet:

> . . . the past of you western European peoples serves us as a lesson and that is all; we do not regard ourselves as the executors of your historical testament.
>
> Your doubts we accept, but your faith does not rouse us. For us you are too religious. We share your hatreds, but we do not understand your devotion to what your forefathers have bequeathed to you; we are too downtrodden, too wretched, to be satisfied with a half-freedom. You are held back by caution, restrained by scruples; we have neither caution nor scruples; all we lack at the moment is strength. . . .[7]

The cultural battle over Russian identity and destiny had been turned since the early 19th century into a bitter conflict between the government bent on suppressing subversion and the intelligentsia bent on freedom. In this struggle the boldest intellectuals turned into determined revolutionaries. By the end of the century the tsarist government had come to resemble a police state, brutally suppressing hardened revolutionaries who did not shrink from terror. Escalating their mutual hostility, both the government and the revolutionaries took their cues from western Europe, the government in adopting the latest techniques of police control, the revolutionaries in promoting the most progressive ideas and advanced conspirational methods.

Typically the government, the intellectuals, and the revolutionary movement were split—rather superficially—into "Westernizers" and "Slavophiles." The former hoped for a Russia capable of overtaking the West on its own terms; the latter copied the German model of glorifying cultural roots, "Slavic" roots in their case. Under Western influence some Slavophiles even turned into pan-Slavs, wishing to unite all Slavs as far away as eastern Germany, the Hapsburg lands, and the Balkans. Correspondingly, some ministers of the tsars and some publicists inspired by the British example hoped for territorial expansion into the Far East, India, or even East Africa.

Whatever the ambitious visions, the hard question remained for all contenders for supreme power in the Russian empire: how could the country be made competitive with the western European countries? By widespread agreement Russia was backward. A continent-spanning multinational empire, it lacked social and political cohesion: a small, outwardly Westernized yet still very Russian elite ruled over illiterate, resentful, and ill-assorted peasant masses; it also lacked industry, let along effective governmental institutions. Virtually all educated Russians admitted, in various disguises, that life in the West was better; few respected the existing order, an "Asiatic" order, as they sneered, placing themselves in their self-esteem higher up on the cultural slope. As for the political future of the country, quite a few, indeed, thought they sat on a volcano of popular discontent ready to blow up any moment.

At the beginning of the 20th century the Russian empire was in a parlous state. One former minister of the interior predicted utter anarchy and collapse should Russia be involved in a major war. The existing order was hollowed out by unending defeat in invidious comparison. What indeed in tsarist Russia was worth defending? And yet many Russians shared a belief, however irrational, in a great future for their country, considering themselves no less worthy than the English and other western Europeans of imprinting their sense of reality upon humanity. Did not the Western ideals of liberty, equality, and fraternity forever incite their ambitions? And was not their huge country, with its untapped potential and impregnable faith in itself, entitled to worldwide preeminence?

In 1914 Vladimir Ilich Ulianov, a radical revolutionary from a privileged family, a well-educated political exile living in western Europe but revealingly self-named "Lenin" after the easternmost Siberian river, the Lena, dreamed of anti-Western world revolution. Lenin was an idea-driven Russian patriot to the marrow of his bones, but in updated form; he cast Herzen's repudiation of the West as well as many reactionary anti-Western sentiments into Marxist language and carried the hope for a European socialist revolution, of which the Russian "proletariat" was to be the vanguard, into the world at large. He envisioned a global counterrevolution to the Western outreach led by its victims—victims half admiring the Western model and half rejecting it—by universalizing its message of freedom, equality, and universal fellowship. Revolutionary Russia's allies were to come from all the colonial and semi-colonial lands. In search of a superior Russia, Lenin imposed Marx's ambiguous vision on the ambiguities both of the Russian intelligentsia and of Russian statecraft.

What awesome problems the rulers—any rulers—of the Russian empire faced! Nowhere else did political ambition based on the potential of a vast country and the quest for superiority born of centuries of humiliation clash so brutally with the sordid realities of political weakness. The tsarist empire was a colossus on clay feet, yet endowed through its intelligentsia with an innate urge to triumph over all handicaps. It was spurred on by the immense promise held out by the tormenting Western model as well as by an inkling of the West's weaknesses. Like all other observers—both Western insiders and non-Western outsiders—Russian intellectuals possessed no insight into the hidden substructures that upheld the West at its core; but they—and especially the Marxists among them—were waiting for a further turn in the world revolution of Westernization to make their bid. They had reasons for hope.

V

At its historic peak before 1914—to return now to a consideration of Europe as it was seen by those who considered themselves part of it—European culture appeared to be capable of unlimited progress in making life more civilized by the standards of its most privileged countries. In world perspective, Europe was a lively cultural whole despite the differences here traced, held together by competition whether in the arts of war or of peace (the latter more conspicuous in an era of limited wars); in finance, industry, commerce, literature, the fine arts, science and technology; all thriving under an ever more integrated system of rapid communication. Science and technology especially scored immense gains, seemingly making man the master of all creation. Their advance tempted the human imagination into extravagant visions of human perfection, through worldwide democracy or through "communism" according to Marx. Such optimism was contagious; it spread downslope together with vague expectations of ultimate solutions in the social order or in political empire—or both combined.

At the same time European life and culture, now broadening out across the North Atlantic, were in a state of rapid flux, recklessly piling up major problems of statehood and social solidarity, precariously balancing between peace and war, between international cooperation and national self-assertion at any price. Power competition had now assumed global proportions, escalated into a struggle for world domination, with incipient American participation. As an American clergyman, Josiah Strong, had argued in 1886, a new stage in world history had been reached, "the final competition of races." The Anglo-Saxon race, "having developed peculiarly aggressive traits calculated to impress its institutions upon mankind, will spread itself over the earth."[8] Some years later, in 1901, a noted English biologist, Karl Pearson, had carried the social Darwinist argument even further, declaring that "History shows . . . one way, and one way only, in which the high state of civilization has been produced, namely, the struggle of race with race, and the survival of the physically and mentally fitter race."[9] These sentiments were but chaff in the crosswinds of contemporary opinion, yet they left their mark.

On the whole, nationalism remained moderate among the pacesetters secure in their worldwide superiority. Downslope it was imitative, deliberately fostered as an antidote to the cultural and political subversion that followed the invasion of innovation from the model countries. Everywhere national exclusiveness grew with greater political participation by the bulk of the population, whose ignorance undermined the faint but widespread cosmopolitanism of liberal middle class culture. At a time when worldwide interdependence deepened by leaps and bounds, nation-centered public opinion narrowed the possibilities for peaceful international adjustment.

Under the liberal ideals of democracy and individual freedom domestic politics now involved the bulk of the population. The state became the state of all people, responsible for their welfare. Thus an ever larger burden fell on the administration of government as well as on the individual citizen. Effective democracy required a heightened sense of civic responsibility and individual self-discipline throughout the body politic. At the same time, the division of labor became more minute, creating social divisions which called for a countervailing sense of solidarity, while in the enlarged contacts around the world and the ever greater complexity of urban-industrial society the tempo of life speeded up. Were the run of people capable of peacefully sustaining the evergrowing burden, attending not only to their own busy lives but also to the needs of their country as a whole (let alone those of other countries with whom they were ever more closely associated)?

The answer depended on where in Europe one looked. In insular England, for instance, optimism prevailed; politics and society seemed reasonably cohesive and harmonious (Ireland excepted), by contrast with the Continent. In France violent attitudes were more commonplace, an inspiration for violent people in countries east and south. Elsewhere one heard even louder talk of revolution in response to sharper political repression. Throughout Europe—and again more toward the periphery—the new mobility and cosmopolitanism, combined with the trend toward mass politics, undermined established elites. The rulers of the great monarchies—Germany, Austria-Hungary, and especially Russia—became dangerously alienated from political reality; but small folk suffered too, as the security of rural or small-town association customary in the past gave way to the insecurities of a society open to a wider world. The dislocations caused a pessimistic penchant for anti-social individual self-assertion or even for violence. At a time of heightened unease over individual and collective identity and status, cultural and racial otherness, as of Jews for instance, constituted a major threat, especially when associated with the suspect sources of change: the banks, the railroads, the press, or an expanded rationality in general. Anti-Semitism flourished, foremost in areas of intense sociocultural conflict, as in Vienna, in parts of Russia, or, more mildly, even in France.

Sharp fissures in the body politic were also revealed where socialism in its many forms was a political force. It pitted the working classes against the "bourgeoisie," the established propertied elements in a society dominated, it was argued, by big banks, big industry, big business, all of them beginning to operate around the world. Yet the manual workers in industry hardly constituted a homogeneous revolutionary class; their state of mind was shaped by many conflicting considerations. And even socialist theory at its best, while extending awareness of the complex tensions in the social order, was markedly deficient in understanding the drift of events.

Yet socialism possessed two major assets endearing it to the victims of the revolution of Westernization everywhere. It was anti-establishment, forever dwelling on the model countries' flaws and thereby giving moral assurance to their victims. Second, it put collective welfare above individual self-interest, thereby justifying an extension of state power with benefits to governments whose subjects did not live by the Western democratic discipline of voluntary cooperation. Thus we find in pre-1914 Europe the seeds of a state-centered socialist nationalism well suited for the needs of the recipients of enforced innovation. That prescription combined the expansive universalism of the model countries with a set of substitutes for the lacking cultural skills of downslope peoples.

By a summary glance, Europe taken as a whole, in its economic, social, and governmental institutions together with its intellectual and cultural life, moved toward an uncertain future. Unfinished, imperfect, tension ridden— its privileged classes too self-indulgent and self-satisfied for their own good, its masses held back by drudgery and ignorance—it stood a divided and brittle center in the interdependent world it was creating. And yet it called the tune by which all other peoples around the world had to march, whether they wanted or not.

The pace was set by those western European countries that could count on the reasonably willing cooperation of their citizens in managing all the activities that undergirded their superior standing in the world. Paradoxically, the peoples most highly disciplined individually and collectively for such cooperation considered themselves also the freest, unaware of the restraints bred into their freedoms and critical of all forms of external compulsion observed in less fortunate countries, which they incited to compete with them.

At the eve of World War I two big questions confronted the leaders of Europe: How could the invisible skills of social cooperation which endowed the model countries with their power be passed on to the rest of the world? And how, more broadly, could that power be shared in a shrunk world with its forever escalating political ambition and material interdependence? These central problems were faced not only by imperial administrators like Lord Lytton, but also by political leaders and their followers in Europe itself.

How little they understood and how poorly they were prepared was revealed when the next phase in the world revolution of Westernization dawned upon them. But the evil that now descended upon Europe and the world was not of men (however much they were blamed), but of uncomprehended circumstance. The Western ascendancy around the world had enlarged the scale and intensity of human interaction beyond all available knowledge and understanding, beyond all resources for peaceful control. The world had grown over the heads of people, especially in those parts of Europe (let alone the world) where industrialization, liberal democracy, and empire had entered as alien but inescapable facts of life. The evil men emerging from World War I (all raised before it) were merely symptoms—or even victims— of the larger impersonal and unpremeditated evil caused by the disparity between the enlarged scale of human interaction and the limited human comprehension of the demands now made upon individuals and their societies.

Had the advance of civilization, so proudly hailed before 1914, made man more civilized?

II

World War I and
the Postwar Twenties

August 1914—to take a long view even beyond the time span of this part—
opened a new phase in the world revolution of Westernization running from
the outbreak of World War I to the end of World War II. This phase witnessed a mighty escalation in all aspects of that revolution. The ambitions of
global power became yet more universal and ruthless, the outreach from the
center and the impact on unprepared peripheral societies more intense.
Those thirty-one years were a time of unprecedented turmoil, brutal experimentation, and testing for ultimate solutions. Which Western country was
destined to stand at the head of the revolution? Could any country among
the victims of Westernization within Europe itself dislodge the Western
model, and at what human price? Beyond Europe World War I started rebellious agitation for independence and self-assertion which drove Western ideals and institutions more deeply into indigenous tradition, causing endless
suffering.

In the same years, global interdependence, perilously disregarded yet
facilitated by relentless progress in the technologies of communication, rose
to a new pitch; it demonstrated its reality above all through war. That confused and calamitous era comprised World War I, its aftermath to the Great
Depression, the triumph of totalitarianism in the 1930s, and World War II.
It ended with a drastic recasting of the global balance of power, with a heightened awareness of a common destiny, and with a new (though short-lived)
mood for peaceful worldwide cooperation. The reassertion of Western ideals
and the ascendancy of the United States prepared a still more far-flung revolutionary thrust into non-Western societies and cultures in the years after
World War II.

The starting point for these momentous trends was World War I. It laid
bare the raw forces at work underneath Europe's civilized prosperity and self-

assurance yet also radiated, through the victory of the Western powers, the Western model still further into the world—with unexpected consequences. In November 1917 Lenin's communists seized power in the Russian empire, determined to counter the Western influences over their country by a superior social order copied from Western experience. In 1922 Mussolini laid the groundwork for a fascist dictatorship, while in Germany Hitler rallied a growing party of militant National Socialists; like Lenin, both men imitated Western achievements in the name of an anti-Western ideology. Meanwhile, the colonial peoples of Asia and Africa, stirred by the war and the victors' call for democratic freedom, agitated for self-determination and independence. The troubled postwar decade ended with the Great Depression, which further undermined the global order built up in the 19th century; yet, like the war and the new dictatorships, it helped to strengthen the ascendancy of the Western model.

5

World War I

The 20th century began in earnest during the four murderous years of World War I. The causes of that war—and of the escalation in the West's revolutionary outreach—were embedded in the dynamics of the competitive European state system which had given Europe its worldwide preeminence. After 1871 the European balance of power, gradually extending around the globe, was agitated by the patent ambition of the newly united Germany, allied with the Austro-Hungarian Empire, to claim its place in the sun. The three other Great Powers of Europe took their precautions in an anti-German orientation which made apprehensive Germans feel encircled.

The strains of the balance of power were aggravated within Europe itself by the tensions resulting from the expansion of the Western nation-states and the attractions of urban-industrial life into unprepared parts of the Continent. Even in the model countries established authorities were threatened, down to the traditional verities of religion; the threats were more serious downslope. The agitation for national self-determination and democracy was a disrupting force especially in the monarchical multinational empires of Austria-Hungary and Russia, which dominated eastern Europe. The pervasive unease heightened the sense of national danger and prompted ever greater reliance on armed force as the ultimate security. In the year 1912 military and psychological preparedness for war was boosted in all major European countries; many observers considered war inevitable.

In 1914 war came. When in late June the heir to the throne of Austria-Hungary was assassinated by a pro-Serbian terrorist, his death set off a predictable chain reaction. The Hapsburg government, fearing the collapse of its multinational empire and counting on German support, soon declared war on neighboring Serbia. That action alarmed the Russian government, threatened at that moment with revolution if it did not aid the Serbs, whose cause excited Russian patriots. The alarm of the Russian government also reached into France, Russia's ally, just as the threat to Austria-Hungary aroused German fears. Were Russia and France to let German power move into the Balkans? But was encircled Germany to let its only ally fall apart?

The political chain reaction leading to war was accelerated by an even more rigidly predetermined mechanism: the long-prepared military mobilization plans. The mobilization of Russia's army, activated by the Serbian crisis, had been conceived in anticipation of war with Germany. Thus threatened, Germany invoked its own emergency plan, premised on an inevitable and desperate two-front conflict. For salvation and survival Germany was to knock out France first by marching through Belgium before the Russians could rally their superior numbers (the German generals never considered the political effects of violating Belgium's neutrality). The German government then took "a leap in the dark,"[1] as the imperial chancellor Bethmann-Hollweg put it, by declaring war on Russia and France. Invading Belgium, it found itself at war, a few days later, with Great Britain as well. Thus it was not design for conquest but the militarists' professional preoccupation with "worst case" eventualities (quickly transformed into self-fulfilling prophecies) which prompted Germany to make the preemptive first moves of declaring war and striking beyond its borders. It hardly seemed to matter that thereby it also saddled itself with the onus of being the aggressor.

Thus Europe began its long-expected war. In the exultation of August 1914, battle was hailed by many prominent figures as a liberation from the intolerable burden of political and social conflict or even of cultural despair, all widespread in contemporary life. If only they had known what lay ahead! Liberation was to come through violence, not on the canvasses of expressionist artists but in mangled bodies, ravaged landscapes, and collapsed systems of government. In fact, there had been no lack of warning. As a few far-sighted observers had pointed out, Europe's prosperity and culture, indeed its place in the world, depended on an expanding network of cooperation in all aspects of life. Disarmament had been a lively cause among a minority at odds with the drift of public opinion toward war. Now the logic of war began to teach its own lessons of interdependence.

From the start the war was a "world war" (Bethmann-Hollweg had used that term already a few days before it broke out).[2] French and English colonies around the world were mobilized for service; the autonomous British dominions across the oceans rallied to their motherland, all of them concerned with the control of the high seas; in eastern Europe, Russia drew on the resources of its Eurasian empire. Within a month the war had spread to the Far East, as Japan took the British side, spoiling to acquire the German bases in its vicinity. In late fall of 1914 the Ottoman Empire, afraid of Russia and distrusting Russia's allies, joined the German side. Thus the war extended into the Middle East along the Russo-Turkish border, into Mesopotamia, and to the approaches of the Suez Canal. Meanwhile, in sub-Saharan Africa German-led Africans, some of them undefeated until the end of the war, started to fight their fellow Africans under English or French command.

It mattered little that in 1915 Italy threw its support to "the Allies" (France, England, and Russia) or in 1916 Bulgaria to "the Central Powers" (Germany, Austria-Hungary, and Turkey). What really counted was the entry, in March 1917, of the United States with its resources and influence around the world. It joined the Allies as an "Associated Power," recruiting also a number of Latin American countries (including Brazil). And finally, later in 1917, two more East Asian countries were enlisted on the Allied side, China and Siam, both sending labor battalions to France and other theaters of war, both hoping to gain in international stature. By the end of the war in 1918 the Central Powers, encircled on all sides, were vastly outnumbered by the Allied and Associated Powers, which, in command of the oceans, drew on the resources of the entire globe. The eyes of the world were riveted on Europe, all energies channeled to sustain its battles for victory and world power.

Among the belligerents expansionist ambitions had quickly outgrown the defensive instincts manifest at the start. By 1915 the Germans were united in their desire to match and outdo the British Empire, partly through expansion on the continent, particularly at the expense of Russia. That country was to be partitioned and dominated politically and economically by Germany, fulfilling long-standing expansionist dreams among anti-Western romantics. Further afield, so German diplomats and generals wondered, might it not be possible to enlist the help of the colonial peoples against their British masters, especially in India and Egypt?[3] Unfortunately, the affirmation of German culture could offer little to the peoples of Asia or Africa (though their African soldiers fought valiantly under German command)—the question of harnessing the colonial peoples' desires for liberation to the power struggles of Europe was left for later experiments under different auspices. In any case, Germans fought in order to make their country a world power after the British model; the ambition sustained simple soldiers through four terrible years in the trenches.

On the Allied side, territorial expansion was a lesser (though not negligible) factor. What counted most was the expansion of Western ideals and ways of life; liberal democracy was pitted against militarist monarchy. Especially after the collapse of the tsarist regime in March 1917 and the entry of the United States, the expansion of democracy became the crux of the Western war aims. By declaring that "the world must be made safe for democracy" President Wilson threw his country's immense prestige into the war and transformed the Western war effort into an ideological crusade. The democratic model, shaped after the American experience, was to radiate the boons of peace and prosperity into the entire world.

The American entry into European power politics was the logical result of the rapid expansion of American power and political vision since the 1890s. By the turn of the century Americans had begun to consider their country a major force in world affairs by virtue of its industrial preeminence, its newly acquired possessions in the Pacific, and its far-flung interests in the Far East, the Western Hemisphere, and the Middle East. Subsequently, a more pacific expansionism by the name of "dollar diplomacy" was formulated, preaching the benefits of peaceful worldwide economic and political cooperation. By offering their most civilized ideals for the service of all mankind, both developed and undeveloped, Americans more than ever globalized their nationalism and self-interest, ready now to back up both by participation in the war.

Paradoxically, the crusade for democracy grew out of a war conducted on all sides with the most illiberal ruthlessness and inhuman determination. Within a short time the war absorbed all human energies. Persistent ammunition shortages, unforeseen by the military experts, forced the mobilization of industry, which soon turned into pervasive control of all economic activities by the state. In Germany, which suffered the greatest shortages, wartime planning led to what was called "war socialism," the subordination of all private economic activities to the insatiable demands of the war. The mobilization of manpower took a similar course; by 1916 even the British had introduced compulsory universal military service for all men. As the men went to battle, the women took their place in industry and agriculture. Both men and women increasingly found themselves under a barrage of war propaganda, their sources of information as well as their private correspondence subject to censorship. Disloyalty or refusal to serve became a criminal offense. Political agitation in any form was suspended for the duration. Religion was drafted for war, and God degraded to a fighting compatriot. As the battles grew fiercer, every source of moral energy was tapped—the war became a contest between good and evil.

At the same time, the weapons used in battle became more ferocious. From the start the machine guns had mowed down human bodies like grass. In 1915 the Germans introduced poison gas in an effort to accomplish the breakthrough envisaged in their mobilization plans. The following year saw the introduction of the tank and the flamethrower. The Germans declared unrestricted submarine warfare, torpedoing all ships in enemy waters for the purpose of starving the British Isles into submission. The Allies had long blockaded Germany for the same purpose; in the winter of 1916–17 hunger struck hard. Thus civilians too became victims of war in another escalation of violence, some most directly when airplanes, another new instrument of war, began dropping bombs on open cities. All of these innovations unleashed more anger, more determination to prevail by redoubled human sacrifice.

Casuality rates had been exorbitant from the start; by 1916 they had become appalling. On the opening day of the battle of the Somme the British alone lost sixty thousand men. The even grimmer and more protracted battle of Verdun, then grinding to a halt, had consumed over ten times that number. What could have been more monstrous than the design of General von Falkenhayn, the planner of that battle, to bleed the French army to death? Or, rather, what could have been more appalling than his desperation, born of the realization that Germany had no other choice if it wanted to escape defeat? Thus the war assumed the proportions of a struggle for ultimate survival. What did the individual count when the fate of the nation was at stake? That logic was implanted into the depths of men's minds for decades to come.

Violence became glorified together with the most modern technology of death. By 1917 the war had become total war. At the same time, its effects began to show.

Inevitably—that word haunts any analysis of the war (as of the entire century)—inevitably not only every citizen but the entire body politic among the European belligerents was strained beyond any previous experience. How soon under the onslaught of hunger, destruction, and death would the bonds of citizenship, so feverishly boosted by war propaganda, snap and governments bent on victory collapse? And what form would the revulsion of the brutalized people take?

By 1917 the war began to claim its toll. In March tsarist rule collapsed, reducing the Russian empire to military impotence, social revolution, and civil war. Mutiny cropped up in the French army; German socialists called for peace negotiations. Early next year crisis was averted in Italy only because of the certitude of an Allied victory with American help. Then, in the fall of 1918, the Central Powers crumbled. Bulgaria dropped out first, the Ottoman Empire next. The Austro-Hungarian Empire was in dissolution before a formal armistice was declared in early November. A few days later, months after the German generals had recognized their failure, a civilian government accepted defeat. Under the strain of war the monarchies which had ruled eastern and central Europe had dissolved, leaving the many peoples in that area to the agonies of finding more effective governments under the victors' guidance.

The victorious Allied and Associated Powers, now—and for the rest of the century—firmly established in ideological and political solidarity as "the West," stepped into the peace under the glowing banner of Wilsonian idealism proclaiming liberal democracy and self-determination. Drawing on prewar progressive thought in western Europe and the United States, Wilson also projected into international relations the ideal of a transnational organization designed to prevent another world war. Called the "League of Nations," that organization represented the first formal recognition of a human community transcending the nation-state. Would that ideal be compatible with the ideals of democracy and self-determination?

The mixed message was heard, with awe and high expectations, throughout Europe and the lands beyond. It made a profound and lasting impression. It redoubled the West's revolutionary outreach, producing innumerable local revolutions in thought and action for individual and national self-determination, for democracy, for political regeneration. Never yet had the superiority of the Western model been so convincing; never before were its ideals so triumphantly projected upon disoriented people, politically aroused as never before, who looked to the victors' political creed for their own salvation. However misunderstood, "democracy" was a key term in all political innovation arising out of the breakup of the prewar order. It became the dominant ideology of the day (and the entire century), stimulating the political involvement of millions of unprepared peoples stirred up by the war and the raised expectations prompted by extravagant war aims. Mass politics required all-inclusive political visions; it required mass-oriented ideologies, all owing a debt to Wilsonian idealism.

The Western victory, however, taught two other lessons at odds with the ideal of democracy; first, that the prestige of universal power amounts to the capacity for reshaping the world in one's own image. Yet, under the Western slogan of equality, should that capacity be a monopoly only of the Americans (or of the West)? What indeed of equality? The second lesson was even more alarming: the moral justification of superior might, however brutal, is victory. Why were the Americans and their allies able to project their ideals upon the entire world? Because of their superior military power. Power is a factor which the powerful are apt to play down in their assessment of the world, though it impresses most poignantly those who are weak; for the latter it seems a precondition of all else. And as for the true nature of that power, the war's monstrous sacrifices were indelibly impressed upon some hardened witnesses as an unavoidable necessity of statecraft. Under these confused auspices the postwar years began.

The end-of-war excitement soon died down amidst universal exhaustion aggravated by a worldwide influenza epidemic. The victors' crusade for liberal democracy subsided; its ideals lingered on as a powerful stimulant, although its credibility had suffered by too many concessions to political expedience in the peace settlements. The League of Nations, the centerpiece of the Wilsonian vision of avoiding war in the future and another article of faith in the liberal legacy, was emasculated from the start by the American refusal to participate. While continuing to extend their economic and cultural influence throughout the world, Americans were not ready to assume a corresponding political responsibility. The other victors, England and France, were materially and spiritually weakened. Their triumph had reaffirmed their democratic and humanitarian traditions, counteracting the brutalization inevitable in war and purging illiberal intellectual trends like social Darwinism popular before 1914. Yet the revulsion against bloodshed also promoted a pacifist caution in foreign policy; a repetition of such inhumanity was to be avoided at all costs. Their prewar self-confidence sapped, their worldwide prestige undermined, and their militancy chastened—what better news for envious opponents eager to match their well-remembered preeminence?

The real gainers were the outsiders, foremost the United States, but also Japan, now the principal power in the Far East. The British dominions, forced to be more self-reliant and demanding recognition for their contribution to the war, were not willing to return to their former dependence. The colonial empires were still intact, but everywhere non-Westerners appalled by the slaughter had lost faith in the moral superiority of Europe. They became more self-assertive, more than ever guided by the Western ideals of self-determination and national power. In the 1920s the colonial masters themselves, under local pressure, were forced to accelerate the Westernizing trend toward emancipation. Had the war not been fought against repressive militarism and for democracy, self-determination, and equality? Had it not stirred up people's yearning for freedom as never before?

Thus the center of political innovation began to shift away from the established liberal democracies, into the fringe lands of Western culture in Europe and beyond. Vastly intensified by the war, the revolution of Westernization developed its own dynamics, paradoxically in the form of anti-Western revolutions. The crusade for democracy was now pressed by the newcomers in their own manner, beginning in Russia.

6

The Communist Counterrevolution

In March 1917 the three-hundred-year-old rule of the Romanov dynasty collapsed. The last tsar was incompetent, his government unsuited for coping with the rapid pace of change in the world, most of his educated subjects disloyal as a result of unfavorable comparison between their country and western Europe (Germany included). In addition, its material resources, organizational skills, and popular endurance were insufficient for the demands of modern war. Under these conditions, the German army, representing the wartime essence of Western power, easily destroyed the old order, precipitating a political and cultural revolution of unimagined ferocity over the next decades.

To start with, the fall of the monarchy set off, under the continued pressure of war, the long-expected volcanic eruption of popular hostility against the old order and the Westernized upper classes. At last freedom came, as the Russian army, which for so long had upheld the old regime, was defeated. Whatever the perils to their country, the masses of peasants and industrial workers (most of them peasants at heart) now could live by their own lights. In the face of such elemental popular revulsion against all state power the liberal-democratic Provisional Government replacing the tsar was doomed. Instead of producing the consensus necessary for coping with the country's crisis, liberation led to fragmentation, separatism, anarchy, and civil war. The disunity which the state-building tsars had tried to overcome, and had thereby tragically perpetuated, broke now into full view.

Was the country now to fall prey to German expansion? Or was it able, at this moment of utter collapse, of evolving a state structure capable of preserving the country from future catastrophe and of satisfying the Russian quest for equality with the always superior West? Was it possible to create, from the human and material resources of a huge but backward country in the throes of utter disorganization, a super-modern polity capable not only of withstanding subversion by the revolution of Westernization but also of serving as a universal countermodel? Russian Marxists, idealists all, set out to prove that the paradoxical feat was possible.

The designer of the most successful global model of a Westernizing anti-Western counterrevolution was Lenin, presenting an exceptional case of a Westernized non-Westerner's "heightened consciousness" applied to politics. He adapted Marxism to Russian use and devised a persuasive prescription for Russian superiority. His Marxism offered many assets. Its stress on universal working-class solidarity allowed an escape from the divisive nationalism threatening the multinational Russian empire; "proletarian" class consciousness supposedly provided stronger bonds. A "socialist" Russia headed for "communism," furthermore, was superior, by Marxist dialectics at least, to the "capitalist" West (just as by the ideology of "official nationality" or of Slavophilism prerevolutionary Russia had been considered superior to decadent Europe). Leninism offered a conviction counteracting the disabling discouragements of backwardness.

And, better yet, Marxist globalism as developed by a revolutionary exile sensitive to world affairs had transformed the "capitalist" world order into a unified stage for social revolution. Even before the war Lenin had called attention to the revolutionary potential of anti-Western nationalism, as, for instance, in the Chinese revolution of 1911. Thereafter, incited by the universalization of war aims, he had developed a theory of imperialism which tied the European "proletariat" exploited by the war to the colonial and semi-colonial peoples under "capitalist"—or Western—domination. In this perspective the Russian "proletariat" was the vanguard and leader of a global revolution by all the exploited peoples of the world. Lenin's occasional recognition of the continued superiority of contemporary "capitalism" subjected his analysis to some strain, but the political potential of worldwide anti-Western anti-colonial rebellion could not be discounted. The bases had been laid before the war; the war itself, which Lenin had watched from neutral Switzerland, augmented the agitation. Here lay a tempting opportunity for a globally oriented revolutionary Russia (following the example just set by the United States) to universalize its self-interest, compensating for its utter weakness and humiliation. Not the West but socialist Russia, with its Marxist vision of universal happiness under communism, was to bless humanity with freedom, equality, and fraternity.

Lenin's globalism provided an intellectual edge over other Russians competing for the succession to the tsars. Another asset was his grasp of the necessities of mass politics in the Russian setting. Drawing on the experience of the revolutionary movement, he advanced its political skills and ruthless determination. In resisting the tsarist police, so he had argued long before the revolution, an effective revolutionary organization could not afford open democratic organization. True revolutionaries in Russia must rely on an elite of hardened professionals under strict obedience to their leaders, in close touch with the masses yet not organizationally linked with them. Thus he had tried to shape his followers in the Russian Social-Democratic Workers Party, best known as Bolsheviks.

According to Lenin, Bolsheviks also needed an effective revolutionary theory providing a clear grasp of the problems encountered through war and peace in the interaction between Russian and Western cultures; it had to be an authoritative guide through the endless confusion of conflicting cultural stimuli. Indispensable for effective political action, that theory had to be comprehensive and systematic, covering a wide range of factors, including international relations in the age of global interdependence, the necessity of large-scale organization supporting the state, and the backwardness of the country. Within the Marxist-Leninist ideology it was a policy-oriented, flexible tool for coping with the tasks that lay ahead.

But what of practical experience? Tsarist tradition combined with exposure to Western society and politics provided a useful fund of abstract knowledge about the mechanics of power. Yet, as for building a state equal to the West out of the anti-modern anarchy unleashed in 1917, Lenin and his followers could draw neither on precedent nor administrative training. They had to prove themselves pragmatically, step by step, in a protracted series of brutal experiments among "unprecedently large masses of proletarians who have just awakened to political life,"[1] as Lenin said in April 1917. The experiments began in November, when the Bolsheviks, in a minor scuffle of the incipient civil war (subsequently much glorified), seized power from the Provisional Government. The experiments continued through the Stalin era, always prompted by the memories of the war and the traditional (and well-founded) Russian dread of foreign invasion, both political and cultural.

The key problem after the seizure of power was like squaring the circle: how to channel the elemental revolutionary anti-modernism of the aroused masses into building a modern state; how to use the Marxist ideology of revolutionary liberation for fitting the anarchic masses into an ever tighter harness of conformity to large-scale civic cooperation; how, in short, to reculture the recalcitrant and widely diverse peoples of the Russian empire—no Marxist "proletariat" by any stretch of the imagination—into citizens as productive, disciplined, and loyal as the citizens of the Western model. Extreme adversity dogged the Bolshevik experiments from the start: humiliation in a lost war, collapse of government, civil war, poverty, famine, endemic violence, a stubborn anarchist narrowness of temperament and outlook among the people, including the Bolsheviks themselves; and perennial defeat in any invidious comparison with the West against which the theoretical assertion of socialist superiority proved ineffectual.

Against these debilitating odds Lenin expanded the tsarist heritage. From the start Bolshevik rule was an avowed dictatorship led by toughened revolutionaries under party discipline. It tolerated no opposition, preached strict conformity and tight organization. "With clenched teeth," Lenin said in March 1918 after concluding a disastrous peace with Germany, he and his party were laying "the firm foundation of socialist society stone by stone," working "with might and main to establish discipline and self-discipline," introducing comprehensive controls for the sake of building up "military might and socialist might."[2] He readily resorted to terror, promising to "cleanse the land of Russia of all sorts of harmful insects, of crook-fleas, or bedbugs"—his terms for the enemies of his regime (among whom he counted all opponents of modernization, including the Orthodox Church and all religion).

His "dictatorship of the proletariat," he maintained, was more democratic than liberal democracy, where elections were decided by money. It was exercised in the name of the liberated masses—including women too—now politically organized in councils (or "soviets"). These soviets had spontaneously sprung up in many communities after the fall of the monarchy; they were now taken in hand by the Bolsheviks as the political foundation of the "soviet system." The Bolsheviks, who soon called themselves communists (following Marx's example), spared no effort to increase their hold over the masses. They had long identified themselves with common people. Now they conformed to them in their roughness of speech and manner, in their callous brutality of political action, and in their stress on working-class social origins for all officials, all the while oversimplifying their Marxism for popular consumption. They also pioneered the essentials of mass politics in rallies, demonstrations, and in the stereotyped language of slogans, posters, and symbols. In order to gain credibility they even fell in with popular prejudices like anti-intellectualism and xenophobia, while stressing proletarian internationalism and the need for mastering revolutionary theory. As revolutionaries they were aware that they could change the masses only by appearing as one with them. For that purpose they always carefully maintained the formalities of democracy, holding elections under constitutions protecting the legal rights of the citizens (but also stressing their duties).

As a "democratic" dictatorship Leninist rule was also socialist. It nationalized trade and industry, trying to put all economic life under central direction, initially in the manner of German "war socialism." All individuals, liberated from past oppression, were to be integrated into larger units, at work, in local communities, in the state, and ultimately in the worldwide proletarian fellowship. Now they would labor for their common gain rather than for self-interest; the rational guidance of the comprehensive economic plan assured them of steady employment and the promotion of their welfare. Marxist socialism as set forth by Lenin and backed up by Leninist organization and theory thus became a far-flung program for the political mobilization of popular energies among an ignorant preindustrial people. Individual wills had to conform, against all spontaneous impulses, to the collective will in all essentials of state power. While Marxism-Leninism assured the unprepared peoples rushed into the modern world that they would be spared the tensions of competitive "capitalism" and their craving for economic security be respected, they were subjected to massive changes in their lives.

Early in his career, when setting the guidelines for an effective revolutionary party, Lenin had formulated the basic rationale for the Soviet regime: "*What is to a great extent automatic in a politically free country must in Russia be done deliberately and systematically by our organizations.*"[3] What in the Western model was accomplished by a spontaneous subconscious consensus so deep as never even to touch the sense of freedom had to be achieved in Russia under a self-conscious world-spanning ideology. Lenin's communism opened the age of political ideology, of deliberate efforts to create a worldview touching all essentials of political existence and comprehensible to simple folk barely able to make ends meet on the ground floor of life. Ideology had to make sense of a world grown over people's heads.

Equipped with such an artificial, theory-based worldview Lenin went one step further. He aimed at setting up a pervasive system of compulsory compliance with the goals projected by his ideology, compliance backed up if necessary by terror. Subliminal internal command was to be replaced by a rationally guided set of external commands; the political revolution of seizing power had to be followed by a never-ending cultural revolution of vast dimensions. Viewed in this manner, Leninism was a comprehensive theory of substitutionism, imposing a deliberate discipline of all-encompassing cooperation in the place of the invisible cultural conditioning for cooperation that undergirded Western power. It held out a vision of a superior society "scientifically" designed, a monument to the human intellect and a powerful stimulant to ambitious state-builders.

In these insights, Lenin found himself, unwittingly, in the company of Lord Lytton. For the sake of "the supreme law—the safety of the state" he too applied "the most refined principles of European society to a vast . . . population in whose history, habits, and traditions they have had no previous existence." To the run of his people his "communism" was just another of those "mysterious formulas of a foreign and more or less uncongenial system of administration, which is scarcely if at all, intelligible to . . . those for whose benefit it is maintained." He differed, however, from Lord Lytton in two significant respects. He was not part of a "conquering race" trying to reculture colonial subjects into proper compatriots, but an enraged patriot determined to do the job himself. Second, he was willing to use extreme measures under conditions far more extreme than those ever experienced by the British in India (or by previous rulers of Russia), and with greater confidence.

Unaware as any Westernized non-Western intellectual of the limitations imposed by cultural realities, Lenin (at least until he fell sick in 1922 and became more cautious) sincerely expected that once the exploited masses were liberated by socialism, their creative energies would blossom; culture-blind like Marx and inexperienced in the world that had outgrown all available experience, he would not recognize the tenacity of tradition. Swayed by the wartime extravagance of vision he even argued in 1917 that already the Russian people were politically ahead of the Western democracies and would soon overtake them economically as well; backwardness would be turned into superiority.[4] Socialist Russia was to set a superior world model; the Union of Soviet Socialist Republics, envisaged in 1918, was to be the core of a socialist world state. Call it whistling in the dark—in 1918 the prospects for the Bolsheviks were dark indeed. But their shrill confidence, stepping up traditional

Russian nationalism to a new global pitch, was more conducive to national mobilization in an age of world wars than the stale creed of old-fashioned patriots in league with "capitalist" foreigners.

And, as we have seen, Lenin possessed an added trump card: world revolution. At the moment of seizing power he appealed to the proletariat in all warring countries as well as to the colonial and semi-colonial peoples to rise against their "capitalist" masters. The message spread with the war news, falling on willing ears. Wilsonian universalism gave Lenin's plea for world revolution a piggyback ride to prominence.

Lenin's plans for world revolution provided a novel thrust to the revolution of Westernization. He hammered the necessity of global interdependence, together with Western ideals and institutions, more deeply into non-Western awareness right down to the masses. He raised expectations by measuring all revolutionary activities by the Western model which, though constituting the target of communist agitation, still provided the universal yardstick. He advanced "proletarian internationalism" and looked forward to a unified world economy under a common plan; he railed against all manifestations or survivals of "feudalism" (the Marxist term for traditional institutions). Calling for radical change in all economically backward countries— they might even skip the "capitalist" stage of development—he warned especially of the petty bourgeois nationalism that glorified indigenous culture.[5] Yet, while claiming leadership, through the Communist International, in this anti-Western world revolution of Westernization, Lenin still acknowledged the realities of backwardness in Russia and elsewhere in the world. Like Dostoyevsky's anti-hero rising above the floorboards of his patriotism, he admitted, especially during the last months of his life, that only the western European proletariat, the true heir of "capitalism," possessed the qualities needed for building a socialist world. In this respect western Europe still retained a residual ascendancy.

What mattered most for practical politics, however, was the active revolutionary outreach from Russia. Viewed in retrospect, the potential for world revolution was virtually nil. But in the pervasive uneasiness following the war, all prewar stability seemed threatened or shattered. Even in the victorious countries statesmen and governments were afraid. Wilsonianism had promised democratic freedom; Leninism outtrumped it by promising social justice as well. Left-wing socialists, sullen soldiers, emancipated women, angry workers and labor union officials were particularly susceptible to communist propaganda; communist parties arose in all Western countries. Thus at a time of virtual impotence in state power, Lenin created two important sources of support abroad at very little cost: first, out of the widespread rebelliousness following the war and, second, out of the fear among "capitalist" governments which forced them to take notice of the Soviet regime.

In regard to basic perspectives, the Western governments indeed had cause for alarm. Lenin's concept of world revolution categorically challenged the global order evolved for centuries under western European auspices; it was the biggest challenge the West had ever faced. Henceforth Soviet Russia claimed the leadership of all non-Western peoples under Western domination. Whatever Lenin's theoretical reservations, Soviet Russia projected a rival global model endowed with an innate magnetism. In order to stop the revolutionary outreach, the Western powers—England, France, and the United States (plus Japan for its own reasons)—tried to overthrow the new regime by sending troops and arms to its enemies in the civil war. Americans especially reacted with nervous sensitivity; Lenin's ideological arrow had hit the heart of their own universalist aspirations, implanting a permanent sense of insecurity. American diplomats—and not only they—suspected communist intrigue ands connivance wherever American policies encountered resistance. The groundwork was laid for defining the American collective identity in outraged contrast to Soviet communism.

Meanwhile, communist rule endured. After the defeat of Germany it was buoyed by the sparks of revolution struck up in the collapse of monarchical rule in central Europe. Aided by the postwar slump among the victors and the susceptibility of some foreign troops to communist propaganda, it survived foreign military intervention. Eventually the communists also routed their domestic rivals, profiting from their party's hold on the industrial centers and their close identification with the popular revulsion against the tsarist regime. They even beat back, at the conclusion of the civil war, a rebellion of the very groups—workers and soldiers—who had raised them to power in 1917 and now resented the Communist Party's dictatorship over the proletariat.

That rebellion, combined with utter exhaustion and widespread famine, led after 1921 to a temporary relaxation of dictatorial control under the New Economic Policy (NEP). Although the West had come territorially closer to the heart of Russia, through the new Western-supported anti-soviet regimes in Poland and the Baltic states carved out of the Russian empire, the external danger receded. The country, officially proclaimed the Union of Soviet Socialist Republics in January 1924, enjoyed a breathing space needed for sorting out basic problems of leadership and the organization of state power. In the 1920s, amidst continuous disorientation in all aspects of life, the Communist Party made considerable progress in harnessing the reluctant peoples of Russia to the task of "building socialism." In the process it carried its message also to its colonial subjects in Central Asia, evolving under trying conditions an administration capable of integrating them into their state and raising them far above the horizons of their Muslim past. While the Communist Party thus began to rebuild the Russian empire in its own image, its revolutionary outreach, although never called off, was reduced. Opportunities grew less propitious; "world capitalism" recovered after the war.

However uncertain the progress toward "communism" in the 1920s, a basic division was permanently established in world affairs. Henceforth two incompatible world models competed in slowly escalating rivalry. The Western model continued in its superiority because of its traditional geographic and cultural assets. It promised freedom because for its own progress it could rely on the invisible hand not only of the market but also of a cohesive culture which made possible voluntary cooperation in all essential civic activities. The other, the countermodel with its theory-based, self-conscious, and defensively aggressive superiority, offered the same accomplisments: power and prestige in the world, but produced by deliberate acts of will and rational design enforced by organization, command, and even terror, in short, by planned reculturation. The latter model, with all its inherent contradictions, appealed to the ambitious Westernized anti-Western intelligentsia around the world because it offered encouraging perspectives and political techniques for getting even with the West by rising above it.

Unfortunately—or rather: tragically—compulsory reculturation, even when plainly in the collective interest, tends to confuse or even stifle the already inadequate popular capacity for creative initiative. Lenin's prescription suffered from the inherent Marxist flaw: surpassing the leading socio-political order with the resources of a presumable superior—or "socialist"—backwardness, the gap in cultural preparedness for global competition in the essentials of power could not be overcome by Marxist dialectics. Beset by contradictions in its ideology, the Leninist vision was marred also by the miseries of its Russian origins. The utter disorganization of the country, together with the massive cultural and moral disorientation prevailing at the end of the war, led to a profound discrepancy between Marxist-Leninist theory and communist practice; even the Communist Party suffered from the pervasive anarchy and ignorance. Communist rule was further discredited by the tide of seemingly senseless brutality rising from traditional callousness to the stark horrors of the civil war; the winner—any winner—in the contests for power had no choice but to outdo all rivals in ferocity. The hardened insensitivity needed for prevailing in such hell became part of communist working equipment; it discredited the humane vision of communist society. Yet, once they had launched their defiant countermodel, could Soviet leaders ever surrender their ambition?

Their ruthlessness certainly raised powerful limits to their revolutionary outreach. Among the intelligentsia of Asia or Africa eager for liberation it might count for little. But in the more privileged and disciplined polities of Europe or North America it aggravated traditional fears of Russia's anti-European messiansim and its lack of civilized restraints in politics as in life generally. Such fears had long simmered throughout the West; now they became an active political force where the anti-Western counterrevolution set off by the war encountered less extreme conditions. Here communism helped to promote fascism.

7

The Fascist
Counterrevolution

I

In central Europe the anti-Western counterrevolution began soon after the Bolsheviks had established themselves in Russia. Here the cultural contrasts with the countries of western Europe were less obvious; the term "backward" hardly applied. Traditional elites, in the church, the army, in government or society, were not too sharply separated from common folk. Yet here too people lacked the culturally conditioned boons of easy conformity. Industrialism and the modern economy in general were latecomers and a source of sharp social conflict, often suspect because of their implicit internationalism. Disorientation over basic directions in life was widespread; indigenous impulses did not harmonize with the foreign influences. Anti-modernism in various forms was widespread among intellectuals but noticeable also in socialist agitation. While the potential for industrial productivity was relatively high (certainly in the case of Germany), human attitudes were not ready for the subconscious large-scale civic-mindedness required for the fusion of industrialism, liberal democracy, and glory in global politics.

The sacrifices of war, followed by disillusionment and disorganization, had weakened these unprepared societies still further. The run of people, however unused to civic responsibility, were now politically more alert than ever and more volatile. No one had escaped the impact of patriotic propaganda. As the popular press, the cinema, and the radio intensified communication around the world, the common people further enlarged their political vision. In the new globalism Wilsonian and Leninist universalism competed with indigenous tradition. In thrusting vast perspectives on excitable unprepared people, the foreign ideologies offered tempting opportunities for political action and individual self-expression without providing the prerequisite collective discipline. The novel call for democracy encouraged a destabilizing political spontaneity, not as explosive as in Russia but likewise releasing anarchic emotions hitherto inarticulate or repressed. Politics became mass politics and turned ugly. The brutality of war carried over into the peace revived the rawest instincts among people barely touched in their depths by the humanitarianism of the Enlightenment. Although less extreme than in Russia, here too violence was harnessed for political use. Communist ruthlessness in revolution and civil war had quickly become part of the radical imagination everywhere.

After 1918 the communist revolution was an ever present provocation polarizing society. On the one hand, its attack on "capitalism" threatened all established classes, heightening their sense of insecurity and undermining their liberal humanitarianism. Its internationalism ran against the grain of national pride, its atheism against religious conviction. On the other hand, it held out hopes for universal social justice or even, with its semi-religious over-tones, of an eventual earthly paradise. Given the balance of social forces, political convictions, and cultural pride, a Leninist revolution was out of the question. But in those agitated years its disciples and their vision cast a pow-erful spell. A small, if militant, minority encouraged landless peasants to seize land from wealthy landlords or workers to occupy factories and set up soviets. Furthermore, communist practice furnished valuable techniques for ambi-tious political leaders trying to master the novel arts of mass politics. Com-munists could teach a trick or two to an anti-communist movement proclaim-ing an indigenous version of freedom and social justice spiced with visibility if not preeminence in the global power struggle.

Under these conditions liberal democracy could offer no satisfactory vision for dealing with postwar politics. It raised high hopes by advertising political freedom in fine words wrenched from their indigenous moorings; yet by giving voice to every persuasion, to every interest group, it aggravated discord in the body politic. Its freedoms, moreover, seemed lifeless, set into complex constitutional and legal formalities above the heads of simple folk. While failing to express the elemental yearnings of the postwar age, the lib-eral ideals also tied their followers to the victors in the war toward whom nationalists felt no loyalty. The saving prescription had to be anti-communist and anti-liberal; it had to offer an all-inclusive ideology that touched hearts and souls. The experiments leading to an anti-communist and anti-Western revolution of Westernization started the moment the war ended; they claimed their first success in Italy.

II

Here the government, encouraged by the victors' ideology, continued the prewar pattern of liberal democracy within the framework of a constitutional monarchy; it conformed to current trends by extending the suffrage to the entire male population in the first postwar election. Parliamentary practice, however, hardly lived up to its promise. More than in the past the required consensus was obtained by deals behind the scenes. In addition, the authority of government was too limited for the tasks at hand. It could not effectively respond to economic dislocation and unemployment, to rising violence in factories and the countryside, and to bitterness over unfulfilled war aims. While the tedious parliamentary bargaining continued, the experiments for more effective leadership took place on the streets among uprooted intellectuals, ex-soldiers, the unemployed, and restless youths. In this setting Benito Mussolini, the originator of the fascist anti-Western revolution of Westernization, rose to prominence.

A schoolteacher's son from a traditionally tempestuous province, he started as a journalist of radical socialist convictions, an Italian Lenin, yet more open to non-Marxist anti-liberal thought from France and Germany, with a bent for dynamic action for its own sake. After the outbreak of war he was carried away by the tide of nationalism, helping, as a left-wing socialist, to enlist his country on the Allied side. After serving for a time as a soldier and being accidentally wounded behind the front, he turned, more ambitious than ever, to politics in search of followers. An outsider of heightened consciousness, his ears tuned to political crosscurrents running from Lenin's communism to fanatical nationalism, he founded his own party in March 1919. It was a small collection of tiny groups called *Fasci italiani di combattimento.* A *fascio* meant a "bundle," a close-knit team of political activists, in this case of militant black-shirted veterans (also known as *squadristi*) determined to bully their way to power. They soon called themselves fascists.

The fascist program at that time was a medley of fashionable causes, ardently socialist in its demands for social justice, stylishly democratic by advocating women's suffrage, defiantly anti-egalitarian in its elitism, fiercely nationalistic in its assertion of Italy's glory and its demands for territorial gain (denied by the peace conference that formally ended the war), and optimistically futuristic. The wide variety of issues and the mix of ideologies gave Mussolini ample freedom for political maneuver. Experimenting for success under the unprecedented conditions of mass politics, he appealed to young men craving action for its own sake, willing to use whatever elements in liberal, socialist, and democratic doctrine still had a living value. From these ingredients—and the memory of the war—Mussolini conjured up an ideology for people yearning to participate in the grand competition of global politics yet also lost in the postwar confusion.

One of his most creative insights was the realization that democratic freedom in the sudden openness—the global openness—of postwar society produced profound anxieties. The old verities that had (barely) held prewar society together no longer worked; more timely ones had to be provided. It was not only that in the postwar turmoil "the nation stood more in need of authority, direction, and of order" than ever; the people themselves "thirsted for authority, direction, and order" as never before.[1] Fascism was a faith for which men were ready to die; as in war, martyrdom was the ultimate proof of truth. For simple people in uncertain times truths are not established by rational argument but by the intensity of conviction and, ultimately, by self-sacrifice. How many middle-class liberals were willing to offer their lives for their creed?

The escape from freedom thus was a flight from a debilitating insecurity that was not only material but preeminently spiritual—or even metaphysical, as in the case of all cultures subverted by the impact of Westernization. The stabilizing subconscious assurances of continuity and security, so essential for a sense of freedom, had dropped away; they needed replacement as a precondition for constructive citizenship. Only uncomprehending Western liberals, whose world had remained intact (reaffirmed by victory in two world wars), would consider the craving for security an escape. "Freedom" is a more complex cultural molecule than the keenest liberal theorists suspect.

What counted foremost in Mussolini's appeal was the optimistic *élan vital,* the elemental affirmation of conviction and psychological wholeness amidst the paltry legalism of parliamentary government and the disorientation of postwar society generally. Part of that elementary vitality was expressed in raw violence, beating up or even killing socialists and communists, driving them out of elected office in towns and cities. Such action ensured the fascists of conservative support in the church, the army, the government, the monarchy, and among the propertied classes generally. But Mussolini always aimed higher than protection of a stale status quo. Attracting more followers, his experiment continued.

In the summer and fall of 1922 fascist *squadristi* stepped up their drive to seize local control in key areas of northern Italy. In October Mussolini threatened to march on Rome itself. He was never put to the test. To forestall trouble, he was offered the prime ministership by the government in power with the connivance of the army and the king, to whom accommodation seemed the better part of political wisdom. His appeal confirmed by the cowardice of the politicians, Mussolini ruled for two years under the facade of the constitutional monarchy. In early 1925, however, six months after unruly fascists had murdered a prominent socialist, he threw off all restraint by imposing dictatorial rule, thereby completing the fascist model.

Its avowed aim was to create a dynamic heroic society raised above class conflict and joining all people in patriotic fervor for national glory.

> Fascism desires man to be active and engaged in action with all his energies; it wants him to be virilly conscious of existing difficulties and ready to meet them. It considers the fact that life is a battle and that it is man's task to conquer for himself that which is really worthy of him. . . . Fascism does not believe in either the possibility or utility of universal peace. It rejects the pacifism which masks surrender and cowardice. War alone brings all human energies to their highest tension and imprints a seal of nobility on the people who have the virtue to face it.[2]

That determination was expressed through the state. "Fascism asserts the right of the State as expressing the real essence of the individual. And if liberty is to be the attribute of living men and not of abstract dummies invented by individualistic liberalism, then Fascism stands for liberty and for the only liberty worth having, the liberty of the State and of the individual within the state."[3] Considered in this fashion, the fascist state was "the purest form of democracy" expressing the will of the people unified by history.

The state, however, was merely an instrument for a yet higher purpose. It led "to the highest expression of human power, which is empire."[4] For fascism "the growth of empire, that is to say the expansion of the nation, is an essential manifestation of vitality, and its opposite a sign of decadence. . . . Fascism is the doctrine best adapted to represent the tendencies and the aspirations of a people, like the people of Italy, who are rising again after many centuries of abasement and foreign servitude."[5] Obviously, in its expansionism fascism testified to the power of the Western model. Like communism, it was a doctrine of liberation from Western ascendancy that carried essential aspects of the Western model into an unprepared polity.

As an empire-minded doctrine of liberation fascism abandoned "individualistic liberalism" for a new imitative collectivism. Like communism, it practiced a pragmatic substitutionism. Like Lenin, Mussolini was trying to square the circle by transforming an unruly populace—or even his own *squadristi*—into a cohesive and docile polity. "Individualistic" liberty was ruled out; the individual fulfilled himself only as far as his interests coincided with those of an all-embracing "totalitarian" state. That state allowed no individuals or groups to stand apart; it outlawed all opposition. It also transcended all class conflict, in its totality representing "the immanent spirit of the nation."

The fascist state was controlled by the Fascist Party and its parliamentary blackshirted units, the political elite. An Italian and more easygoing version of the Soviet communist one-party state, Mussolini's dictatorship served the same purpose: imposing a single will, if necessary by terror; continuing demonstrations of force were needed for discouraging or even eliminating opponents and asserting the fascist *élan,* its showpiece. The party advertised the common will of all Italians by organizing massive rallies, national festivals, and demonstrations of state power, all staged with the operatic theatricality beloved by the people. In the center at these occasions always stood *Il Duce,* Mussolini himself, the embodiment of the collective vitality, the sole source of inspiration and energy (yet secretly suffering from ulcers), his voice booming from the festooned grandstands into the alleyways of the country's ancient towns and cities. From the start fascism paid much more attention to the personality factor than did Soviet communism (at least until after the death of Lenin). In their simple ways individualistic people shared in the glories and the macho self-assertion of their leader, the supreme individualist, thereby legitimizing his authority and submitting to it.

Forever in search of impressive phrases Mussolini was the first to call his political system "totalitarian." The term highlighted the contrast with liberal democracy, in which the state played only a limited part and contributed, through its competing parties, to civic discord and individual disorientation. Expressing the totality of the people's aspirations, totalitarian fascism promised to rescue them from fragmentation and make them again whole human beings fulfilling themselves heroically in their service to the community embodied in the state. The prescription met with some success. From its inception well into the 1930s the fascist dictatorship enjoyed considerable popularity. The term "totalitarianism," however, was soon turned against Mussolini by his socialist opponents as the epitome of a repressive regime. In that sense it was subsequently employed by the adversaries of both fascism and communism in the age of Hitler and Stalin.

The intensity of Italian totalitarianism, however, remained limited. Admittedly Mussolini pleaded for a stricter collective discipline:

> The Fascist virtues are tenacity at work; the extreme parsimony of gesture and work; physical and moral courage; absolute loyalty in personal relations; firmness in decisions; affection for comrades; hatred for enemies of the Revolution and Fatherland; unlimited faithfulness to an oath that has been taken; respect for tradition, and at the same time the desire of accomplishment for the morrow.[6]

Obviously, even as moderate a form of anti-Western Westernization as Italian fascism required a novel code of individual conduct; traditional Italian spontaneity in "gesture and word" was not compatible with fascist aims. But the impact of such pleading remained limited. Mussolini's regime covered up much inefficiency and continued disunity by a formal show of common purpose. Its chief function was psychological, to endow the country with a common will for self-assertion in foreign affairs, to make Italians proud of their country in the manner of the Western model. Regimentation for political unity required no social revolution. The fascist dictatorship came to terms with the monarchy as well as the Catholic church. The existing social relationships and the class structure persisted, except for making room for the new fascist elite. In economic affairs it introduced the "corporate state." Outlawing socialist labor unions and imposing a few pro-labor restraints on the employers, it stressed the common interests of all participants in economic production, leaving the country's economic order essentially undisturbed. The regime boasted a few projects of economic development, such as draining the Pontine marshes and increasing the wheat crop, but otherwise accepted the inherited "capitalist" order. Italy was not as economically backward as Russia; it required no drastic reculturation.

Nor was it, as one of the lesser victors in the war, under extreme pressure in foreign affairs. Its existence was not threatened and its ambition remained focused on the Mediterranean basin rather than the world as a whole. Yet in ideological competition with the egalitarian universalism of Wilson and Lenin, Mussolini's fascism aspired to an equal visibility. Admittedly, as an expression of Italian national genius and a deliberate counterideology to its universalistic rivals, fascism was not "for export." Eager to call attention to itself, however, it set forth a model of national regeneration that could also be applied elsewhere. (Could any political force rise to significance in the 20th century without arousing universal interest?) Soon an informal fascist international came into existence.

The conditions that had spawned fascism in Italy were at work elsewhere as well. The postwar expansion of Western democracy into central, south-eastern, and eastern Europe, into lands formerly ruled by monarchies, led to division and disunity at a time when the formation of new nation-states demanded effective government. Within a decade authoritarian or even dictatorial regimes, all afraid of Soviet communism, arose under various guises. Many of them eventually conformed to the fascist model; only Czechoslovakia escaped the trend. The smallness of the new states and their preoccupation with their own problems ruled out any independent dreams of universal significance. It was otherwise in the case of Germany. Here the war and the Western victory produced the most potent anti-communist anti-Western revolution of Westernization. That counterrevolution profited from Italian fascism, but grew on its own out of the accumulated tensions in the country's past aggravated by defeat and the overthrow of the monarchy.

III

The defeat of imperial Germany, previously the major challenger of Western ascendancy within Europe, subjected that country to direct Western domination; Germany's defeat and revolution constituted a victory of the world revolution of Westernization. Disarmed, shrunk in territory, its lands west of the Rhine occupied by French and English garrisons for ten years, burdened with reparations which would have lasted (according to the final settlement in 1930) until 1984, postwar Germany presented a humiliating spectacle to a generation expecting global preeminence. The peace forced a drastic change also in the country's political culture. The Weimar constitution was not the product of a spontaneous internal consensus matured over time, but an extension of the victors' political creed, welcomed in the West but flavored with the bitterness of defeat for the Germans. The first election held under its auspices proved its lack of majority support.

Inevitably, the revolution of Westernization took its course, fomenting disunity and rebellion, propelling a layer of the population hitherto silent and passive into politics. The "revolt of the masses," prompted by the impact of the war and democratic ideology, activated the raw underside of German society hitherto covered up by the cultural achievements of an upper stratum which had established the image of a civilized Germany at home and abroad (an underside generally ignored in the histories of modern Germany). In the fervor of democratic postwar politics some people proclaimed separatism, others class war; communists struck at the Weimar regime from the left, more extreme nationalists from the right. Political parties proliferated across the political spectrum. Coalition governments succeeded each other with alarming rapidity, failing for lack of common ground.

Before long, Weimar politics were dominated by four incompatible constituencies, each sporting its own paramilitary organization: the communists under red banners; the Social-Democrats flying the Weimar colors of black, red, and gold; the nationalists sticking to the black, white, and red of the imperial era; and the swastika-bearing National-Socialists. What was lacking was a strong liberal democratic and humanitarian center. The German middle classes, never strong politically, divided by religion and political creeds, shaken by the war, scared now by communism, lost not only their economic security and firmness of their convictions during the disastrous postwar inflation, but also their civilizing and moderating popular appeal.

In any case, what relevance did the Western liberal democratic experience have under German conditions? The Western model furnished no skills for coping with the novel phenomenon of postwar mass politics. Consider the long string of adverse factors at work: the disorientation inherited from the past; the violence of the war and its passionate nationalism; the sudden fall from expected victory to utter defeat; the influx of two powerful alien ideologies: Wilsonian democracy and Leninism; the accentuated globalism promoted by these ideologies and by the new technologies of communication; the political and economic dislocations produced by the war at home, in Europe, and around the world; and the utter incapacity of the newly politicized bulk of the population to cope with this huge overload.

Also consider the people excitedly drawn into this wide-open arena: peasants and ex-peasants, small craftsmen and shopkeepers, white-collar employees, teachers, small-town folk and rootless big-city people, ex-soldiers, and unsettled men and women from all walks of life. Marxist social democracy (but not communism) provided the industrial workers with a reasonably integrated worldview. Conservatives likewise remained loyal to their convictions, as did religious people—especially Catholics. Yet nobody was free of doubt. The great majority, and particularly the young, keenly felt the loss of meaning; hungering for wholeness, they avidly searched for a unifying all-absorbing purpose in an unhinged world.

A cultural earthquake of this magnitude was bound to aggravate all unresolved tensions of the past, including the romantic anti-modernism so prominent before the war, together with anti-Semitism, its eternal companion; from even deeper layers it revived the pagan instincts embodied in myth and folklore. Encased in the trappings of modern science and technology, repudiated on one level of consciousness and absorbed on another, these unsettling impulses produced atrocious hybrids of pseudo-science and inhuman technology (the latter already demonstrated in the war). Enriched with the social Darwinism popular in prewar England and the universalism of globe-spanning power-hungry millenarian ideologies, the heady mix was offered to minds illiterate in the impersonal civic discipline of Western democracy— minds imprisoned in parochial perspectives and fiercely determined to recapture their lost bearings in life. From that utterly discordant yet inevitable combination arose the apocalyptic penchant for final solutions that is the essence of the German variety of fascism.

In that setting Adolf Hitler, an ambitiously imaginative and keen-witted outsider, fashioned an effective political force. Barely one generation removed from peasant life and all through his career alluding to his peasant instincts, a failure at school and an unsuccessful artist yet gifted with a marginal man's heightened consciousness, Hitler was apprenticed to the tensions of urban society in Vienna. The city was a hothouse of unresolved ethnic and cultural conflict, where—to use a Freudian metaphor—Western civilization met its discontent, its baroque pre-Enlightenment and anti-Western denial. Perhaps more a symptom of his time and environment than an independent source of evil, he absorbed there the fanatical nationalism and anti-Semitism that became his hallmark. Forsaking the moribund Austro-Hungarian Empire on the eve of the war, he moved to Munich, where he, pan-German nationalist and inveterate reader of popular political literature, steeped his convictions in the most extreme manifestations of British imperialism as reported in Germany. At the outbreak of the war in August 1914 he was carried away by the spontaneous outburst of patriotic fervor, the only upwelling of civic community recorded in German history—for war rather than civic freedom. Then he served four years in the trenches on the western front, ready to sacrifice his life. In 1918 the cause that had upheld him in battle suddenly and shamefully collapsed.

The shock propelled him into a political career. Soon he became a master in the novel arts of mass politics, taking his cues from many sources, including communist and fascist practice and (most likely) a prophetic book entitled *The Psychology of Crowds* written in 1895 by the Frenchman Gustav LeBon. Suddenly he discovered that by his words he could transform an indifferent audience into "a surging mass full of the holiest indignation and boundless wrath" over Germany's humiliation.[7] In the impersonal vastness of postwar politics his impassioned voice added a personal touch; his brilliant staging of rallies and demonstrations radiated firmness of purpose as well as an assurance of community to all participants. Throughout he capitalized on the common experience of the war and on the German tradition of militarism; his storm troopers were more disciplined than Mussolini's *squadristi*. And more than Mussolini he made fanatical will (toughened by four years in the trenches) into a political asset; it was his ultimate proof for the truth of his message.

Adjusting to the common people (as good liberals never did), he clairvoyantly catered to their deepest yearnings, comparing himself to the saints he had seen in the baroque churches of Austria: "When human hearts break and human souls despair, then from the twilight of the past the great conquerors of disgrace and misery, of spiritual slavery and physical compulsion, look down on them and hold out their eternal hands to the despairing mortals."[8] A virtuoso at psychopolitics, he sensed the widespread mood of uncertainty, of great expectations mixed with profound anxieties. Disregarding the rational arguments of middle-class liberal politicians, he concentrated on the emotional needs of an irresolute postwar generation overburdened with political problems beyond its ken. His patent success as *der Führer* proved the eagerness among the masses for submerging their identity in his. Inevitably, his ego swelled proportionally. Yet in his depths he never overcame his nagging doubt about his qualifications as he stepped on the vast stage of the postwar world. In case of failure at any point he was ready to shoot himself, his personal version of final solutions.

Like Mussolini, Hitler freely borrowed attractive ingredients from current ideological trends. He proclaimed himself a socialist, appealing to the masses and stressing the fact that the welfare of the individual depends on the solidarity of the community. Realizing that the bulk of the German people craved security, he knew that he must come to power nonviolently; he rejected the Marxist call for social revolution. Yet, sensing the popularity of the term "revolution," he also called for a spiritual revolutionizing of the German people, for a transformation at least as gigantic as a transformation in the Marxist sense. He held out vast perspectives, the creation of "a completely new world" that combined socialism with nationalism.

The chief prerequisite for such grand accomplishments (so much in keeping with the war-induced expansion of horizons) was a dynamic national commitment, a patriotic fervor like that of August 1914, but institutionalized and perpetuated until the goal was achieved. For that reason he wanted to "win the masses for a national resurrection," whatever the costs of "the social sacrifice."[9] He admired the victors in the war for their "herd instinct," their sense of unity in time of national danger, hoping to overcome all disorienting division in Germany by a similar "herd solidarity." As a descendent of cattle-breeding peasants, he preferred to think of a national unity in terms of race ("race" in any case was a common term at the time; Englishmen, for instance, spoke of their people as "the British race"; in 1919 their colonial secretary, Lord Milner, described himself as "a British race patriot").[10] According to Hitler, a follower of English social Darwinism and its call for a "conscious race culture," unity could be achieved only by eliminating the spiritual sources of weakness, above all the Jews, the chief purveyors of the debilitating spiritual and cultural disorientation that came with the influx of modern ideas and institutions; anti-Semitism was part of the anti-Western counterrevolution. As a biological determinist—as a racist—he called for a biological cleansing of the German people, eliminating Jews as well as physical misfits. In the typical ambiguity of anti-Western Westernizing intellectuals, his anti-Semitism covered a secret admiration for Jews, for their intellectual abilities as well as for the resilience and sense of righteousness that in the past had carried them through centuries of persecution. Would that the Germans possessed the same ingrained resilience!

His vision of the new world order was set out, like his entire political philosophy, in *Mein Kampf*, a fascinating book if viewed in the contexts of the new mass politics. Its conclusions were shaped by the imperialist expansion of the late 19th century led by Great Britain. In this light he saw the postwar world as "a world of great power states in process of formation."[11] Opportunities were open, the competition keen. "As guardians of the highest humanity on this earth," the Germans had to hold their own, or, as he phrased it in the most extreme Disraelian alternatives: "Germany will either be a world power or there will be no Germany";[12] that was the essence of "the German question." The concept of national living space, already aired in prewar Britain, held a particularly powerful hold over his mind. When he declared that "the most sacred right on this earth is a man's right to have earth to till with his own hands,"[13] he translated not only the national-liberal dream for a German place in the sun but also global politics generally into peasant terms, which appealed also to city dwellers with a romantic attachment to the soil.

Forceful though it was, Hitler's program abounded with the contradictions characteristic of all anti-Western revolutions of Westernization. It combined authoritarianism with mass politics, peasant instincts with global politics, and anti-modernism with a preference for the latest technology (Hitler preferred to be driven in the fastest Mercedes-Benz car and showed a keen interest in the sophisticated weapons developed in World War I). He was a Westernizer of sorts in trying to imitate the fullness of prewar British imperial power, a Wilsonian in coping with the impact of democracy; he was a Marxist of sorts in proclaiming the ideal of social justice and fellowship while holding out the prospect of a revolutionary regeneration. At the same time, he was forced to advocate—and glorify—an anti-Western substitutionism, using the techniques of totalitarianism as outlined by Mussolini and Lenin (though with a peculiarly German twist of efficiency) for creating the "herd instinct" that had provided the model countries with their patent sense of common purpose.

In the case of Germany substitutionism did not have to be as extensive as in Soviet Russia; by contrast with Lenin's Russia, it could count on many assets. The population was more homogeneous; patriotism more deeply entrenched and combined with a strong work ethic and a preference for public order. The country's industrial capacity was high; despite all anti-modernism, the captains of industry enjoyed unusual popularity. Reculturation through substitutionism in the form of national-socialist totalitarianism was limited to the creation of a dynamic collective will deliberately hardened by the vehemence of extremism. The traditional structure of society was to remain largely untouched; the Western models were to be matched through mobilizing existing resources.

Yet Hitler's variety of fascism, though less total than Soviet communism, was more extreme than Mussolini's. The accumulated tensions of centuries of political disunity aggravated by the persistent alienation through foreign influences throbbed more sharply than in Italy; political ambition was more extravagant, the outcome of the war more humiliating, and the expectations for the country's future based on bigger potential resources. Hitler wanted to raise Germany to global preeminence not through a universal ideal as in Wilsonianism or Leninism, but, fascist style, through conquest in the name of a superior national identity.

In the 1920s, to be sure, Hitler's movement remained in the wings. The Weimar Republic survived all postwar crises and, with the help of American investments, seemingly thrived until the beginning of the Great Depression. The pervasive disorientation prompted the "heightened consciousness" expressed in artistic and literary creativity at the edge of Western society. Yet "Weimar culture," often praised in retrospect, contributed little to curing the cultural malaise; indeed it promoted it.

Meanwhile fascism, especially Mussolini's variety, found many admirers even in the more secure countries of the West. It appealed to patriots afraid of communism and of ineffectual government. Its nationalism hardly seemed a threat; its expansionism was turned eastward, most prominently so by Hitler, who revived prewar dreams among German romantics for colonizing the "culturally inferior" Slavic peoples, crushing communism into the bargain.

IV

In this fashion the 1920s witnessed the rise of two Westernizing counterrevolutions based on newly created worldviews (or ideologies) trying to provide a sense of meaning to simple people overwhelmed by the confusions of the postwar age. The communist counterrevolutionary ideology was the product of Russian Eurasia, its fascist counterpart of countries nearer in geography and cultural conditioning to the West, both the products of the revolution of Westernization escalated by the war. The first one, deliberately universalist, appealed to other countries in similar straits, where indigenous culture was unprepared for matching the Western resources for state-building and therefore required drastic reculturation under the leadership of theory-oriented Westernized intellectuals. The other, romantically anti-intellectual, stressed the competitive viability and even superiority of indigenous culture as mobilized by militant nationalists; it offered a hardly less universal model to countries similarly constituted. Both were quintessentially counterrevolutions, shaped differently only because they arose under sharply different conditions. By their ideology they were mutually exclusive, though both anti-Western and always ready to learn from each other as they shared common problems; both tried to match the Western spontaneity of civic cooperation by indoctrination, organization, and command.

They also shared a common willingness to use brutal compulsion for making their recalcitrant subjects more docile, thereby tragically inhibiting civic creativity in the body politic. Their leaders, ignorant of the true sources of Western power and necessarily in a hurry, were always tempted to substitute will power and violence for the lacking civic skills. With the Western model constantly before their eyes, prompted by the Western ideals of equality and self-determination, incited by the arrogance of Western power as well as their own pride, did they have much choice? Letting their confused and divided people drift aimlessly would spell further humiliation. What was needed was not a democratic laissez-faire policy, but a firm hand to restore order and purpose for the sake of building up their countries' strength. Was not the models' external security and preeminence in the world the source of their prosperity and comparative domestic happiness? The spokesmen of the countermodels raised troubling questions about justice and equity in the emerging world order.

8

The Effects of the War
Outside Europe

While the war left its heaviest and most ominous imprint on Europe, its reper-
cussions were felt around the world. What perspectives it opened among
Westernized non-Europeans even at the start may be seen from the words of
a Sudanese Egyptian associated with pan-Africanism and the Universal Races
Congress. The day after the British government declared war on Germany he
wrote, in London:

> We can only watch and pray. Unarmed, undisciplined, disunited we cannot
> strike a blow, we can only wait the event. But whatever that may be, all the
> combatants, the conquerors and the conquered alike, will be exhausted by
> the struggle, and will require years for their recovery, and during that time
> much may be done. Watch and wait! It may be that the non-European races
> will profit by the European disaster. God's ways are mysterious.[1]

Already in the course of the war much was done for the "non-European
races," although not quite in the sense intended by this writer. The Allies
brought men from all parts of the world to serve under their colors, unwittingly Westernizing them and, through their propaganda, teaching them
advanced political views. Paris particularly turned into a center of political
education for Africans and Asians, many of whom became leaders of anti-
Western revolutions. In addition, the war news carried the message of freedom and self-determination around the world, a message subsequently reaffirmed by the promise of the League of Nations to assist in the advance of all
colonial peoples toward self-government. The impact of the war took different forms in different cultural settings, yet everywhere it hastened the revolutionary reculturation under way since the previous century.

I

As already stated, the chief non-Western winner of the war was a thoroughly
Westernized Japan, Britain's ally and now principal heir of Germany's bases

in the Far East. In 1919 its territorial claims and special rights in China, though not its plea for including racial equality among the aims of the League of Nations, were recognized by the Versailles Peace Conference. Three years later, in the Washington Treaty of 1922, it gained equality with British and American naval strength in the Far East (but not around the world). Now Japan stood on its own, the first non-Western power to rise to an established if suspect regional preeminence, responding to Western expansionism with its own ambitions and backed by Western instruments of power. Self-consciously nationalist in its imperial tradition, it could not, however, aspire to a global universalism.

For a time indeed Japan continued under the Western spell. Within the framework of its imperial constitution Wilsonian ideals of universal suffrage and liberal democracy gained ground. As early as February 1919 agitation began in the Japanese Diet over universal suffrage; public demonstrations followed on behalf of the excluded agricultural and industrial workers. Bitter controversy and political riots over this issue continued until the mid-1920s, when at least universal male suffrage was enacted, more than quadrupling the electorate and giving new opportunities to labor unions and widening the spectrum of political groupings. The agitation for universal suffrage coincided with a marked internationalism facilitating participation in world trade and the quiet consolidation of Japan's new global significance; its relations with the Soviet Union and China improved. The young emperor who in 1926 succeeded to the throne was known for his modern convictions; he hoped his reign would bring "enlightened peace."

In the latter twenties, however, the pro-Western trends grew stale as the country struggled with their unsettling consequences. From the early postwar years Western entertainment and lifestyle introduced to Japanese youth, especially from the United States, had shocked their elders by their frivolity and unseemly self-indulgence; they undermined the traditional discipline of Japanese life. Marxism, popular among university students, was also considered subversive. The extension of the suffrage, moreover, rather than improve parliamentary practice, degraded it; voters were shocked by reports of corruption among their representatives. Simultaneously, a banking crisis discredited the big financial concerns, another source of unsettling foreign influence. Meanwhile, ominous clouds gathered on the horizon: at home a rapid population growth and abroad increasing anti-Japanese discrimination. In the United States the Alien Exclusion Act of 1924 and in China anti-Japanese nationalism limited Japanese foreign trade. The country did not share the widespread economic boom of the later 1920s. Cultural disorientation and the anxieties of the times encouraged an anti-Western updating of tradition (as had already happened in Italy).

In this atmosphere a Japanese variant of fascism arose called the "Showa Restoration." It advocated a "revolutionary empire," both national and socialist, under the dictatorial leadership of the emperor's military advisors, dedicated to expansionism. Big business and political parties would be swept away in favor of the restored sovereignty of the emperor. By the end of the decade this vision found much support among young military officers who considered themselves guardians of the national heritage. They needed no new political party, putting their hopes on the army instead. When the Great Depression destroyed the economic foundations of international cooperation on which Japan's future depended, their hour came. In 1931 they began their outward thrust into China, taking advantage of their neighbor's weakness.

II

How exceedingly different and difficult by comparison with Japan was the condition of China at the end of the war! Consider the realities: a huge territory, landlocked and land-oriented for the most part; its huge population (ten times as large as that of Japan, the largest in the world) desperately poor and intensely soil-bound, trying to make ends meet with the help of family, clan, and the petty deities of traditional culture; lacking adequate means of transportation and communication, with virtually no access to, or knowledge of, the outside world that determined their future. Tragically, the country was isolated by the incompatibility between Chinese and Western cultures; the refinements of Chinese culture were of no use in the modern world. Added to that incapacity was a widespread unwillingness to change. Subliminally, the Chinese stuck to their tradition: their country was the Middle Kingdom, all others barbarians—the burden of change fell on them. This arrogance was part of the racial identity cherished by all patriots and perpetuated in defense against the persistent humiliations in all contacts with the barbarian world.

That world—European, American, Russian, Japanese—had succeeded in discrediting the age-old imperial regime by taking away territory, sovereign rights, and "face." It had established, mostly along the country's seaboard, a small minority of merchants, professional experts (including soldiers), intellectuals, all halfway people, ardent patriots, and yet agents of Westernization. With the help of other disaffected elements, these people had overthrown the last emperor in 1911. Yet the republic, proclaimed soon after, represented a central authority only in name; government fell into the hands of local warlords ruling by armed power in traditional style, supported by traditional elites and conducting their own foreign policy. The "real" China was but a product of the imagination among scattered and uprooted intellectuals desperately trying to build bridges between China's past greatness and an uncertain future in a Westernized world. They wielded no power; yet they felt themselves to be the country's conscience.

Soon after the outbreak of war this disjointed China was confronted with Japanese claims to the German-held territory in China and to additional privileges, the famous Twenty-One Demands of 1915. The Chinese response was to proclaim the day of the Japanese ultimatum as National Humiliation Day (all the same, for their own gain some powerful Chinese cooperated with the Japanese). In 1917 China officially joined the war against the Central Powers in the hope of regaining its lost rights. Labor battalions were sent to France numbering over two hundred thousand men—men with minds responsive to all the political currents abroad at the end of the war. They (like other Asians transplanted by the war) joined their compatriots at home in cheering when Woodrow Wilson spoke out in favor of national self-determination.

Meanwhile, the drumbeats of war resounding from Europe had prepared the minds of Chinese intellectuals for a more radical nationalism. Sensing the heightened danger to their country, one of their most articulate spokesmen, Chen Duxiu, had issued, as early as 1915, an alarmed "Call to Youth":

> Considered in the light of the evolution of human affairs, it is plain that those races that cling to antiquated ways are declining, or disappearing, day by day, and the people who seek progress and advancement are just beginning to ascend in power and strength. It is possible to predict which of these will survive and which will not. . . . Our people will be turned out of this 20th century world and be lodged in the dark ditches fit only for slaves, cattle, and horses. . . . I would much rather see the past culture of our nation disappear than see our race die out . . . because of its unfitness for living in the modern world.[2]

In this spirit his *New Youth* magazine pioneered a simplified literary style for use by a less erudite readership, at the same time preaching an anti-traditional nationalism promising China a dignified place in the world. In its pages a vague recognition dawned that what was needed first was a profound cultural transformation—essentially a cultural revolution of Westernization—to allow "the Chinese race" to survive in the modern world. These sentiments found much response among students and intellectuals in Beijing and the seaboard cities who kept up with the news from Europe.

Preparing for the fourth anniversary of National Humiliation Day in May 1919, these excitable patriots heard that the Versailles Peace Conference had agreed to the Japanese demands. On May 4 the news set off instant protest. Demonstrations started in Beijing and spread to other cities (even to Tokyo, where many Chinese lived), initiated by students and backed by their professors but soon attracting businessmen and professional people, workers and common folk, women too, all in ever larger numbers. The authorities proved powerless to stop the demonstrations, which led to strikes and a boycott of Japanese goods as well. The agitation, which lasted for several weeks, represented the first stirring of patriotic mass politics in China, a turning point in its history. Under the impact of war and renewed national humiliation a sudden sense of national community had flared up. It was limited to those parts of the country and those segments of the population most open to foreign influence, yet inspiring the future leaders who set the course of the entire country for the rest of the century.

Suddenly, after the vague talk about cultural transformation, politics moved center stage. *New Youth* declared politics an important aspect of life, advocating mass movement and social reconstruction. "We believe that in a genuine democracy political rights must be distributed to all people."[3] Politics now meant mass politics. It had come to China at the conjunction of three major events: the Western victory under Wilsonian auspices, the stinging humiliation of China at the Peace Conference in violation of Wilsonian principles of self-determination, and the Bolshevik revolution, the latter brought to the full attention of Chinese intellectuals only in the wake of the May Fourth Movement.

The postwar years were thus a time of high excitement interpreted, in the social Darwinist terms then popular even in China, as a time of ultimate solutions. Echoing Chen Duxiu, Sun Yatsen, foremost among China's revolutionary leaders, said in 1924 that the country was in danger of becoming dismembered, a "hypo-colony" of the world's power nations; "another century will see our country gone and our race destroyed." He concluded by saying (in rather traditional terms): "Heaven has placed great responsibility upon us Chinese; if we do not love ourselves we are rebels against Heaven."[4] But how were China's leaders to translate that responsibility into practice among the highly resistant and brittle Chinese realities? So in China too a protracted and painful series of state-building experiments began, mixing political reorganization with cultural transformation. It added up to one of this century's most heroic achievements and greatest human tragedies, its end not in sight even now.

After 1919 several visions competed, all concerned with basic perspectives of national development expressed in political theory—reculturation had to begin in human minds and human wills. For a time, the Wilsonian message stood in the foreground, reinforced by the visits of two distinguished Western philosophers, John Dewey and Bertrand Russell, and seconded by Chen Duxiu's call for a new pragmatism under the guidance of "Mr. Democracy" and "Mr. Science." Such outright pro-Western orientation, however, quickly ran into trouble. The war's brutality and the cynical violation of Wilsonian ideals in the peace settlement had diminished the West's moral prestige. Moreover, the Western prescription was gradualist and consensual: the transformation would require a long time just when time was of the essence. Individual freedom too, so central to the Western model, posed profound problems among the Chinese people, who resembled, in Sun Yatsen's phrase, "a sheet of loose sand."[5] Where was the cement to be found for binding the granular multitudes to a common will for radical change? In the face of these problems the Western model lost its lustre, though it kept beckoning in the distance; in its own orbit, obviously, it worked, putting all else to shame.

In any case, the basic problem confronting Westernizing Chinese patriots was how to reach into the masses. In this respect initial experiments had been made under Sun Yatsen's revolutionary leadership since the 1890s, though hardly successfully; the republic which he had helped to found had been superseded by the rule of the warlords. As head of the Nationalist party, better known as the Guomindang, he strongly endorsed the May Fourth Movement, though lacking the means to advance it. Leadership required an effective mass following. Yet how was it to be organized in a country that had never known systematic political participation among the run of people?

At this point, the Soviet model became relevant, with support from Moscow itself. Lenin had watched the rise of Chinese nationalism since 1911. In his anti-imperialist crusade after 1917 he had made it Soviet policy to support national liberation movements everywhere, promising in the same breath to renounce all privileges enjoyed by tsarist Russia. He did so ostentatiously in the case of China, where the tsarist government had claimed special rights (subsequently the promise was not fully carried out). In addition, he sent out Soviet agents to teach the inexperienced Chinese intellectuals political skills for times of violence and civil war. One result was the founding of a military academy under Guomindang auspices to provide officers for a modern army, followed by improvement of the Guomindang organization; a third result—more indirect—was the vigorous propagation by Sun Yatsen of his political ideology suited to Chinese conditions.

The ideology, evolved early in his revolutionary career, was presented under the title *The Three Principles of the People, San Min Chu I*. It constituted a curious mix of Chinese tradition with Western influences absorbed either directly or through Soviet Russia and Japan. "Nationalism," Sun Yatsen said, "is that precious possession which enables a state to aspire to progress and a nation to perpetuate its existence,"[6] stressing throughout that "the Chinese people are of the Han or Chinese race with common blood, common religion, and common customs—a single pure race." He did not mention Confucius, but urged the preservation of "our ancient morality," including filial devotion, kindness, and love. He admitted that the Chinese would have to match the West's scientific learning, although without surrendering the old wisdom of China, which was superior to the wisdom of the West. He denounced cosmopolitanism as a tool of Western imperialism, but also struck a note of universal mission: "Let us today, before China's development begins, pledge ourselves to lift up the fallen and aid the weak; then when we become strong and look back upon our own sufferings under the political and economic domination of the Powers and see weaker and smaller peoples undergoing similar treatment, we will rise and smite that imperialism. Then will we be truly 'governing the state and pacifying the world.' "[7]

As for democracy, it meant popular sovereignty, traceable in Chinese history for over 2,000 years (in Europe for only 150 years) and indispensable in "the age of people's power." It did not mean, however, the assertion of individual liberty. On the contrary, the Chinese needed to "become pressed together into an unyielding body like the firm rock which is formed by the addition of cement to sand."[8] Equality, likewise, was not suited to Chinese conditions, nor was Western democracy generally. Sun Yatsen's democracy was advertised as a superior Chinese creation, remaking "China into a nation under completely popular rule, ahead of Europe and America."[9]

The third principle was "People's Livelihood," dealing with the economy and social justice. Sun Yatsen rejected any identification of this principle with socialism, but agreed that mankind would enjoy the greatest blessings only when "the social problem" was solved. He advocated economic mobilization and social justice as suited to Chinese conditions where feudalism had been abolished over 2,000 years earlier and Marxism, designed for European conditions, did not apply (it did not apply in Russia either, he hinted). He favored

the equalization of landownership, an income tax, and control of land speculation, attacked landlords for their inefficiency, and called for peasant liberation. At the same time, he also recognized the need for attracting foreign
capital in promoting, under state guidance, further industrialization, which
would help all the people. He would do away with vagabonds and parasites,
saying that "the government should force them by law to work and try to
convert them into honorable laborers." He concluded with a vision no doubt
inspired more by Confucian tradition than by Marx: "when the loafers are
eliminated and all men have a share in production, then there will be enough
to eat and to wear; homes will be comfortable and the people content . . . the
problem of livelihood will have been solved."[10]

Put into our context, Sun Yatsen's *Three Principles* appear as yet another
major effort to create in a disjointed, disoriented, and humiliated polity an
integrated worldview designed for political mobilization. It represented the
currents of the age: the rise of mass politics under the slogan of democracy,
the assertion of national superiority in a competitive world, the need for
catching up, and the impatience for speedy and decisive action befitting an
age of final reckoning. It also reflected a syncretist mix of discordant inputs,
the contradictions born of the combination of backwardness with a proud or
even universalist affirmation of tradition. Yet it remained disappointingly
vague on the practice of political organization. Who was to provide the vital
cement, who to pour it? Sun Yatsen obviously assumed it would be done by
the Guomindang, but as a traditionalist he offered no practical advice on how
to weld the Chinese people together "under completely popular rule."

In this respect the tiny Chinese Communist Party, founded in 1921, was
potentially better equipped by adhering more closely to the Soviet model.
Among radical patriots the Soviet emissaries had found greater receptivity
than in the Guomindang, as much for their militant anti-imperialism as their
Marxism-Leninism. Even before the May Fourth Movement, Li Dazhao, a
restless follower of Chen Duxiu and chief librarian at the Beijing National
University, had begun to look toward Moscow. He soon convinced himself
that the victory of Bolshevism was inevitable: "the bell of humanitarianism is
sounding. The dawn of freedom has arrived."[11] Looking at the new Russia,
he considered backwardness an asset, because it created "surplus energy for
development."[12]

That comforting thought was accompanied by admiration for Bolshevism
as a universal creed capable of transforming the entire world. "The mass
movement of the twentieth century," he wrote, "combines the whole of mankind into one great mass. The efforts of each individual within this great
mass . . . will then be concentrated and become a great irresistible social
force. . . ."[13] The Bolshevik revolution, obviously, struck deeper chords
among Chinese radicals than Wilsonianism, releasing more massive energy
eventually applied to the un-Confucian task of organizing the masses.

Inescapably, the impact of the Bolshevik revolution and of Marxism-Leninism remained abstract as in all cases of cross-cultural transfers. The substance of the news from Soviet Russia dealt with ideals and theory, both
devoid of the tough political sense of Bolsehviks trained in a long revolution-

ary tradition. No wonder Chinese Marxism assumed a different form as its proponents wrestled with the theoretical and practical questions raised by Chinese conditions. China resembled a "capitalist" country on the verge of a socialist revolution even less than Russia. It was essentially a country of peasants and landowning gentry, with but a tiny class of "capitalists" and even fewer "proletarians" in the seaboard cities. Lenin, to be sure, had argued that in Asia the revolutionary thrust would come from peasant soviets, but under proletarian leadership. Were the Chinese workers ready to assume that role?

What leadership, moreover, could come from Chinese intellectuals still raised, regardless of their political creed, in the Confucian ethic? Li Dazhao, for instance, would not swallow Marxist dialectical materialism. He continued to believe in the primacy of consciousness and will (as did Lenin, without making an issue of it). He also had little use for the concept of class struggle. Lumping the discredited Chinese tradition together with Western bourgeois values, he considered contemporary China a proletarian country, part of the non-Western world exploited by imperialism and capable, through its revolutionary élan, of skipping the "capitalist" phase of development.

More than its Russian prototype, Chinese communism in its formative years was awash in extravagant and vague ideas drawn from poorly understood Western experience thinly stretched over poorly understood indigenous conditions. Above all, despite their professed attachment to the Soviet cause, Chinese communists remained patriots. Li Dazhao, for instance, while talking about a new Asianism or dreaming of transforming the life of the whole world into the life of a single family, still assumed that China rather than Soviet Russia would lead mankind to socialism. One can understand the dismay with which the emissaries of the Communist International watched their comrades in the Chinese Communist Party.

Soviet policy in those years aimed above all at strengthening national liberation movements which promised to have greater political influence than incipient communist parties. For that reason the Chinese Communist Party was advised to form a united front with the Guomindang, which lasted until Sun Yatsen's death in 1925. Thereafter the divergent aims became irreconcilable. The Guomindang under the Soviet-trained General Jiang Jieshi (Chiang Kaishek), more strongly wedded to Chinese tradition and the social classes that upheld it, allied itself with the anti-imperialist and moderately progressive but otherwise conservative monied interests in the seaboard cities. It evolved in a sinified fascist pattern and, with the help of its new army (a conventional instrument of power), set out to reunify the country, reducing the warlords and drowning all communist agitation in bloodshed. The surviving communists had to turn to the peasants as their only hope for national revival through a socialist revolution.

While Lenin had held out the possibility in Asian countries of peasant soviets, no Marxist-Leninist experience for working with peasants had yet evolved. Chinese communists willing to go to the peasants had to start from scratch, transforming themselves, intellectuals cradled in Confusian gentleness, into ruthless organizers capable of handling the most unpromising human raw material for realizing the Marxist vision. The pioneer in this excruciating transformation was Mao Zedong, a youthful follower of Chen Duxiu and Li Dazhao risen from a comparatively prosperous peasant background. When in 1927 General Jiang Jieshi wiped out all communist organizations in the seaboard cities, Mao turned to the peasants of his native Hunan Province, to study their revolutionary potential. His report began with an inexperienced theoretician's flourish: "Within a short time, hundreds of millions of peasants will arise in Central, South, and North China, with the fury of a hurricane: no power, however strong, can restrain them."[14] How little he knew what lay ahead! His ascent as a Marxist peasant leader was slow and painful, the result of long and bloody experiments in reculturation, all tied to the mounting world tensions in the 1930s.

III

History took a different course in India, where the people were better prepared—and conditions more propitious—for absorbing the new shock waves of Westernizaton set off by World War I. Under the British *raj* India had been open to Western influences for over a century, its political institutions and incipient nationalism a Western product. Its attachment to British rule was proven by the support volunteered after the outbreak of war; men and money rallied to the British cause.

Yet here too the war left its mark. As its costs became apparent and its aims more ideological, patriotic pride advanced; agitation for home rule increased beyond its prewar intensity. Sensing the drift of opinion, the British government in 1917 promised an extension of Indian self-rule, which, together with the news from Russia, mightily raised Indian expectations for liberation from colonialism and set off public unrest. Alarmed, the British authorities turned repressive. In April 1919, after a handful of Englishmen had been killed, a British general at Amritsar ordered his Gurkha troops to fire indiscriminately into an unarmed crowd down to the last round of ammunition; he wanted to teach a lesson: there would be no revolution in India. The "Amritsar massacre" set Indians ablaze with indignation, even though later that year the British extended Indian self-government, giving Indians control over less essential services. But few Indian nationalists were in a mood for compromise.

Violence also flared up from another source. The heightened nationalist agitation of the war had mobilized Indian Muslims against their Hindu neighbors (Indian Muslims had refused to fight against Muslim Turks). Now militant Muslims were determined to safeguard their own rights against the Hindu majority. The rise of a tiny Indian communist party combining class struggle with anti-imperialism added to the turmoil of the postwar years. In India too, then, the crusade for freedom and political rights, for democracy, brought political strife and, more quickly than in China, extended it down to the bulk of the population, to the peasants.

In this setting Gandhi proved his mettle as the pioneer of Indian mass politics and charismatic synthesizer of conflicting ideologies and political ambitions. Of slight figure, bespectacled, and seemingly self-deprecating,

Gandhi stirred the Indian people to their depths, himself rising to a commanding position in the Indian nationalist movement. In 1921, sensing the necessity of identifying himself with the masses, he shed his European clothes, appearing ever thereafter in a cotton dhoti and a shawl covering his bare shoulders, as lean as any peasant, poor in everything but spirit, admired for his "great soul." A cosmopolitan intellectual, he knew how to appeal to peasants; a widely read Westernized non-Westerner with experience of life in England and South Africa, he could adjust ancient traditions for modern use. Unconcerned with the contradictions in his role or views, he was an inspired transcender.

He preached and practiced peaceful cooperation between Hindus and Muslims, searching for common ground beyond their differences. He leavened Hindu custom with Western egalitarianism by advocating political rights for the untouchables, whom he called "sons of God," or *harijans*. His practice of "soul-force" *(satyagraha)* combined Christian pacifism with Hindu (or, more precisely, Jain) teachings. Although his political ideal was a society of self-supporting peasants living in simplicity and purity and his vision of an Indian state a loose federation of village republics, he created a mass movement in favor of an independent nation-state. He denounced Western industrialism because it undermined the ascesticism essential for *satyagraha,* but gladly took financial contributions from leading Indian industrialists. Above all, he opposed the British authorities by nonviolent resistance, invariably offering his own life, confronting British police power with the disciplined spirituality of his followers.

Always close to the British model, he was anti-Western in the moral superiority of his pacifist appeal. Yet he played, with superb skill, the arts of modern mass politics in the Indian setting, a saint and a wily politician, all along rejoicing in setting a universal model for nonviolent political change (in the tradition of Vivekananda, who had wanted India to conquer the world spiritually). Gandhi did not stem the tide of Westernization in Indian industry, or education, or life generally, but infused it with an indigenous spirit which, reminiscent of Japanese Westernization, allowed a constructive subliminal fusion. A contemporary of Lenin, Mussolini, Hitler, and Sun Yatsen, he built for his compatriots an experimental bridge, spiritual, mental, and political, into the future. The bridge, admittedly, remained highly fragile, more a vision than a reality. The struggle for independence continued through a series of crises and accommodations with the British masters—on a subcontinent mercifully sheltered from the direct impact of global power politics and its glorification of violence.

IV

In postwar Africa Egypt trod yet another path in the revolution of Westernization. Here Britain, in charge of the country before the war, tightened its control and installed a loyal king after the Ottoman Empire had joined the Central Powers and moved troops toward the Suez Canal. In fighting the Turks the British freely drew on Egypt's resources, reaping hostility and strengthening prewar sentiments in favor of independence; in 1919 British troops suppressed a full-scale anti-British insurrection. Nonetheless, as in India, the British government was ready for moderate concessions in the spirit of the times, offering limited independence; the offer was met with resistance by the nationalists. The formal granting of independence in 1922, with due consideration for British needs, was followed by the introduction of representative government and universal suffrage, with dubious results. The first election led to an overwhelming victory for the nationalists and subsequently to endless turmoil and some violence (the British high commissioner was assassinated). Finally, in 1928, the Egyptian parliament was dissolved for three years pending renegotiation of the British presence—a troublesome issue left unresolved for several decades. Meanwhile, the country had made some progress toward sovereign statehood, acquiring the trappings of democratic government under a constitutional monarchy, beginning to regulate the flow of the Nile with the help of Western technology, and introducing its people more fully to the ambiguous benefits of Western culture.

Beyond Egypt, in the vast African continent stretching to the south and west, the pattern of postwar Westernization was far more complex and diversified. Indigenous society was splintered into innumerable small-scale linguistic and ethnic units, most of them still untouched by European influence, although virtually all under European colonial rule. Britain and France through their own political vitality had made the deepest inroads into indigenous society, dealing with it from overwhelming strength.

Far more than in China, indigenous resources were no match for European power; not even the preconditions for communism or fascism applied. For Europeans, as for Russians or Chinese—all judging other peoples by their own standards—Africans were "primitives." Even the run of Westernized Africans from the French or English colonies along the west coast stood in awe of the white man, claiming equality on grounds of their often considerable personal achievements rather than of their cultural heritage. Eager to protect themselves and their peoples from European encroachment, they nevertheless sought quick access to the sources of Western power, preferably through education in the West or through Western schools in their own lands (most of them pioneered by missionaries). Undeterred by the vast gap in cultural resources, they were quick to use Western ideals for their own advantage, aided by fellow blacks in the diaspora of the West Indies or the United States. Indeed, the most dynamic advocate of Africanism came from the pan-African movement originating among assimilated Negroes in the Western Hemisphere. Better educated than the run of whites, they felt most keenly the discrimination based on cultural stereotypes.

The war revived, under W.E.B. Du Bois' leadership, the flagging pan-African agitation. Already in 1915, carried away by the battles then raging, Du Bois had threatened:

> The colored peoples still will not always submit to foreign domination. . . .
> These nations and races, composing as they do the vast majority of humanity,
> are going to endure this treatment as long as they must and not a moment

longer. Then they are going to fight and the War of the Color Line will outdo in savage inhumanity any war this world has yet seen. For colored folk have much to remember and they will not forget.[15]

While the Versailles Peace Conference was in progress, he masterminded the second Pan-African Congress in Paris, determined "to have Africa in some way voice its complaints to the world."[16] On that occasion, however, his aims were quite moderate: Africans must have not independence but "the right to participate" in colonial government. At the third Pan-African Congress two years later he called for "the establishment of political institutions among suppressed peoples. The habit of democracy must be made to encircle the world."[17] The aims of Du Bois' pan-Africanism remained moderate throughout the 1920s, but its political pragmatism was marred after his first visit to Africa (1923) by a romantic attachment to African culture: " . . . Different, Immense, Menacing, Alluring . . . a great black bosom where the spirit longs to die."[18] Thus he contributed to the Back to Africa movement popular among alienated black intellectuals in the metropolitan centers of the West.

Bigger contradictions flared in Du Bois' rival, Marcus Garvey, a product of Jamaica and London. On the day the war broke out in Europe he founded in Jamaica a "Universal Negro Improvement Association" with a lengthy agenda. It was to create a "Universal Confraternity" among the black race; promote "the spirit of race pride and love . . . ; assist in civilizing the backward tribes of Africa . . . ; establish a central nation for the race"; promote "a conscientious spiritual worship smong the native tribes of Africa"; and set up "universities, colleges, academies and schools for the racial education and culture of the people."[19] In 1915 he moved to New York's Harlem, where after the war he rose to prominence, working (within the security of American state and society) to fuse globalism, nationalism, the poltiical excitement born of the war, equality, democracy, and the impulses of uprooted, oppressed, but lively American Negroes into a meaningful creed. Marcus Garvey was a modernizing black messiah in the revival of black culture centered in New York City.

Raised as a Catholic, he founded the African Orthodox Church worshipping a Black Christ and a Black Madonna; he formed a Universal African Legion, a Universal African Motor Corps. and the Black Star Line for ocean-going ships. He even appointed himself "Provisional President of Africa," with the avowed intention of liberating the continent regardless of the wishes of the Africans; a delegation he sent to Liberia was arrested on arrival. Yet he had an enthusiastic following among American Negroes and impressed African students studying in the United States (decades later one of them, Kwame Nkrumah, adopted the flag of the Black Star Line as the National flag of the new state of Ghana). Garvey's cry "We shall now organize the 400 million Negroes of the world into a vast organization and plant the banner of freedom on the great continent of Africa"[20] aroused deep emotions, without, unfortunately, supplying the necessary skill and discipline of enduring political organization.

More practical English-speaking African nationalists meanwhile emerged from West Africa, where agitation for indigenous representations in the colonial administraton had surfaced in the local African-edited press before the war. Not surprisingly, considering the news from Paris, in March 1919 a deputation of leading Western-educated Africans from the Gold Coast (the most Westernized of the British colonies in West Africa) petitioned the British governor that Africans be consulted on all matters of importance to them through freely elected institutions. The following year the West African National Congress was formed, representing the chief British colonies in the area and demanding greater African representation in the colonial administration. When in 1923 and 1925 Nigeria and the Gold Coast received new constitutions, their demands seemed met; the Congress' agitation had also produced another result: the extension of Western schools, culminating in the foundation of Achimota School in the Gold Coast. Paradoxically, the English authorities wanted to promote the skills most suited to African life, including African languages and agriculture; the Africans (untouched by the intellectual "Back to Africa" trend) wanted the classics, the learning that distinguished British gentlemen.

African nationalism, meanwhile, also grew among West Africans studying in England, who in 1925 founded the West African Student Union. It counted among its members all the future leaders of anglophone African independence movements. It was in touch with all other African political groups in the United States, the West Indies, France, and francophone Africa. It produced Westernized anti-Western African patriots, pan-African in sentiment yet bound through family and lineage to their local origins, all of them determined to gain credibility by success in their professions. Influenced by their British training, they remained moderates.

Political Westernization took a somewhat different turn among Negroes under French influence. Unlike England, France had recruited colonial subjects from northern and western Africa into frontline service, broadening their perspectives and assimilating them into French culture. Assimilation had long been part of French colonial policy, with some notable successes and many failures. Negroes from the francophone West Indies also gathered in France. Like most non-Europeans (the Asians included), they fell under the spell of socialism, welcomed as fellow victims of "capitalism" by French socialist and soon even communist leaders. Among the blacks, Africans generally were more radical than the West Indians, more easily carried away by the promise of extreme creeds. All of them fell prey to the fragmentation common in French politics. Yet some excelled in literary protest, culminating in the celebration of *négritude*, promoted by Aimé Césaire from Martinique and Leopold Senghor from Senegal. These writers drew on the romantic worship of indigenous roots as well as on the work of Western ethnographers studying African culture. Responding to the widespread anti-modernism of the times, Senghor even briefly sympathized with German national socialism.[21] What Negroes craved above all in the face of African backwardness and Western cultural arrogance was racial pride, combined under socialist (or communist) influence with an affirmation of a common humanity—a concept far beyond the comprehension of simple villagers guarding the African roots.

In the late 1920s even the Communist International discovered the opportunities hidden in the agitation for African liberation. Lenin's anti-imperialism set the basic appeal. In response a few Africans and Afro-Americans went to the Soviet Union to study at the University of the Toilers of the East, taken in tow by the Comintern and sent back to unite blacks in France, England, and the United States under the banner of the "League Against Imperialism." That League attracted widespread support at its first congress in 1927, which pledged itself to "liberate the Negro race throughout the world."[22] Liberation included equality with other races; African control of Africa; freedom of speech, press, assembly, and communications—goals, in short, drawn from European precedents and, as any ethnographer could prove, incomprehensible to most Africans. All of them testified to the progress of reculturation starting first in human minds.

Meanwhile—and taken for granted by the black intellectuals—the Westernization of their continent proceeded apace after the war. For the benefit of the mother countries (and less clearly so for the colonial peoples), economic development began in earnest, occasionally even with a new compassion born out of the war's suffering. Health services improved, education advanced, often with the help of white missionaries or black churches in the United States (especially the African Methodist Episcopal Zion Church); slavery was suppressed. In addition, harbors were enlarged, railways built into the interior, communication by road and telegraph extended, exports advanced, and local markets expanded. At the same time, Western ethnographers (foremost among them R. S. Rattray in the Gold Coast) helped colonial administrators understand—or even appreciate—their subjects better, while imparting to Africans a Westernized version of African identity.

Given the vast disparity between African and Western cultural resources, progress was bound to be slow and beset by controversy. The Europeans, preoccupied with their postwar problems at home, were reluctant to make sacrifices for Africans. The Africans impatiently or even rebelliously complained about the insufficient attention given to their needs, yet also eager, again with Western encouragement, to preserve their culture—a culture unobtrusively cleansed in their minds of all embarrassing barbarisms. What began to emerge in these years was a Westernized version of African culture, carrying into Africa the intellectual and emotional conflicts of Westernized non-Westerners familiar in European romanticism. Glorifying the African vitalist (sensualist) immediacy of life, African intellectuals rebelled against the ascetic self-discipline required for the voluntary coordination of human wills under a modern state. Yet their defense of *négritude* was bound to prove a hindrance rather than a help on the endless and infinitely cruel road to independence and equality in an interdependent competitive world.

In sum, wherever one looked outside Europe, outside the West, whether into Japan, China, India, or Africa, the revolution of Westernization relentlessly advanced, causing more troubles than it resolved, piling up ominous problems even for the leading countries of the West.

9

The Great Depression

Meanwhile, the countries at the center of global "capitalism" entered a major crisis, the Great Depression. That depression, linking the postwar 1920s with the prewar 1930s, highlighted unresolved tensions within the West itself, thereby providing new opportunities for anti-Western Westernizers.

Economic crises and times of economic hardship had been long-standing phenomena in European history. They grew bigger as the European economy expanded into a world economy, their causes embedded in the combination of competitive diversity with cooperative unity that characterized all of European society and culture. A multiplicity of producers and consumers were linked by a multiplicity of commercial and financial intermediaries, all living in different localities and states yet tied to each other by a thousand threads of interaction, all of them overtaken at times by greed and speculation. They competed with each other in all aspects of economic life (as in all other fields of human accomplishment). The marketplace therefore was at the mercy of the uncoordinated actions of individuals and collective bodies ranging from business firms to local associations and states. Unrelated events like natural catastrophes, civil commotions, or wars also affected it. The inclusion of non-European peoples and their economies as a result of European expansion added further complexity. Through the centuries good years and bad alternated with seeming cyclical regularity.

In this vast uncoordinated interaction popular preference and government policy had come to encourage individual initiative. The diversity within the cultural unity of the major parts of Europe allowed private enterprise increasing freedom in the pursuit of wealth, fame, and power. Private initiative, however, was always contained within institutional frameworks, formal in the earlier centuries and, starting in the eighteenth century, increasingly informal through internalized civic restraints. The collective bearers of the collective discipline of liberal "individualism" were the emerging urban middle classes of western Europe, the "bourgeoisie," open toward the landed gentry above and toward the workers and peasants below; especially in countries enjoying exceptional external security (foremost the United States) class divisions were remarkably fluid. By the end of the 19th century certainly, middle-class ways of life with their economic base of private enterprise linked in complex networks of interaction—the so-called capitalist system—had become the norm throughout the countries bordering on the North Atlantic which furnished the models of statehood to the rest of the world.

While middle-class society ensured a measure of social, political, and cultural stability, the open market perpetuated its risks to all participants, though in unequal proportion. Propertied and educated people were reasonably well equipped to cope with uncertainty. They had accepted it as a legitimate goal for further exertion, drawing from their religion a set of intellectual and emotional skills capable of making the most of adversity. Inevitably, people lacking such cultural resources and more vulnerable to starvation, the poor and uneducated, like the newcomers from outside, whether from central Europe or the rest of the world, felt lost in that dangerously "open society." Even its beneficiaries tried to limit the damage, searching for the causes of economic crises and advancing the study of economics.

Among the economic theorists Adam Smith, representing the self-confident new middle classes of the late 18th century, defended the benefits of the self-regulating market. In the long run, he argued, the enlightened self-interest of the competitors led to constructive adjustments for the common advantage (including a better grasp of economic processes). He discouraged interference by special interests whether of monopolies or governments; it limited rather than enhanced the collective capacity for creative adjustment. A contrary view emerged in the 19th century when socialists, representing the victims, deplored the injustice of the unregulated market and the high social costs of economic depression. Their recommendations included replacing the irrational anarchy of the market by rational control.

Yet, considering the obstacles which socialists encountered in organizing even their limited following, how could they rationally control the vastness of the market in the emerging world economy? Would any central planning agency possess the necessary information, knowledge, and flexibility? Would the myriad actors in the global market conform to the necessary regulations? Control through central planning certainly overtaxed all existing human resources; it could hardly be called a rational enterprise. It also raised a big question: would such control maintain the high level of cultural creativity fostered by the open market? And at what point would the loss of creativity out balance the human suffering thus contained? Admittedly, under special circumstances, as in Russia after World War I or subsequently in developing countries generally, socialism proved a handy tool for rapid political mobilization, but at a high price. Besides, a planned economy in one country—or even several countries—hardly created the desired harmony of the world economy. A rationally planned world economy obviously required a world state endowed with superhuman wisdom among both the governors and the governed.

In the imperfect and fragmented world of the 19th century, at any rate, the expanding economic marketplace continued its painfully creative anarchy. Hardship taught its lessons, giving all participants more insight and control over their destinies. Certainly, by widespread agreement, the competitive economy had led to progress in all fields of human achievement (including moral sensibility) and to superiority over all other ways of life. It had also led to legislation limiting, at least within separate countries, the worst effects of the business cycle. As the European economy spread around the world, it even evolved a fragile network of economic controls extending beyond nation-states. Yet understanding and control always lagged behind events, while the consequences of economic downturns grew more ominous.

Take the depression of 1873, extending with ups and downs to almost the end of the century, the greatest before 1929. It was sustained, in part, by the appearance of cheap overseas grain on the European market, depressing the rural economy everywhere, worst perhaps in central and eastern Europe including Russia, and jarring state and society. In the United States the depression brought aroused farmers and workers into politics, strengthening American democracy. In central and eastern Europe it sharpened social and political conflict, hardening the reactionary instincts of vested landed interests and even of peasants, promoting the advance of industry and labor unions as well as the rise of militant anti-Semitism (depression, like prosperity, promoted the revolution of Westernization).

Before World War I, however, the world economy, then centered in London, had attained a measure of stability based on the universal acceptance of the gold standard. At a time when national banks had risen as useful, though still limited, agencies of economic control and gold was the common currency, the Bank of England acted as "the banker of the last resort." Used to taking a global overview sufficient for the times, it stepped into the breach whenever a financial crash seemed imminent, thereby softening the impact of dislocations in the world market. World War I, of course, reduced that system to shambles. The international private-enterprise-oriented economy gave place to nationally controlled war economies; the financial center of the world gravitated to New York, the source of war loans to the Allies; farmers and industrialists around the world increased production while the Europeans squandered their resources on destruction. After the war no effort by international conferences could revive the past; the war had wrought too many changes.

Soviet Russia had virtually dropped out of the world market, repudiating the foreign loans of the tsars. Germany, formerly a major economic force, was exhausted, wracked by internal strife and saddled with a huge bill for repairing the damage its soldiers had caused the Allies. France was impoverished and indebted to the United States, its eastern regions devastated by battle. England too had suffered grievously, likewise indebted to the United States; it had lost its financial preeminence as well as vital markets for its industries. All Europeans felt the loss of productive manpower. In order to restore normalcy, wartime economic controls were ended, private enterprise revived. Yet internationalism favored tariff protection, especially hurting the English, who stood by their traditional policy of free trade despite their high rate of unemployment.

By the later 1920s, admittedly, the American credits extended to Germany for the larger purpose of resolving the tangle of reparations and war loans helped to restore private enterprise, international trade, and European over-all prosperity to prewar levels (while the United States enjoyed a boom largely based on the attractions of the automobile). As a result, the world enjoyed a fleeting moment of hope for peaceful cooperation also in the relations between states, highlighted by the Kellogg-Briand pact of 1928, which out-lawed war as an instrument of national policy (signed by all countries which, a dozen years later, found themselves at war). Yet despite the optimism, the economic upswing was superficial. Throughout the postwar years agricultural producers around the world suffered from overproduction and low prices, which limited their purchases of industrial goods. For want of prosperous consumers, the industrial boom in the United States and Germany petered out in 1928, although stock market speculation mounted until its dramatic collapse in October 1929.

By itself that traumatic conclusion to excessive financial speculation would have been of minor importance; tied to the world economy, it brought down the fragile postwar economic order in a resounding crash. The financial superstructure of Allied war loans and German reparations, seemingly secured by the Young Plan of 1928, suffered fatally when American banks recalled their German loans. How could it have stood up in any case? The Smoot-Hawley tariff, the highest in American history (and one of the highest in the world), ready for congressional approval even before the stock market crash, prevented Germans and other European debtors from earning dollars for repaying their loans. By 1930 overextended industries everywhere cut back production, laying off workers and reducing purchasing power all around in a vicious downward spiral. In 1931 bank failures began, starting in Vienna and extending to Germany and England; they reached the United States in early 1933. Meanwhile, unemployment mounted to record figures in all the industrial countries.

What stands out in retrospect is not so much the economic misery as the incapacity of economic experts and political leaders, let alone public opinion, to rise to a full comprehension of the factors at work. As for domestic policy in the face of unprecedented crisis, consensus advised balanced budgets, cur-tailing government expenses and thereby further aggravating the depression. Worse yet, no one in a position of influence after the war seemed to recognize that national economies operated in an interdependent global network (its interdependence intensified in the 1920s by even faster means of telecom-munication). The spirit of nationalism, always lurking in the background and stimulated by the war and the depression, had dangerously shrunk the per-spectives needed for peaceful cooperation. Even the United States, which after vigorously contributing to the postwar economic revival should have taken the lead in stemming the tide of protectionism, instead promoted it; least dependent on international trade, it concentrated on its domestic woes. After 1929 all governments scrambled to protect their own economies. Yet by beggaring their neighbors they again hurt themselves, forced into yet more drastic defensive measures. By 1933 the international currency based on gold had collapsed and with it not only the financial superstructure of war loans, reparations, and American payment plans, but also any hope for international economic cooperation.

In the shrinkage of economic internationalism realistic remedies now had to be nation-centered. Governments had to deal within their own countries with unemployment and impoverished people still disoriented by the aftereffects of the war. Were the governments prepared? Especially in countries with fragile political systems the Great Depression was more than an economic disaster; it deepened the postwar metaphysical crisis of confidence in the existing order of state and society in the worldwide "capitalist" system. It not only put the fabric of economic and political cooperation under exceptional strain, but also unleashed unprecedented hostility toward international cooperation in any form. The interdependent world created in the past hundred years had grown over the heads of people recently organized for war and now faced with prolonged unemployment and economic insecurity.

As peaceful economic cooperation among countries had proved impossible after the war, the necessity—or even the ideal—of self-sufficient national power at any price rose correspondingly. The Great Depression raised the global power struggle, somewhat abated after World War I, to a deadly climax in the 1930s; it prepared for World War II.

III

The Prewar Thirties and World War II

Under the impact of the Great Depression the world revolution of Western-ization progressed further, prompted more by raw considerations of military might and political competition than by peaceful cooperation; it turned ugly as it forced accelerated change on unprepared societies. Paradoxically, it pro-ceeded from a weakened center. The ascendancy of the Western countries that had served as world models, briefly enhanced at the end of the war and then reduced by its after effects, was further diminished by the Great Depre-sion. It aggravated their social tensions, heightening the contrast between the promise of liberal democracy and social reality; it encouraged close scrutiny by eager critics pointing out the internal contradictions of "capitalism." Yet, compared with the victims of the revolution of Westernization, the model countries retained their overall advantage of social cohesion and cultural con-tinuity, though with dimmed lustre.

After 1934 France, which had weathered the first years of the depression rather well, suffered from sharp internal tensions and lack of conviction in external relations. The British fabric of parliamentary government withstood more successfully the economic crises, but again at the price of reduced vig-ilance in foreign policy. In the United States, Roosevelt's New Deal absorbed and resolved the tensions of the depression by concentrating on domestic affairs. Under these conditions the League of Nations, already deprived of American backing, was reduced to futility as a peace-making agency. The ini-tiative in world politics fell to the anti-Western counterrevolutionaries press-ing their own forms of Westernization.

In pursuit of power, the militant upstarts were forced to globalize their sights far beyond their past perspectives. In mobilizing for war, they had to develop their industrial potential and rationalize their economies; they also had to force their peoples into more intense discipline of cooperation and docility in large-scale organizations. The logic of their ambition enforced continuous imitation of the sources of strength among the models they wanted to replace. It meant either copying the instruments of Western power or defeat.

Westernization by imitative substitutionism took different forms in different cultural settings. In the Soviet Union, where the challenge had originated in 1917, it rose to the fury of the Stalinist revolution. Starting already before the outbreak of the Great Depression, that revolution proceeded independently, yet cast its shadow over the entire decade. Japan's attack on Manchuria in 1931 started the 1930s' open season for aggressive expansionism, leading to further, if disguised, Westernization in both Japan and China. In 1933 Hitler came to power in Germany, determined to challenge the British Empire as the source of Western superiority. Although affected or prompted by the Great Depression, these counterrevolutions were the product of the fierce political dynamics released by World War I. It was so most clearly in the case of Hitler, who wanted to bring that war, which in his mind had not yet ended, to its conclusion in a German victory. In Stalin, too, the pulse of the war still throbbed ominously.

Both Stalin and Hitler, key figures in the 1930s, have become arch symbols of evil in the modern world. Yet in high-altitude perspective they, like their less notorious fellow dictators, appear as symptoms of their age, as much its victims as its villains. Only by viewing them as integral parts of the tensions of their times can we understand the causes of their actions and thereby dismantle the incomprehension and irrationality they have bequeathed to posterity. Not they alone, but all those who in their ignorance have shaped the era—as well as their successors who perpetuate that ignorance embalmed in righteous indignation—deserve our moral concern.

10

Stalinism

To look, then, at the grim developments of the 1930s in chronological order, beginning with the Stalinist revolution, the most harrowing phenomenon among the counterrevolutions of Westernization. Even in retrospect it stifles rational analysis, breeding violence in thought and action. How can anybody remain reasonable and tolerant in the face of Stalin's terror devouring, by extreme estimates, more lives than two world wars combined? On the other hand, how can we prevent a repetition; how can we calm the violence-producing emotions aroused even by the memory of Stalin's brutality; how can we understand the twentieth century, if we do not fathom the causes of Stalinism? The symbolic figure of Stalin, epitomizing the cruelest features of totalitarianism, constitutes a test case for insight into the age in which we live. Were his "crimes" his personal responsibility or the responsibility of *all* forces at work in his lifetime? Let us recall the relevant contexts.

Stalin, fifty years old in 1929, was a product of the tsarist empire in its final decades and of its collapse. That empire, long weakened by Western cultural and political subversion, was defeated by a superior German army, its very survival in question. At that point a group of revolutionaries under Lenin's leadership, and driven by an updated (or internationalized) version of the intelligentsia's traditionally intense patriotism, seized power. The Western model was constantly in the forefront of their minds, although of course they would not admit that fact openly (if they admitted it even to themselves); it drove them to relentless emulation, both in military might and cultural achievement. Seen in this light the Leninist regime was not an indigenous Russian creation but a counterrevolutionary response to Western ascendance, theoretically superior to the West in a mirror image of Western superiority. Its originality was limited to its prescription of how to achieve its goals under the conditions of Russian backwardness.

The essence of that Leninist prescription, admittedly a remarkable achievement, was an imitative substitutionism. What was done automatically in a politically free country was replaced by organization, command, and compulsion enforced by terror, thereby achieving the reculturation designed for matching the Western model. Stalin inherited this prescription, more as a formula than as a working sytem. It lacked, even while Lenin was in command, not only effective central administration in party and state but also experience and time for raising the institutional infrastructure of substitutionism to the required dimensions. After Lenin fell sick, and especially after his death in January 1924, the Communist Party, the core of the Soviet experiment, more than ever lacked cohesion and a central will. Who was to replace him, who to restore that powerful driving force, with what directives? And how?

Bolshevik practice, shot through with intrigue and factionalism despite all emphasis on discipline, furnished neither precedent nor institutional machinery for consensus-based transfer of leadership. As a would-be democratic organization the party held periodic conferences and congresses; but it could not overcome its debilitating personal feuds or administrative wrangles; it did not live up to its Leninist ideal. What was needed to offset these weaknesses was a towering personality endowed with superior energy and willpower, and with the political savvy capable of pulling together unruly, ambitious, temperamental rivals and their restless followers, all inexperienced in the individual self-control required for large-scale organization, all products of the collapsing Russian empire. Yet effective leadership involved more than establishing ascendancy over an unruly Communist Party; it called for legitimizing the communist regime among the population at large in the postwar era of mass politics. True leadership had to be steeped in traditional popular political culture, familiar with the human raw material destined for reculturation.

The raw recesses of popular culture in the vast Eurasian landmass extending deep into central Europe is an unexplored and often deliberately avoided subject (like the hidden callousness in German society that produced Hitler). It embarrassed or exasperated the intelligentsia; it has never been realistically explored by Western analysts, who perhaps too readily followed the lead of Westernized Russian authors. A glimpse into the lower depths of Russian society takes us into a petty world of intense localism and clannishness where otherness in physical appearance or cultural traits was an act of aggression against God's order where tradition had to be defended at all costs. The pervasive insecurity of existence made human relations tense and merciless. Nature was harsh and unforgiving; so were human beings in their cramped quarters. Violence abounded, often rising to senseless or even refined cruelty.

As for Russian statecraft, the available evidence—from prerevolutionary court records, firsthand accounts of war, revolution, and civil war by Russian writers or German prisoners of war—inspires grim conclusions. Dealing with such coarse human raw material, Russian rulers had been traditionally brutish in their relentless efforts to create an effective polity. The concept of compromise, so essential in Western life and politics, was not part of Russian popular culture. Admittedly, one observes also extraordinary kindliness, even saintliness, but to no political effect. Foreign invasion, a cruel climate, poor soil added to the pervasive callousness in life, manners, and human relations generally. These realities fitted neither into the patriotic and Western-oriented self-image projected by the tsars or the prerevolutionary intelligentsia; they were anathema to any Marxist characterization of the Russian "proletariat." Why then explore them? All the same, they determined the style and dynamics of Soviet policy in the postwar age of mass politics.

Who, then, was able to provide leadership at a time when the revolutionary regime was reasonably secured and its grand objectives had at last to be addressed? Could the revolutionary intellectuals who had spent crucial years in western Europe fathom the depths of Russian mass politics? Did Lenin grasp the hard political realities confronting his regime, especially after he fell sick in 1922? His most prominent lieutenants were preoccupied with questions of basic direction, whether to expand the New Economic Policy or to strike out boldly for rapid industrialization; they hesitated when confronted with the costly consequences of each course. None were concerned with the prior necessity and the nitty-gritty of constructing a monolithic party organization without which no consistent policy, no matter how cleverly conceived, stood a chance.

Considered realistically in the context of the times, Lenin's most likely successors lacked essential qualities: physical energy, single-minded concentration on top priorities, domineering will power, and a feel for the ground-floor people with whom they must work. Foremost among the contenders was Leon Trotsky, "fiery, vibrant, filled with the zeal of his historical mission,"[1] ruthlessly brutal, to judge by his civil war record, but lacking physical vitality at the crucial time; besides, he also considered himself above the mundane chores of administration. And, like the other party intellectuals, he failed to recognize the rules of the succession struggle or of mass politics generally. By contrast, Stalin, although no match for the articulate intellectuals on their own turf, played the rough game with ruthless virtuosity. In the words of one anti-Stalinist Soviet writer, "he saw several stages further ahead in the various possible lines of play than did his rivals."[2] In the open race for the succession Stalin won hands down, on merit.

And now that thoroughly Russified Georgian, Stalin himself. What was he like? His detractors ascribe his nasty character to his illegitimate birth; his father supposedly was a high Russian official stationed in Georgia. Whatever the facts, he grew up in an uncouth cobbler's household amidst violent public commotions deepened by fervid local patriotism. Prompted by an innate sense of protest, he was first drawn to anti-Russian Georgian nationalism and steeped in the lore of Caucasus mountain feuds. At school—a seminary of the Orthodox Church where, it is said, he learned a cynical attitude toward doctrine—he turned revolutionary Marxist, fired subsequently by the furies of beginning industrialism in the oil center of Baku and similar cities of southern Russia. Soon he became submerged in the Russian revolutionary tradition which had long combined idealistic ends with morally questionable, or even criminal, means. In the revolutionary underground he alternated between political agitation and time in jail or Siberian exile. Among his comrades he was respected for his willingness to stay on the job when softer revolutionaries preferred foreign exile; thus he endeared himself to Lenin.

As a frequent victim of the tsarist prison system, he experienced firsthand the roughest realities of popular political culture. In the tsarist prison system, so Leo Tolstoy has written,[3] "all sorts of violence, cruelty, and inhumanity are not only tolerated but even sanctioned by Government when it suits its purpose." That inhumanity, Tolstoy continued, was impressed upon all political prisoners thrust together with thieves and criminals. The entire tsarist prison system "seemed purposely devised for the production of depravity and vice, and for spreading this condensed depravity and vice broadcast among the whole population." It produced people "who excelled [in] Nietzsche's newest teaching holding everything allowable and nothing forbidden, and spreading this teaching first among the convicts and then among the people in general." Life in prison among convicts was part of Stalin's political schooling during his formative years as a Leninist revolutionary. Its ethics became part of his character and of his working equipment for the tasks set by the Bolshevik revolution.

After the seizure of power he was given important assignments, which according to all accounts he handled badly, in part because they were intrinsically unmanageable. A prominent figure in the civil war, he excelled in ruthless brutality rather than strategic skill. Judged a mediocrity, he was appointed the party's general secretary in 1922, a job which the leading party intellectuals considered beneath their dignity. In this role he paid little attention to the discussions of high policy, devoting himself instead to what mattered most: party organization and discipline in the rough manner of popular political culture. In the process he built up, with an ex-convict's finesse, a personal following bound to him by self-interest, temperament, and experience. With its help he mustered the majorities at party congresses that enabled him to triumph over his rivals. As in other countries hard hit by the revolution of Westernization, so in Eurasian Russia: the most successful leaders rose from below, from the masses, conforming closely to their temper and habits. These qualities enabled Stalin to step into Lenin's position and to use a more tightly knit party as an instrument for drastic reculturation. While Lenin had supplied a theory of substitutionism with little comprehension of what it held in store and no opportunity to put it into effect, Stalin was ready for the far more difficult task of making it work. Thus the stage was set for the second great trauma descending within one generation on the people of the Soviet empire, the "Stalin revolution."

By 1929 the Communist Party, its Soviet subjects, and the world at large had to take Stalin as he was, shaped by a troubled childhood, by the nationalist struggle within the multinational Russian empire, by the revolutionary underground and the tsarist police, by World War I, by the Bolshevik seizure of power, the civil war, and the succession struggle. Last, but not least, he was also shaped by the Western impact that had brought catastrophe to a backward country of heightened political sensitivity at the edge of the European state system, imposing upon it the inhuman necessities of drastic recculturation. The scope of his notorious ambition was not set by himself, but by the globalism of the times first formulated in the West; the dimensions of his terror were foreshadowed by the savagery of World War I. What did the individual count when the fate of a country or the construction of a superior social order was at stake? In sum, Stalin and Stalinism were symptoms of the age of the global confluence, its vast scope and its ignorance. And more significant, considering the moral outrage directed at them, they were ultimately the responsibility of the model countries uncomprehendingly shaping that age.

Stalin was—to press the point—an uncouth Hobbesian Leviathan trying, within the confines of the Russian empire, to create order, authority, and meaning from a condition of anarchy and violence which offered no guarantee of the security needed for civilized existence. The world at large had just emerged from World War I; within that framework the Soviet Union hardly constituted an orderly polity even in the mid-1920s. Its revolutionary elite, riven by bitter feuds, lacked legitimacy amidst an unruly people suspended between dark backwardness and the most progressive imported views. In the absence of any institutions or bonds of cohesion strong enough to hold state and society together, the only effective unifying force was the will of an individual; and looking at Stalin, we should add, of an individual not only inarticulate by nature but also hamstrung by the inadequacies of Marxism-Leninism for dealing honestly with the given realities. Like the tsars, the Soviet rulers developed no political theory growing organically out of state practice; the ideological facade covered up many inadmissible realities. Lacking theoretical guidance and time for analysis, action arose out of wordless and often sordid spontaneity. More than Hitler, Stalin acted by political instinct guided by sheer will. The revolutionary accent rested on will, particularly at the core of government. Here monstrous elemental willpower was the locomotive of history.

The task of that locomotive was to "overtake and surpass" the "capitalist" model, to make the Soviet Union as powerful and prosperous as any Western state—with the United States looming ever more clearly as the universal model—and to endow the country with a superior social order capable of serving as an inspiration in its own right. That huge job was made even bigger by the demand that it be achieved in the shortest time possible. Stalin, temperamentally and politically predisposed toward extremes, took as his starting point not the superficial pacifism of the late 1920s but the political catastrophe of 1917–18; he too imagined himself to live in an age of ultimate solutions. Moribund "capitalism," especially in its "fascist" variant, he assumed, was locked in deadly battle with the communism of the future; Soviet Russia had to be ready for all eventualities. "We are fifty or a hundred years behind the advanced countries," he said in 1931, "We must make good this distance in ten years. Either we do it or we shall be crushed." Drawing on deeper resources of Russian patriotism, he cited the setbacks caused by Mother Russia's weakness in the past, concluding: "Such is the law of the exploiters—to beat the backward, the weak. . . . You are backward, you are weak—therefore you are wrong; hence you can be beaten and enslaved. You are mighty—therefore you are right; hence we must be wary of you."[4] In these resounding words he voiced the moral challenge and the defensive innocence stirring in all victims of the world revolution of Westernization, though with a characteristic trace of hypocrisy. Bolshevik theory and practice, continuing tsarist foreign policies, had contributed their share to the tensions in international relations from the start. (Yet was Stalin's hypocrisy worse than the Western hypocrisy in not recognizing the consequences of world-wide Westernization?)

In any case, catching up in ten years meant concentrating all energies on domestic mobilization while observing the utmost caution in foreign policy. And domestic mobilization foremost meant all-out industrialization. Already in 1927 the party, agreeing on this point, had decreed a first Five-Year Plan for the implementation of that policy; Five-Year plans for economic development have been the rule ever since, highlighting the Marxist contention that rational planning can prevent the inhumanities of the "capitalist" market economy. At the start of forced-tempo industrialization it became clear that industrial planning also required strict control over agriculture, which was in part still based on traditional communalism or in its more productive aspects managed by private ownership. The next stop, therefore, was collectivization, the forcible transformation of peasant agriculture into large-scale farms subject to the Five-Year Plan and progressive mechanization. Together industrialization and collectivization called for the drastic recasting of the socio-cultural order, imposing the complex and subtle discipline of urban-industrial society upon utterly unprepared peoples through organization, instruction, and command. Such drastic reculturation finally required a huge administrative machinery of substitutionism capable of translating what was done automatically in the West into deliberate words, symbols, and inescapable compulsions. Such all-inclusive substitutionism called for immense creative energy, imagination, intelligence, and cunning of a kind not required in the Western model. And not wanted.

The most obvious and crucial form of deliberate substitutionism, to be mentioned at the outset, was practiced by the secret police responsible for internal security. Already the tsars had relied on a special police force, with agents even abroad, to guard themselves against civic commotion or assassination. In the exceptionally hazardous postrevolutionary era Lenin expanded that tsarist institution with the help of its experienced victims. What in the West was provided freely by a reasonably peaceful and loyal citizenry had to be replaced, under a controversial revolutionary regime in a notoriously unstable country, by an elaborate network of secret agents with power to enforce public docility by fear and terror. With Stalin's help that secret police, under various names, virtually became a state within the state, all-powerful and omnipresent.

It was linked to another form of substitutionism strenghtening state security by alleging universal preeminence for the Soviet system. The Western sense of superiority generating immunity against all subversion from without had to be artificially reproduced in the vulnerable Soviet Union. It was done by the endlessly reiterated claim that "socialism" rather than "capitalism" showed the way to the future. This dogmatic assertion was backed up by the prevention of all harmful contacts with the outside world; the Communist Party made sure that the Soviet Union emerged victoriously in all invidious comparison. Thus the invisible cultural barriers keeping the West immune to outside subversion were replaced in the Soviet Union by a tight security fence designed to achieve the same effect.

The most obvious form of substitutionism, of course, was the Marxist-Leninist ideology itself. A substitute intellectual culture in its own right, it took the place of the Western self-assured, closely integrated, and largely subliminal sense of collective identity and place in the world, spelled out in ever greater detail. Already in the 1920s Soviet intellectuals under party direction had created a Soviet worldview of encyclopedic proportions, covering all fields of knowledge and all controversies between the Soviet Union and the "capitalist" West. Such mental and spiritual framework—impressive proof of Soviet substitutionalist emancipation, though highly artificial from Western perspectives—was an indispensable precondition for reculturation.

In the 1930s substitutionism began to reshape Soviet state and society at top speed. First came the implementation in the fields that counted most, in economics and politics. Implementation, however, suffered throughout from the lack of prior experience among both leaders and followers in the party and among the run of the people. Admittedly, some technical knowledge existed, inherited from the past; additional know-how was imported. But the real challenge lay in fitting technical expertise into the human contexts, in drawing, under untrained leaders, the untrained masses into large-scale organization in industrial and agricultural production, and in political cooperation as well.

Not surprisingly, the first Five-Year Plan unleashed a militant effort among party activists not only to recruit workers and peasants into managerial positions for shaping a true "proletarian" society, but also to liberate popular creativity through art and literature appropriate to the all-out effort cast in the doctrinal form of "socialist realism." All human creativity was to be commandeered for the gigantic mobilization. Given the overriding need for organization, however, it was no wonder that the original spontaneity of that miniature "cultural revolution" was soon lost in the ever-growing bureaucratic apparatus needed to guide the all-inclusive cultural revolution of Westernization. That revolution in its Stalinist variant demanded total control under the facade of Marxist-Leninist rationality. Examined closely, however, it was a process of confused trial and massive error driven by a ruthless will with methods learned in the revolutionary underground and the civil war, brutal yet unavoidable considering the circumstances. The standard of living, barely up to its prewar level in 1927, fell sharply; industrialization required concentration on capital construction, not on consumer goods. Economic planning, however logical on paper, covered glaring confusion, mismanagement, and appalling waste; the first Five-Year Plan was scrapped before its conclusion. Only gradually did the planners learn the tricks of their trade, though never living up to their promise.

Collectivization brought worse calamities. Pushed by Stalin in 1929 with vindictive fervor, it led to a virtual "war against the nation." Uncomprehending, stubborn, and frequently violent peasants were herded into collective farms at gunpoint; resistance was sometimes punished with volleys fired into helpless crowds including women and children. The worst sufferers were the most enterprising peasants, the kulaks; if not killed outright, they were separated from their families and driven off to forced labor. The casualties ran into millions. In part because of peasant resistance, famine broke out in key agricultural areas, adding to the human toll. And yet, despite the famine, the country exported grain in order to buy foreign industrial equipment (under highly adverse terms of trade during the depression). Under these conditions collectivization, even more than industrialization, crushed the recalcitrant and suspicious anti-modernism entrenched among the backward masses. Drastic mobilization could never succeed except by breaking the will of the peasant majority. And if, in addition, the punishment could settle old scores, such as the Russian resentment against Ukrainian separatism, it seemed all the more deserved, whatever the human price.

Part of that inhuman price was charged to Stalin. Was such brutishness really necessary? Was it in keeping with the Marxist promise? The challenge to his leadership which surfaced in 1934 not only touched his suspicious ego but also raised basic questions. Who else among the top party cadre had the will and ruthlessness to carry the unprecedented experiment to its conclusion? And more: how could the inevitable elemental resistance to the drastic and basically uncomprehended change, and to its unending confusion, be disarmed? Here again Stalin showed his mettle. Top priority, obviously, belonged to strengthening the central will, the guiding authority conducting the perilous experiment. That will, obviously, had to be the more imposing the greater the confusion. It required an unassailable image of the supreme leader.

Stalin had taken the lead in exalting Lenin after his death, drawing on popular instincts deeply embedded in tradition. Now he deified himself as "the Great Leader of the Soviet People" or even the "Greatest Genius of All Times and Peoples" (in the exaltation of "the leader" one finds little difference between fascists and communists). In part the leaders, forever insecure, needed psychological crutches for their egos (one wonders whether the crutches were really effective in their innermost psyche). More important, they needed powerful—or even magic—symbols for gaining the elemental legitimacy needed for leadership.

Stalin's "cult of personality" should therefore be viewed not necessarily as a description of an organizational reality—there is good evidence that his intentions did not automatically prevail in all party decisions. It rather served as a deliberate political instrument, an essential part of total mobilization; it filled an obvious need for metaphysical security amidst frightening disorientation and helplessness. And did he not have strong justification? Having won in open competition for leadership, he could well claim, Leviathan fashion, that *his* self-interest represented the collective interest as well. In the absence of any legitimation by established institutions, *he* was the state; *his* security was state security; *his* will constituted sovereignty; *his* power created the distinctions between right and wrong; *his* personality set the style for the heroic Soviet experiment that was to complete Lenin's vision.

Considering his training, it comes as no surprise that when challenged he defended his leadership also by outright terror, striking at known or potential rivals with the devious cunning he had learned in tsarist prisons. Yet again, his terror was more than an instrument of self-aggrandizement. There was reason in his ferocity. Hidden in political instinct and censored even at the source, it deserves rational analysis as we touch the core problem of enforced reculturation.

How could rooted convictions be adjusted and established perspectives be enlarged; how could hardened human wills—especially the obstinate wills of Russians—be made fluid so that they would readily fuse with other human wills in large-scale cooperation? How could the diverse and self-assertive peoples of the Soviet Union be taught the docility needed for common submission to the myriad demands made by men, machines, and organizations in an industrial society? In the Western model such civic docility had been evolved from within, under highly favorable circumstances, in the bloody wrangles of nation-building through centuries, to be buried eventually under a collective consciousness of freedom. In Soviet Russia—as in any non-Western country bent on catching up fast—that externally imposed cultural transformation had to come overnight, a deliberate revolution of drastic reculturation engineered by the Leviathan and his henchmen.

It had been Lenin's ambition from the start to create a new type of men and women, "a new man" capable of building a socialist, and eventually a communist, society. The "new man" was to be hardworking, enterprising, dedicated to civic cooperation, technically competent—a superior replica of the Yankee capitalist. According to Lenin, communists were to be "engineers of souls," shaping that new type of citizen through education in the schools, in youth organizations and professional associations, through literature and art, through the media. Under Stalin that effort was consolidated and intensified. Stalin, however, was not content to be merely an "engineer of souls"; he aimed at shaping human wills. Souls cannot be engineered as long as the human will remains inflexible. His all-inclusive cultural revolution was to start at the innermost core of human volition. But how?

On a small scale, familiar to Stalin from home and prison, the prescription was to slap an obstinate child or to beat an unrepentant revolutionary in order to soften—or break—their wills. In the maximalist reculturation of a vast population, the prescription—equally reasonable—was large-scale unmitigated blind terror. The gradual chastisement of recalcitrant human wills achieved in the West through the calamities of war and civil war combined with strong incentives for voluntary cooperation had to be accomplished in Russia artifically, by a Leviathan's elementary body blow against established habits, administered without distinction of rank or person. Certain suspect categories of people offered special targets: old Bolsheviks with authority to criticize, assertive nationalists among the national minorities, refractory dissidents drawing unfavorable comparisons with the West. Yet what mattered most was the impact of raw terror on the people as a whole: to make them malleable like clay in the hands of the grand master of reculturation. Grandly staged showtrials of once prominent party leaders demonstrated, with hardly a hitch, their submission to Stalin as an example for all the others.

In this gruesome enterprise Stalin found a ready response from popular political culture, its callousness enhanced by the moral confusion caused by enforced, uncomprehended rapid change. All accounts by surviving victims of Stalin's terror testify to the spontaneous or even refined inhumanities practiced by a multitude of interrogators or camp guards. Many years later one of the most heroic among the survivors, Alexander Solzhenitsyn, asked a fateful question about his tormentors. "Where did this wolftribe appear from among our people? Does it really stem from our own roots? Our own blood?" Devastatingly he answered: "It is our own."[5] Stalin merely made the wolfishness of popular culture an instrument of high policy (with ominous precedents in the Russian past), while trying to meet a challenge never before faced in history by any people. And for what gain?

The costs of forced reculturation in the 1930s were indeed appalling. There was the brazen cynicism of covering inhumanities with the optimistic phrases of Marxist-Leninist ideology—Stalin declared "socialism" achieved in 1935. Worse was the endless squandering of material resources because of mismanagement, carelessness, fatigue, or indifference; worst, of course, the extermination of tens of millions of human lives with their talents, some of them irreplaceable. Not all wills were broken; some survivors carried into the future a steely anger matching Stalin's ruthlessness.

Yet the gains of that cruel decade were also impressive. Industrialization and collectivization had established a reasonably productive economy sufficient for basic needs. A system of tight political controls was also in place holding together a population disciplined (if artificially) as never before in Russian history. In addition, Stalin, the Russian nationalist, had revived Russian cultural and political tradition, drawing on the country's past to provide a badly needed cement for social cohesion at a time of over-rapid change. Marxist depreciation of the feudal past gave way to identificaiton with past glories, proving that reculturation did not imply a loss of traditional identity. At the same time, sexual morality, much relaxed after the revolution, was tightened. Both in cultural outlook and social mores Stalin's Russia turned conventional or even conservative.

A more palpable boon was the new guarantee of minimal social security. There was substance in the claim that Soviet socialism served the needs of the masses, especially those employed in industry. The uncertainties frightening simple people in the vastness of urban-industrial society, especially unemployment, were minimized. Soviet popular culture did not furnish the psychospiritual props of individual self-reliance found in the more affluent Western model. It was up to the Soviet state to supply that security (another instance of substitition), at a high price of inefficiency in productivity perhaps, but as a public service legitimizing the regime. Opportunities for advancement lay open to workers and peasants; the rigid social distinctions of tsarist days were gone. Contemporaries reported an air of buoyancy emerging toward the end of the thirties.

Within this framework Stalin, the product of a preindustrial age, apprenticed his people to the urgency of mastering technology, working in large organizations, engaging in constant improvisation for improvement under conditions of utter uncertainty, all under extreme threat of arrest, forced labor, and annihilation. He bullied them into the docility of civic cooperation—thanks to him Soviet Russia never suffered from peasant rebellion or revolutionary terror. He also instilled into them an involuntary heroism, a capacity for enduring intolerable hardships, unwittingly preparing them for the climax of sacrifice that came after Hitler attacked in 1941. No people in the West has passed through such extreme—such demanding—suffering as the people of the Soviet Union in the throes of precipitous reculturation. In the process culture-blind Marxism-Leninism had at last come to grips with the hard realities of backwardness previously hidden from sight, its optimism interlaced with a pessimistic but realistic penchant for violence and compulsion, its ignorance paid for in human lives. Stalin's successors—and indeed all the Soviet peoples, one might argue—have benefited from the socially disciplining catastrophe he engineered. But to this day the extent, the cause, and the methods of compulsion in that transformation remain an intellectually and morally unresolved burden on the Soviet conscience. Had, then, the Soviet Union made progress in catching up to the West or had the distance from the West even been widened?

And the people of the West? How could they understand what happened, lacking all relevant experience? How could they apply their sense of rationality—or even of humanity—to a condition entirely outside their own historic evolution? How could they judge morally what lay so totally beyond their grasp, especially as they were so wrapped up in their own problems?

Among the Western democracies—to remind ourselves of the global contexts—the Great Depression lasted virtually through the 1930s. Unemployment remained high, giving Stalin an unexpected propaganda advantage. During the early Five-Year plans, a few adventurous young men from "capitalist" countries took jobs in the Soviet Union. Others back home were favorably impressed by Soviet planning. The depression also increased the voltage of international tension, flaunting not only the red flag of Soviet ambition and ruthlessness in political mobilization but also the banners of nationalism raised by the breakdown in international trade and interaction. The trouble-makers bent on upsetting the postwar settlement for their own benefit found their appetites stimulated as the depression opened tempting opportunities. The first country to take advantage was distant Japan.

11

Japanese Expansion

In Japan, as we have seen, disillusionment with Western institutions and influence had fostered a nationalist revival with fascist overtones, strong particularly among younger army officers. The depression, hitting with special vengeance a tightly populated country dependent on international trade, played into their hands: their people needed more living space, their national economy more secure access to raw materials and markets. The obvious field for expansion lay in nearby China, especially in Manchuria with its considerable industrial resources. Why not seize it while China was feeble and the Soviet Union preoccupied? In September 1931, the commanding officers of the Japanese troops stationed in Manchuria, protecting Japan's interests, seized control of the country, with support from their headquarters in Tokyo. They forced their embarrassed government's hand, unleashing a wave of patriotic support which legitimized their insubordination. By early 1932 Manchuria was a Japanese puppet state.

The Manchurian adventure had far-reaching consequences. At home it soon ended government by the political parties. Through a series of assassinations and military plots against civilian leaders the army gradually strengthened its hold on the government, all the while cultivating anti-capitalist and anti-foreign sentiments. In the "Showa Restoration" political freedom and Western ideas were replaced by the ancient "Way of the Gods," by Shintoism (with the help, admittedly, of modern means of communication). The emperor, although uneasy about the ascendancy of the army, felt unwilling to face the public unrest bound to follow from reducing the military to its constitutional place; besides, his hands were tied by the rising nationalism in his country as well as the continuing expansion of military operations in China. The times were not meant for "enlightened peace," as was obvious also in the international reaction to Japanese expansionism.

In 1931 strong protest over Manchuria were launched by the League of
Nations, backed up also by the United States, but never enforced by effective
sanctions; seen from depression-ridden western Europe or Washington, Manchuria was too far away. In China the nationalist government under Jiang
Jieshi resisted as best it could, providing a further challenge to Japanese
aggression. Starting in the north and spreading out from Shanghai to the
south and inland, the Japanese army began open war on China in 1937. In
the early operations an American and a British warship were bombed, with
no adverse reactions from their governments except in words. Unchecked,
the Japanese campaigns in China continued, the Far East prologue to World
War II.

Whatever the professions of traditionalism, Japanese expansionism was
now set into the advanced perspectives of the new globalism. Its most
extreme vision was sketched by Prime Minister Tanaka as early as 1927.

> The way to gain actual rights in Manchuria and Mongolia is to use this region
> as a base and under the pretense of trade and commerce penetrate the rest
> of China. Armed by the rights already secured we shall seize the resources
> all over the country. Having China's entire resources at our disposal we shall
> proceed to conquer India, the Archipelago, Asia Minor, Central Asia, and
> even Europe.[1]

Army policy in the 1930s was more cautious. Japan was to create a "New
Order in East Asia," combining its own resources with those of Manchuria
and China, yet not without reference to Europe.

The more the Europeans warred among themselves, the freer the hands
of the Japanese warlords. In 1936 Japan began openly to take sides, concluding the anti-Comintern pact with Hitler's Germany. German Nazis and Japanese warlords held certain goals in common. As anti-communists, both eyed
the Soviet Union as a target for expansionism; for their largest vision they
anticipated a new world order placing the nationalist (or fascist) counterrevolutionaries in command. As Westernizing anti-Westerners, they also struggled with common problems in trying to combine their ideological traditionalism with the necessities of winning power, with the latest military technology
as well as the skills of large-scale organization, both attempted to put willpower and valor above technical expertise (to their ultimate undoing).

Committed to proving the merit of their experiments of national aggran-
dizements in war, both challenged an interdependent world too big and complex for comprehension by people so ardently attached to their roots. The
uncertainties of the new globalism were rudely impressed upon the Japanese
leaders when Hitler and Stalin concluded a nonaggression pact in August
1939; European and Far Eastern Politics did not necessarily run in the same
groove. Japan's attention remained concentrated on China and on political
developments there.

12

China—Toward the "Yanan Way"

I

The chief experiment in East Asian anti-Western Westernization was not taking place in Japan (a reasonably Westernized country despite the Showa Restoration), but in China. There it ran in two parallel and competing channels, through the Guomindang and the Chinese Communist Party, the former the most prominent at the time, the latter holding the keys to the future.

The Guomindang established itself as the official government in China in 1928; it enjoyed its best years before 1937. Its hold over the country, however, was always limited; even in the areas which it dominated its writ was often disputed or disregarded. All the same, it represented China to the outside world, considering itself a nationalist revolutionary force aware of its precarious existence under the shadow of the historic either/or, of disappearance or rebirth, as formulated by Sun Yatsen. In its formal organization the Guomindang combined the unity of a political party, originally shaped under Soviet influence, with the military discipline of its army (now advised by German experts). Yet it suffered from the factionalism common to all regimes in the throes of Westernization: which of the innumerable yet mutually exclusive options should it follow? How much tradition should be mixed with how much Western innovation? Jiang Jieshi, a military man rather than a politician, was hard put to maintain unity among his followers, even though he followed current trends by building up his image as "the leader."

His chief problem—he faced it in common with all anti-Western Westernizing leaders—was how to evolve a convincing fusion of the old and the new sufficiently dynamic to release the popular energies needed for the hoped-for national revival. Influenced by developments in Italy, Germany, and Japan, Jiang leaned toward the past. Hard pressed to compete with the com-

munists, he and his wife started in 1934 a "New Life Movement," an ideological drive to rally the country to its ancient virtues. "The Chinese of today
seem to have forgotten the old source of China's greatness in their urge to
acquire material gain, but, obviously, if the national spirit is to be revived,
there must be recourse to stable foundations." Those foundations were the
four hallowed principles of 'Li" (courtesy), "I" (service to self and others),
"Lien" (honesty and respect for others), and "Chih" (high-mindedness and
honor). Through their proper application, "the standard of official honesty
would become so high that corruption could not exist. . . . " It "would
become shameless to stoop to anything mean or underhanded; . . . the moral
character of the nation" would "attain its highest standard."[1] At the same
time, however, officials of this tradition-oriented organization worked with
foreign missionaries for the treatment of opium addicts or the abolition of
footbinding. Subsequently, courting ridicule, the New Life Movement
preached petty aspects of Western civilization: clean fingernails, personal
hygiene, and the extermination of rats and flies.

Conviction-sapping contradictions showed up at every turn in Jiang's
faintly fascist ideological pronouncements. When he addressed his people in
1943, saying: "we hope that you will practice with deep compassion the teachings of our sages and 'not limit your filial piety and parental love to your own
parents and children," he added: "we must direct the interests of all individuals, families, villages, and regions toward the welfare of the whole nation."[2]
Unfortunately, the concept of "the whole nation" had never carried much
weight in traditional political culture. More important perhaps was Jiang's
concept of race. As he emphasized in a tract on "revolutionary education,"
when the old moral principles had been restored, "the splendid innate qualities of the Chinese race will shine forth . . . all we need to do is to recover
the ancient character of our race and we shall be equal to, nay, even superior
to other races."[3]

Yet how did industrialization, another of the Guomindang's official goals,
fit into the ancient virtues? Whatever the answer, much was done in economic
modernization. More railways were built, and more highways. The currency
was reformed, financial security promoted, new banks founded, all in close
touch with Western firms and countries. The majority of top officials in the
government boasted Western academic degrees; sympathetic Western
experts lent a helping hand. Yet the anti-capitalist and anti-foreign elements
in the government succeeded in declaring Confucius' birthday a national hol-
iday. Could the well-meant efforts of modernization, mixed with the simultaneous affirmation of tradition, bridge the profound cultural gap between
China and the West?

Consider the obstacles, starting with the Chinese language, "a singularly
intractable medium" both for modern mass politics and the expression of
Western ideas.[4] Compared with Western languages it was deficient in nota-
tions of number, tense, gender, and relationships; many of its complex characters (too complex for essential modern instruments of communication)
conveyed a wide range of meanings depending upon contexts set by the sub-

tle ground-floor details of Chinese life and therefore a poor resource for expressing the abstractions and general classifications so essential in Western life. Missionaries trying to translate the Bible into Chinese had long despaired of the results—its central idea found no equivalents in Chinese thought patterns. Similar problems beset the translators of scientific works or political theory. There was no guarantee in translation that Western ideas would be transmitted in any precision even in their narrow meaning, let alone the broader cultural contexts of state-building.

Take another example, the operation of a Western factory in a Chinese setting.[5] The Western-trained engineers would neither soil their Mandarin hands by tinkering with the machines or associate with the workers except in traditional terms, as revered authorities. The illiterate workers were splintered among themselves by regional origin, craft training, age, and function. An integrated technical process was thrust into a highly granular human context in which any friction generated excessive or even destructive emotional heat. How to overcome the traditional granularity accentuated by culture clash within the confines of a factory? How to overcome it within society at large in a country so huge, diversified, and illiterate?

Of divided mind in its own ranks, the Guomindang hardly represented Chinese society. Based in the seaboard cities in close contact with the outside world and allied with the defenders of gentry ascendancy in the countryside, it never reached into the peasant depths of the population. In economics as in politics, city and country, the Guomindang elite and the masses remained cultures apart. The Guomindang's well-intentioned and often reasonable reforms (even on behalf of the peasants) remained largely on paper, resisting or defying translation into popular practice. Challenged by the "communist bandits" and, more alarmingly, by Japanese aggression, its shortcomings grew. In the unbridged interstices between tradition and foreign influence disillusioned followers pursued their naked self-interest, their energies diverted from the national mobilization needed ever more urgently if the country was to survive. Its failure gave a lift to the other experiment, conducted by the Chinese communists led by Mao Zedong.

II

There was no end of discouraging obstacles on the communists' road to power, beginning with the kaleidoscopic nature of the countryside where they had chosen to work. Rural society—that almost powdery "sheet of sand"— was marked by no clear class boundaries as in Russia. Poor, middle, and rich peasants, together with poor gentry and even rich gentry, were interlaced in factions and fleeting dependence and interdependence, their relations also affected by secret society, "local bullies" (toughs retained by landlords or self-employed), officials of the Guomindang, and local warloads with their own troops. Intense clannishness and localism prevailed and, even within the smallest units, a self-contained individualism as uptight as the tight-walled courtyards crowding villages and towns. Relations between the hardened people suffered from an appalling callousness and brutality. Unwanted children, especially girls, were sold into virtual slavery. Mass executions were much-attended spectacles, cutoff heads or hands piled up high for public display. In a densely populated, backward country under an almost perpetual state of civil war human lives were dirt cheap; the prevailing mood mixed passivity and fatalism with an elemental urge to survive by customary ways. How under these conditions could there arise the civil security, the social accommodation, and the initiative needed for voluntary large-scale organization, for socialism, for national greatness?

The material preconditions for such civic advance were likewise lacking, including the means of communication, the mechanical contrivances freeing human energy for work in the superstructure of social responsibility, let alone schools overcoming the pervasive illiteracy. Only a small minority of the people had access to education, frequently taught by radical intellectuals (sometimes of communist views)—who else was willing to serve in the boondocks? And there was always the dehumanizing pressure of abject poverty, overcrowding, and hunger, often aggravated by natural disaster or civil commotion. Into this society the Chinese communists were to implant not only the concepts of an all-inclusive Chinese nationalism but also the mind-boggling universals of Marxism-Leninism lifting the country into the global world. In that setting not only state and society but also the individual mind down to its psychic depths had to be restructured.

Progress along that interminable road could come only through endless and forever unsettling experimentation, through a radical willingness not only to learn from incessant failure but also to dismantle all lessons of the past, all personal and collective pride. Chinese Marxists had to accommodate themselves to peasant perspectives, to local need as interpreted by peasants; they had to win peasant goodwill on the peasants' terms. A way of life preserved unchallenged in its essentials for millennia and hardened by recent and current trouble was not to be recultured overnight to match the social cohesion and power of the Soviet model, let alone of the West. Chinese society was not even ready for Stalinist all-out substitutionism. All the preconditions were lacking: an effective party apparatus, an industrial base, a politically unified country, experience with mass politics. The Chinese communists were starting their substitutionism virtually from scratch, under extreme duress, always preoccupied with power for the sake of sheer survival, aware, as Mao observed midway along the revolutionary road, that "power grows out of the barrel of a gun."[6] The Chinese communist experiments in these years were military experiments set into peasant unrest, civil war, and the war with Japan.

Painful failures dogged the Chinese communists from the start, forcing, under immense risks and sacrifices, constant refinement of political analysis, revolutionary theory, and adjustment to peasant mentality. After the collapse of the United Front with the Guomindang in 1927, Mao's prediction of a hurricane-like peasant revolution was not borne out. Although peasant uprising in Hunan did take place, a Red Army recruited, land redistributed, and soviets formed to take the place of landowners and Guomindang officials, these beginnings were wiped out within a short time by Jiang's army. Subsequently, in the border region between Hunan and the neighboring province of Jiangxi, a communist stronghold arose which in 1931 Mao proclaimed the "Chinese Soviet Republic" with himself as chairman. Under continuous Guomindang attack he presided over an uncertain experiment marred by violence, confusion, and ignorance of how to raise political power out of self-willed peasants, to no gain.

Facing annihilation, Mao and his army of some 130,000 followers fled in October 1934, turning first southwest, then west, crossing the gorge of the upper Yangtse River, heading eventually through Muslim territory to the foresaken loess hills on the border of three rather remote provinces, Shaanxi, Gansu, and Ningxia, where communist guerillas had operated for years. Here they arrived a year later, barely 30,000 strong, physically exhausted from their 8,000-mile march, with Mao now their acknowledged leader, all mentally toughened and more open to reality. Along their way the peasants had not risen; the communists came as armed outsiders, sometimes given local support, sometimes fought off, never welcomed as liberators. Reasonably secure at last, Mao's army in 1936 entered the town of Yanan, which became the heart of the communist-dominated area, numbering more than forty million inhabitants.

Yanan became the laboratory of the next batch of communist experiments in sociopolitical mobilization culminating in "the Yanan Way," the symbolic term for the assorted skills that were to make the communists the masters of China—but only after searing tribulations and reconsideration of many preconceived notions. At first (in 1937), communist rule in its new out-of-the-way base was favored by a second united front with the Guomindang; submerging their differences, the rivals promised to cooperate against Japanese aggression. Two years later the truce broke down, while Japanese armies increased their pressure with more sophisticated strategies and weapons. When in 1941 the Guomindang decided to block the communist stronghold, Mao's communists found themselves in dire straits. On the brink of annihilation, they were forced to concentrate more than ever on practices and policies that produced military and political power by an improvised combination of Marxist-Leninist doctrine and Chinese tradition.

That pragmatic experimentalism and radical readiness to learn was the source of Mao's superior strength demonstrated in these years. Driven by patriotism, he was more prepared than Jiang Jieshi, or even some of his fellow communists, to disregard the Chinese heritage. Responsible for government in a remote area under siege by merciless Japanese armies and the Guomindang, and a resolute patriot in a country in utter peril, he (like Lord Lytton) ventured to apply "the most refined principles of European society," expressed in Marxist terms and mediated through Soviet experience, "to a vast . . . population in whose history, habits, and traditions they have had no previous existence." He pioneered mass politics on the most primitive, most resistant ground floor of life yet hit by the revolution of Westernization. In the process he was forced to make many practical concessions to the custom-bound peasants without whose cooperation he was powerless. Yet he was not leading a peasant revolution, as is sometimes assumed, but a Marxist-inspired national liberation movement, a counterrevolution against the "capitalist" West which set the ineluctable universal model.

Mao's conceptual framework for communist success was premised essentially on the worldview propagated by Marxism-Leninism. Communist China was part of the proletarian revolt against imperialism, building, in alliance with other victims of imperialism, a peaceful world society of socialism headed for the blessings of communism. The relations between Communist China and the Soviet Union remained open for discussion; Mao carefully preserved his distance from Stalin. What mattered was the cultural translation of the communist worldview into Chinese conditions, first of all in the organization of the Communist Party and the training of effective cadre.

Given the fact, repeatedly emphasized by Mao, that China was a backward "semi-colonial country with a vast territory and rich resources that has gone through a revolution and is unevenly developed politically and economically,"[7] suffering also from an underdeveloped culture—Mao in these years was a realist—the Communist Party, the key to success, had to be recruited from all layers of the population. Admittedly, ideology required that preference be given to the "proletariat," a class, as Mao once observed, "largely made up of bankrupt peasants."[8] But in practice "the proletariat" was defined rather loosely; Mao openly admitted that "among the proletariat there are still many who cling to petty-bourgeois ideas, many peasants and petty-bourgeois who have backward ideas." The most effective source of communist cadre, particularly in the higher ranks, remained (as always) the intelligentsia, a group "of special value" despite its "bourgeois" or "petty-bourgeois" origin and its penchant for cultural arrogance.[9] What counted above all was the transcendent "consciousness," which was a matter of training and conditioning. In this way the Chinese Communist Party, even more than its Soviet model, was a classless elite corps of believers committed to a common language and a common discipline for a common goal.

The first problem was how to create an effective language. Chinese communists deriving their inspiration and mental equipment from Marxism-Leninism, faced as many difficulties of translation as the Christian missionaries. Not only in making available key Soviet documents but also in their daily communications they were crossing boundaries between incompatible cultures, a burden rarely appreciated by Westerners reading their smooth texts of Mao's writing. Every day the communist leaders confronted a monstrous intellectual challenge when trying to translate the foreign concepts with their perspective-enlarging abstractions into political practice among intellectuals as well as illiterate peasants.

The next problem was how to enforce discipline and proper consciousness among the brittle and heterogeneous human raw material available. Party membership had grown rapidly after it became clear that Mao's communists, rather than Jiang Jieshi's nationalists, offered the most determined resistance to Japanese aggression. How could the newcomers be transformed into effective communists? Mao's guiding principles, crucial in the evolution of Chinese communism, were set forth in the "Rectification Campaign" launched in 1942.

Ruthlessly castigating the traditional failings of the Chinese intelligentsia, its formalism, subjectivism, and dogmatism, these principles tried to create the pragmatic, flexible, humble, and transcendence-oriented open-mindedness needed for a political organization bent on uniting the country and claiming its place in the world. Never before had the narrowness of the Mandarin mind-set, even as applied to Marxism-Leninism, been denounced in such deliberately down-to-earth language. "Those who regard Marxism-Leninism as religious dogma," Mao observed to an amused audience, "show this type of blind ignorance. We must tell them openly, 'Your dogma is of no use,' or to use an impolite phrase, 'Your dogma is less useful than shit.' We see that dog shit can fertilize the fields, and man's can feed the dog. And dogmas? They can't fertilize the field, nor can they feed a dog. Of what use are they?"[10] Throughout Mao demanded that theory be permeated with practical experience among common folk. At the same time, he insisted that practical work be leavened with theory and intellectual curiosity. "CP members must ask 'why?' about everything, examine it carefully in their minds, and ask whether it conforms with reality"—and always with a sense of concern for the whole, taking "the interests of the whole party as a point of departure in every specific task and every time they speak, write, or act. Absolutely no opposition to this principle is to be tolerated."[11]

The overriding need, obviously, was for liberating purposeful individual initiative (especially wanted in guerilla warfare) by breaking down the tight sociocultural walls fragmenting Chinese society—all the defenses of pride, status, and learning which had prevented cooperation in the past; all the distinctions between manual labor and intellectual labor, city folk and peasants, old and young, insiders and outsiders in clans, secret societies, religious cults, men and women. Special attention was paid to women, hitherto left outside politics. "Women constitute one-half of the population of China. Without women's participation in the revolution the revolution cannot succeed. The number of women in the Party is too small. . . . We must oppose the excuse offered . . . that there is little chance for contact between men and women in Chinese society."[12] Whatever the issues, the radical dismantling of custom brought endless and troubling conflict. In order to contain it, Mao prescribed the quintessential wisdom of peaceful transcendence: "each [side] must make allowances for the other, and both must engage in exacting self-criticism."[13] What could have been more humiliating considering the ancient habit of "saving face"?

The widening-out process was not limited to battering down the barriers within the country; it also extended to the relations with the outside world, again with a touch of humility. While, for instance, praising the Chinese language for its richness and vigor, Mao admitted that it was "not sufficient for our use" and that it was necessary not only to absorb progressive theories from abroad but also "fresh terminology" (like the term "cadre"—*gen bu* in Chinese). And, more generally, "All-in-all we must absorb a great many good elements from foreign countries."[14] But—in the transcendence-promoting dialectical approach preached by Mao there was always the other side—"It should never be indiscriminately and uncritically absorbed. . . . These foreign materials we must treat as we treat our food. We submit our food to the mouth for chewing and to the stomach and intestines for digestion, add to it saliva, pepsin, and other secretions of the intestines to separate it into the essence and the residue, and then absorb the essence of our nourishment and pass off the residue."[15] Superior foreign accomplishments, like Marxism-Leninism itself, had to be "sinified," ingested into the special conditions of China and mixed with the good in native tradition (hopefully without loss of effectiveness), always experimentally, dialectically, expansively, beyond the traditional resources of the Chinese language and the ways of thinking embodied in it.

Marxist dialectics indeed served as a handy instrument for promoting inclusiveness of thought and perception; Mao's use of that foreign import provides unique insight into the problems of reculturation. As he observed, "With the continuation of man's social practice, the sensations and images of a thing are repeated innumerable times, and then a sudden change in the cognitive process takes place, resulting in the formation of concepts. Concepts as such no longer represent the external aspect of things, their individual aspects, or their external relations. Through concepts man comes to grasp a thing in its entirety, its essence, and its internal relations. . . . In the complete process of knowing a thing, this stage of conception, judgment, and inference is more important than the first stage. It is the stage of rational knowledge"[16]—the abstract knowledge required for power in a Westernized world.

The core of the cultural transformation, however, remained in the innermost consciousness of the individual party member who had to acquire a model character as a superbly socialized communist, as a member of a huge organization with a single purpose and will. How, then, to restructure a party member's character and innermost motivation? In this respect Mao developed an ingenious technique of substitutionism. The social discipline and humble conformity acquired spontaneously in the West out of the hidden promptings of culture had to be achieved in China deliberately, through intensive education in small study groups forever practicing criticism and self-criticism.

In these group sessions individuals were put under extreme psychological stress, leaving them psychologically naked in their shortcomings until inwardly reformed. "The first step," Mao said, "is to give the patient a powerful stimulus: yell at him, 'You are sick!' so the patient will have a fright and break out in an overall sweat; then he can actually be started on the road to recovery."[17] The communist technique of "thought reform" aimed at no less than reshaping individuals in the depths of their wills, the source of cultural change. Relentlessly applied in the Rectification Campaign (and never completed), the process did not kill, unlike Stalin's terror; but given the peculiarities of the Chinese mentality, it was as brutal in its application. Yet who could deny its justification considering the changes needed to save the country from the Japanese then pounding at the doors of the communist base area?

Equipped with more effective cadre and a more cohesive party, Mao's task of organizing the social, economic, political, and military aspects of the Yanan Way was made easier. His emphasis throughout rested on promoting military power and all the nonmilitary activities on which it rested: agricultural and industrial productivity, effective administration, concentration on enlarging human resources and energies while minimizing human conflict. In organizing, reshuffling, and streamlining administration, Mao learned how to run a wartime government in a sizable area with minimal resources and maximal reliance on peasant energies.

Considering the need for unity in the war effort, the communists, although pledged to peasant revolution, practiced moderation in their treatment of landlords, rich peasants, and business people in general. As Mao admitted in 1942: "the capitalist mode of production is the more progressive method in present-day China . . . the bourgeoisie, particularly the petty bourgeoisie and national bourgeoisie, represent the comparatively more progressive social elements and political forces. . . . " He considered the rich peasants "the capitalists in the rural areas and an indispensable force in the anti-Japanese war and in the battle for production."[18] At the same time, he promoted social justice for the poorest peasants through rent reduction and lowering their financial burdens; in the battle for food, disorder in the countryside was out of the question. Besides, he gained militant supporters from the most marginal elements in the countryside.

Always commanding "go to the masses," Mao went even further on their behalf by organizing a massive "to the peasant" movement, sending intellectuals, teachers, artists, party cadre, and administrators into the villages, all ordered to practice humility in learning from the peasants. It was another experiment in transcending traditional barriers, leading to endless unsettling encounters on the way toward greater mutual trust. The communists' chief innovation in the countryside, however, was the rural "mutual aid movement," peasant cooperatives sustained by peasant initiative as the basic unit of production.[19] Without abandoning the private base of the peasant household, its land and trading habits, the mutual aid movement was designed to bring together the peasants of a village in a new spirit of cooperation, aided by resourceful outsiders helping to modernize agricultural operations and widening peasant horizons—such, at least, were the expectations in experiments that eventually culminated in the establishment of vast rural communes.

Another major need was industrial production, a relatively easy task in textiles but a more difficult one in metal products and weapons. Here Soviet experience helped as industry was concentrated in public ownership, subject to central planning even when decentralized. Through the textile mills peasant women were drawn into the labor force; in the metal industries workers gained a new sense of human mastery over nature. In order to raise the prestige of the workers and boost productivity Mao, again borrowing from Soviet practice, introduced a new authority figure: the "labor hero," an honored activist dispelling the fatalist lack of motivation among Chinese laborers.

Whatever needed to be done in agriculture, industry, administration, or the recruitment and training of soldiers—it was an endless list of innovations—required a literate and politically alert work force. Mao thus relentlessly pressed for the expansion of schools and reform of traditional modes of learning and teaching. Schools had to serve the needs of the plain people; education had to be practical, teachers willing to learn even from the poorest villagers in order to serve them properly, all pressed to ask "why?" and thereby emancipating themselves from superstition, custom, tradition, or plain thoughtlessness. The communist revolution as part of the revolution of Westernization plowed deep.

Such cultural deep-plowing, indeed, was the essence of the "mass line," the high point of the Yanan experiment. "Going to the masses" had been an old slogan among Chinese revolutionaries. It needed constant repetition and reapplication because the bulk of the population resisted being recultured; large-perspective policies never effectively reached down into the village. However eager for improving their lot, the bulk of the people were unable to initiate the progressive cultural transformation that could make China into a powerful nation; they merely wanted to have their way by their own time-tested insights; they would resist, like the peasants in the Soviet Union, any effort to subject them to unwanted and uncomprehended change. Yet, in the last analysis, effective change, although initiated from without and aiming at goals far above peasant comprehension, could come only with their help and therefore, at least partially, on their own terms.

Let Mao speak for himself on this central theme in the "mass line." In 1943 he wrote:

> All correct leadership is necessarily from the masses, to the masses. This means: take the ideas of the masses (scattered and unsystematic ideas) and concentrate them (through study turn them into concentrated and systematic ideas), then go to the masses and propagate and explain these ideas until the masses embrace them as their own, hold fast to them and translate them into action, and test the correctness of these ideas in such action. Then once again concentrate ideas from the masses and once again take them to the masses, so that the ideas are preserved and carried through. And so on, over and over again in an endless spiral, with the ideas becoming more correct, more vital, and richer each time.[20]

In this classic formulation (much quoted in later years by his admirers), Mao, however, got hung up in a characteristic contradiction. His Marxism-Leninism, an exotic import even if sinified, could hardly be conceived as a product of "scattered and unsystematic ideas" gleaned from the masses. But utterly dependent as Mao was on peasant support—he had none of Stalin's armed force to wage "war against the nation"—he needed to win peasant goodwill; catering to the masses by tracing the revolutionary ideals to them was a trick of Leninist mass politics (revealing an ingrained aversion to foreign influences, especially strong in China). Unfortunately, this propaganda fiction obscured essential realities. The internal dialectics of the Chinese revolution as outlined by Mao were still set into the larger dialectics between revolutionary China and the fast-moving outside world. His revolution, however closely tied to peasant support, was more than an internal Chinese revolution. It aimed at the mobilization of peasant China against the powerful imperialists, the Japanese on Chinese soil and the West as the ultimate source of China's humiliation.

To drive home the point in non-Marxist language: using the Chinese people (and especially the peasants) as the only available vehicle of change—and their tradition (in a characteristic contradiction) as the source of the new communist patriotism—constituted a tragic limiation on the extent of change. Pulling themselves up by their sovietized Chinese bootstraps, the Chinese communists could go only the lengths of those straps in a world still moving by seven-league "capitalist" boots. Advancing by "learning from the masses" meant holding back the experiment of catching up with the "advanced" countries. Yet what else could anyone in Mao's position have done?

The Maoist revolution of Westernization culminated for the time being in the Yanan Way. Its intensified "mass line" sufficed, using the levers of peasant self-interest combined with war and civil war, to yank the people from their millennia-old moorings. Mao had achieved what had never been done before: establishing a close link between the peasant village and an organization claiming to represent the country as a whole (and eventually, because of that link, building the most effective state in Chinese history). He had poured cement into the "sheet of sand," laying the groundwork for transforming peasants into citizens. A seemingly impossible achievement (and more messy in practice than in theory or historical analysis), it was certainly the most heroic and incisive experiment in reculturation yet undertaken. It saved the communist base from Japanese conquest and prepared the way for a sovereign China under communist rule—in an out-of-the-way place, unwatched and unappreciated, while world attention was focused on events in Europe.

13

Hitler's Germany

Back now from the periphery to the center, back to the common woe affecting all the world and Europe foremost, the Great Depression; and forward to its most striking result: the rise of Nazi Germany, the experiment of anti-Western Westernization holding center stage after January 1933. Within a decade that experiment produced the most dangerous challenge to Western ascendancy in the next great upheaval common to all humanity: World War II.

While under favorable circumstances the Weimar Republic might have become a legitimate form of government for Germany, the Great Depression destroyed what little credibility that Western implant had acquired since 1919. Briefly revived by American loans now suddenly recalled, heavily dependent on exports and foreign trade now disastrously shrunk, requiring stability and international goodwill for growing roots in German tradition, postwar German democracy was found wanting in all respects. The moderate parties which had combined in the fragile coalitions sustaining Weimar democracy failed to agree on the proper remedies for catastrophic unemployment, bankruptcy, and public demoralization; their political convictions offered no remedies against the growing disillusionment aggravated by the memories of the lost war.

By 1930 the German government operated under emergency decree, while the parties opposed to the Weimar constitution—the Communists and the National Socialists—scored massive election victories; by 1932 they easily outnumbered all other parties combined, with the latter impressively in the lead. Under parliamentary rules Hitler, commanding the largest contingent in the Reichstag, was entitled to claim the top position in the government. The president, the aging Marshal von Hindenburg, advised by conservatives who thought they could dominate Hitler, appointed him as chancellor. After three years of mounting unemployment and near civil war between communists and Nazi storm troopers, he seemed to be the best choice. What other political leaders with proven capacity for handling mass politics in times of crisis were available? Scheming conservatives without popular support like Papen, Schleicher, or Hugenberg? The Catholics under the scholarly and colorless Brüning? Or the Social Democrats—under whom? The communists obviously were not in the running; their relative success at the polls propelled many scared voters into the Nazi ranks. So Hitler peacefully took control of the Weimar Republic, preparing (with a subsequent minor show of repressive violence) the grand bid for world power he had planned since the early 1920s.

Nobody was aware of Hitler's true aims in 1933 as he set out to reshape Germany according to his own principles of leadership. Keeping silent about his ultimate plans (revealed in *Mein Kampf* but now glossed over), he succeeded with surprising ease. Within a half a year he had dissolved all political parties except his own; within another year he transformed the Weimar Republic into a *Führerstaat*, a one-party state held together by his own undisputed will. The speed of the transformation proved the hollowness of German middle-class liberalism in all its varieties and of the susceptibility of Germany's traditional political culture to Hitler's psychopolitics.

The German middle classes, never politically minded, lacked political views reaching beyond their social, economic, professional, or regional interests; they had always taken government for granted while attending to their own business. The cultural disorientation of the preceding century, the lost war, and the Great Depression had hardly increased their humanitarianism or enlarged their horizons except for rousing their patriotism. Hitler easily assuaged their misgivings about Nazi crudities by his nationalism; he overwhelmed their residual moral qualms by the massive show of common will and idealistic purpose so superbly staged at party rallies. What resistance might be left was held in check by the terror of the early concentration camps or the rumor thereof. How many "good" Germans held convictions for which they were ready to give their lives in the manner of militant storm troopers or communists fighting in the the streets at the height of the depression? Making the supreme sacrifice for a democratic constitution was not a plank in the political programs imported from the West; in German minds that sacrifice was reserved for the fatherland in the case of war. The roster of dissenters thus remained rather short; the strong faith of martyrs was found only among radical Christians or Marxists. The great majority were patriotic trimmers, some with more, most with less civic courage as they compromised their integrity. (But let the trimmers be judged only by people who have acted more bravely under similar circumstances.)

In any case, throughout the 1930s Nazi totalitarianism, though sharply contrasting with Western liberal-democratic practice, sat rather lightly on the German people. Conformity under the Nazi one-party dictatorship affected the political surface of life more than its depths. Little substitutionism was required to assure the civic cooperation and economic momentum needed for reducing unemployment (the depression was overcome, incidentally, before rearmament got under way.). As for labor discipline and industrial productivity, Hitler relied on the proverbial German efficiency; he merely supplied political directives, gradually gearing the economy for war. The industrial labor force, deprived of communist or social-democratic leaders but no longer worried about unemployment, was held under firm political control, appeased by a show of social solidarity—Hitler, after all, proclaimed himself a national *socialist*.

As for nationalism, achieving for once in German history a truly united and unitary Germany was now made easy; the regional differences responsible in the past for federal decentralization ceased to matter in the celebration of common resolve. Nazi racism, on the whole, was not taken seriously; the Nuremberg laws barring Jews from government employment touched, like anti-semitism generally, only a small minority of the population. The majority, especially the youths, were captivated by the emotions released at the annual party rallies with their solemn affirmation of common will embodied in Hitler himself. Compared with the hesitancy and doubt—or outright disunity—troubling the Western democracies in the depression years, Nazi Germany appeared a model polity; Hitler had followers in France, England, and the United States. In invidious comparison with the rest of the world many Germans felt they now could hold their own.

From the start Hitler proved his skill in taking advantage of Western irresolution by showing the world that he meant to wipe out German inferiority. Soon after coming to power he violated the Treaty of Versailles by repudiating the unilateral disarmament of Germany; in 1936, more boldly, he sent his army into the demilitarized zone of the Rhineland. And in the spring of 1938 he united Austria with Germany, making a cherished national dream come true and releasing an outburst of national pride almost as exalted as in August 1914. If only he had stopped there! Instead, his successes released his deepest ambition. Now he held the German people in the palm of his hand, the docile instrument for the realization of his true design, of which after November 1937 he no longer made a secret in his inner circle. Confessing that he wanted to wage his war before he was fifty, he was ready to step forward for the ultimate solution of "the German question," with the full force of his personal convictions that went far beyond the ambitions of most Germans.

In 1939, when, at age fifty, he took the plunge, Hitler dropped some remarks that allow a clear view of the dynamics behind his plans. While the imposing new chancellery in Berlin which he was building for his seat of power was under construction, he addressed the workers, justifying the extravagant structure. "Why always the biggest?" he asked, instantly providing the crucial answer: "I do this to restore to each individual German his self-respect. In a hundred areas I want to say to the individual: We are not inferior; on the contrary, we are the complete equals of every other nation."[1] We are not inferior; we are the complete equals with (it went without saying) the model countries that by their superiority had always oppressed the German ego, as well as his own—such was his innermost motivation.

In the same year he gave himself away even more revealingly. "You would hardly believe," he told his architect, Albert Speer, "what power a small mind acquires over the people around him when he is able to show himself in such imposing circumstances."[2] Though he confessed at that time that he would rather live in simple surroundings—he was thinking, he said, of the smaller minds of his likely successors who needed an imposing backdrop—he stated a basic personal insight: little people can be made big by the grandeur of the organized space with which they surround themselves. His insight, of course, was not original. Inspired by the Hapsburg palaces in Vienna echoing the spendors of Versailles, Hitler stated a truth known to rulers for millennia. But his ambition was now couched in the global dimensions of mass politics, when the ego even of little people craved to feel big. The highest point in all areas of human achievement was imperial domination.

In the grand game of world politics, where the proper spatial framework for proud human beings was empire, Hitler was now ready to cast his dice. As he had announced, with a steely ring in his voice, at this conference with his generals in November 1937: "it was his unalterable resolve to solve the question of German space between 1943–1945 at the latest."[3] A chancellery second to none for a thousand-year *Reich* was the proper edifice for curing the humiliated German ego in an age of inflamed worldwide invidious comparison. Symbolically, the new chancellery, "the greatest building in the world" according to Hitler, was to be crowned by the German eagle spreading his wings over the globe.[4] As Goebbels, mirroring Hitler's globalism, put it as late as 1943: "Whoever rules Europe will be able to seize the leadership of the world."[5]

Examined in global perspective, there was little originality in that vision. It was imitative, shaped by expectations introduced from without, above all from the British Empire, toward which Hitler felt a characteristic love-hate attachment. The imitation was mixed with defiant pride: equality was to be achieved through indigenous resources (improved, admittedly, by some skills derived from further imitation). In this manner Hitler voiced, with deep moral fervor, the universal ambition among all victims of Westernization: "We are not inferior; we are the complete equals," and on our own terms! As a protestor in the fascist mold, he admittedly perpetuated inequality by assigning to his country a superiority of its own, unlike the communists, who claimed to be the voice of *all* the humiliated. Yet the source of protest was the same.

Hitler, as much as any non-Western undergroundling, had lived much of his life in the humiliating crevasses between high expectations and personal misery—as a school dropout inspired by the grandiloquence of a Richard Wagner opera; as a marginal loner in Vienna studying Hapsburg architecture and Vienna politics; as an unemployed artist in Munich impressed by the grandeur of the British Empire; as a soldier offering his life in a glorious war ending in defeat. Now, as German chancellor, he saw his historic opportunity to prove that he who had risen from obscurity was equal to the greatest and that his country, long suffering from similar inferiority, could become a world power. From this layer of psychopolitics he tried to create a new patriotic outburst of self-sacrifice and shape it to his vision.

But, viewed in proper perspective, there was an ominous darkness in his vision, the darkness of a world too large for human understanding and control. It projected into the future the deadly misery of all the consequences of sudden modernization, of the revolutionary Western impact on the unprepared peoples of central Europe. It propagated the fierce struggle for national superiority born of the clash of nationalities and ethnic cultures in Vienna, that many-faceted outpost of the West in the Danubian borderlands. It was surcharged with the primitive ruthlessness of small folk defending themselves, through anti-Semitism, against the relentless inroads of railways, banks, big business, advanced technology—against forces far beyond their resources and therefore murderous for their economic and spiritual security. It magnified the widespread *fin de siècle* disorientation in German life, the explosive anti-modern expressionist search for depth and soul. What an appalling mix of incompatible cultural stimuli and resources, of modern science and ancient myth (as in race biology), of global perspective and ground-floor craftiness, of reason and instinct, of humaneness and bestiality, of peasant culture and world politics! In addition, that confused vision was surcharged with the inhumanity of four years of trench warfare in World War I; it breathed the battle-hardened political millenarianism of ultimate solutions. Through Hitler, more drastically than through Stalin, an uncomprehending, stunned West was given a demonstration of the dynamite piled up by the world revolution of Westernization—and in a relatively "civilized" part of the world at that!

The demonstration got under way in 1938. Wanting war, Hitler thought he could have it on his own terms. Judging the Western powers by the spinelessness of the German liberals, he noted the prevailing aversion to war. It was obvious in the failure of collective security after Mussolini invaded Abyssinia and after he successfully defied the Treaty of Versailles, not to mention the unchecked Japanese expansion in the Far East. The political trends of the times, as he saw them, were on his side. The democratic model countries were degenerating. Dictatorship was on the rise, as shown by General Franco's victory in the Spanish Civil War or the collapse of parliamentary regimes in eastern Europe and in Japan. Thus buoyed, he expected to build up German power through a series of small wars, defeating his victims one by one, first Czechoslovakia, next Poland, eventually the Soviet Union, scaring the timid Western powers by his victories and, in his attack on Soviet communism, perhaps even gaining their support.

Yet in crossing the threshold to war he entered a stage larger than German politics, where thus far he had triumphed. He now put himself at the mercy of factors beyond his expertise. The war he had wanted in the fall of 1938 over Czechoslovakia was snatched from him by his ally Mussolini, who, in the Munich agreement, arranged a peaceful compromise. Frustrated in foreign policy, he was disappointed also at home. The prospect of war dimmed his popularity, persuading him in November to jolt his unresponsive people into bloody militancy by an anti-Jewish pogrom, the infamous *Kristallnacht*. Beginning to doubt the resolve of his people, he was even prepared, if necessary, to crush popular protest by force of arms.[6] In his new exultation nothing would stop him from reaching for his ultimate goal.

Preparing to attack Poland, he succeeded in covering his rear by a non-aggression pact with Stalin, a sharp surprise for Poland's Western allies, designed to discourage them from going to war on Poland's behalf. Yet, to his own surprise and profound dismay, England and France did declare war on Germany after his invasion of Poland. While Stalin rejoiced (prematurely) over the prospects of war between the "capitalist" powers, Hitler suddenly realized that his plans had gone awry; the war had expanded beyond his control. But the new uncertainty (temporarily relieved by the German victories of 1940) merely sharpened his appetite for "the final solution of the German question." There would never be a negotiated peace; Hitler had moved past the realism of *realpolitik*.

Now he, the rustic from lower Austria risen, in his own eyes and those of many Germans, to be "the greatest German in history," was faced with a vast world beyond his experience. Although he had long let his imagination roam over the entire globe, he had no intellectual grasp of the realities outside his limited cultural conditioning. The product of preindustrial society, he took German industrial productivity for granted; he had no sense for scientific inquiry, no regrets over the emigration of Germany's leading scientists. And administrative efficiency, so essential in war, he treated with amateurish disdain. Even more crucial, his understanding of alien countries, above all the Soviet Union and the United States, was shaped by political and cultural cliches; except for war service in eastern France he had never been abroad. At his core he was a self-made primitive carried by his astounding ascent far beyond his depths, driving his limited and twisted political insights to their ultimate conclusion with fanatical resolve, psychologically always on the knife's edge between triumph and utter defeat. The national-socialist experiment, far more than Lenin's or Stalin's, was a personal one, built around an individual psyche that was driven by indomitable will and endowed with immense power over people in a highly unstable society craving a savior.

It was another indication of the extraordinary nature of this unsettled age that a person like Hitler could rise in breathtaking swiftness to a position of supreme command over untold millions of lives. Characteristically, as the battles began to take their human toll, the domestic (or biological) campaign for the final solution also got under way, starting with the mercy killings of cripples and misfits. Only in war did Hitler dare to reveal his raw thoughts and unbridled imagination, both trained by the furies of the times in his youth and early manhood, climaxed by World War I. He dreamt of a new German race to emerge from the frenzy of a war that, even more than the last one, knew no limits.

In looking at Hitler at the outbreak of the war, one is reminded of Dostoyevsky's underground man (quoted earlier), who observed: "I for one would not be in the least surprised if in that future age of reason, there suddenly appeared a gentleman with an ungrateful, or shall we say, regressive smirk, who, arms akimbo, would say: 'what do you think, folks, let's send all this reason to hell.'" Here we might redraw that image and say: There Hitler stood, defiantly recasting the Western ideals of freedom, equality, fraternity, together with national unity, the selection of the fittest, and world power— all the Western ideals of greatness—into his own unreasonable, primitive, and distorted mold. He had taken the reasonable ideals from their natural habitate into the underworld of a disoriented and humiliated mind raised in a disoriented and humiliated society. In 1939 he, who until then had been a moderate dictator compared with Stalin, was ready to send the reasonable world of the West to hell and destruction in the frenzy of a war which in his imagination was the culmination of World War I.

14

World War II

Hitler's war brought the peoples of the world, hitherto still existing in fairly loose and abstract interdependence, into accelerating interaction. It was like a funnel into which was poured the turmoil of the inter-war years and all human diversity too. It was the most powerful common experience yet imposed upon humanity. Admittedly, the intensity of the impact varied, but few people anywhere escaped the expansion of perspectives that accompanied the news of the fighting and the discussion of the issues at stake.

The challenge came from the fascist counterrevolutions led, under the banner of an exalted and exclusive nationalism, by Hitler's Germany, assisted in the Far East by a like-minded Japan. Their opponents defended both the world order established over centuries and the universalism of their liberal-democratic culture. In the crucial phases of the war communist universalism, muted during the Five-Year plans and further weakened by Hitler's attack in 1941, was allied with Western universalism. Fighting for sheer survival, the Soviet Union reemerged as a separate force only toward the end, when victory was certain. By that time parts of its revolutionary universalism had been co-opted by its archrival, the United States.

The central issue of the far-flung contest, therefore, was reasonably clear. It was particularist nationalism against global universalism. Would the world be dominated by self-consciously superior national cultures (or even races) with rather little concern for the rest of humanity? Or would Western internationalism, which submerged nationalism in a more inclusive vision, continue its ascendancy? Never had the universalist West been so crassly challenged in all-out war. Could it still command the resources needed to win and to continue its revolutionary outreach? The contest thus was not only over patterns of world order but also over the cultural resources sustaining them. Which culture could prove its superior vitality?

In any case, the essential contest was limited to elites. Although profoundly stirred by the war, the bulk of humanity was at the mercy of events to be dominated either way, yet sensing the differences between the rivals. The aggressive and inexperienced newcomers set out to prove their superiority by cultural exclusiveness based on brute force. Their opponents, the pioneers of globalism defending and reasserting their preeminence, seasoned their armed might, however incongruously, with the universally appealing ideals of freedom, equality, and human fellowship. As a cultural resource these ideals proved their merit even in the war, releasing energies and building support crucial for the outcome.

II

World War II began gradually. At its start in September 1939 two regional wars of unequal proportions ran parallel, one radiating from Europe, the other contained in the Far East. Linked by the logic of the interdependent global state system, both wars gravitated toward each other, with the initiative in Hitler's hands until the linkage was completed in December 1941.

In Europe's first battles Hitler's armies quickly overran Poland, absorbing the larger part of the country while Stalin took back the eastern regions seized in 1920 from Soviet Russia. True to their pledge to Poland, France and England, together with their respective associates and dependents scattered over the world, declared war on Germany, benefiting from the benevolent partisanship of their close partner in World War I, the United States. The other war, mercilessly pressed by Japan against China, had been under way since 1931, escalated in 1937, and fought doggedly if inconclusively thereafter. The two wars were geographically separated by the Soviet Union and, over even larger expanses (mostly of oceans), by the United States. Powerful states, both observed a closely watched neutrality and resisted direct involvement until attacked.

In April 1940 the European war, in abeyance after the conquest of Poland, exploded in dramatic action. First the Germans overran Denmark and Norway, acquiring wide access to the North Atlantic. Next month they swept through Belgium and Holland into France, dominating most of western Europe and further broadening their Atlantic front. Hitler's astounding success encouraged his empire-hungry ally Mussolini to join the war. Italian troops moved into Albania and Greece, and also from Italy's Libyan colony against Egypt, carrying the war into the Mideast and imperiling British communications through the Suez Canal. At the same time, the fall of France brought the war to all French possessions, and foremost those in North Africa, forcing them to take sides for or against the German-dominated French government at Vichy.

In the wake of the French collapse Hitler toyed with the invasion of England—halfheartedly, revealing again his ambivalence toward that country. In the face of determined British resistance, he soon turned to the east, the ultimate target for raising the German eagle above the globe. While preparing to invade the Soviet Union he was temporarily sidetracked into rescuing Mussolini's troops in the Balkans and North Africa. In the spring of 1941 German armies moved into Yugoslavia and Greece down to the island of Crete, while replacing the hapless Italians in the drive on Egypt. Finally, on June 22 1941, Hitler surprised Stalin by crashing into the Soviet Union with a massive army reinforced by contingents from other European countries under his sway. Knowingly tempting fate, he escalated the European war to deadly intensity.

While his armies were overrunning the industrial centers of the Ukraine, encircling Leningrad, and approaching Moscow, Hitler had cause to worry about his other front against England and its ever closer ties to the United States. His navy was reaching out into the Atlantic, threatening British and American shipping and thereby cementing Anglo-American military and political cooperation. Obviously the United States had to be kept out of the European war. What better strategy than letting Japan divert American resources to the Pacific? While his armies in the Soviet Union advanced, he urged the hesitant Japanese leaders to go to war with the United States, clearing the way for Japanese expansion into southeastern Asia. There beckoned the Philippines, the Dutch East Indies, British Malaya, Burma, possibly even India—all targets of Japanese ambition since the 1920s. Japan's military leaders, like Hitler, expecting the imminent collapse of the Soviet Union, could safely transfer their troops southward. The British were preoccupied by the war in Europe; and the United States, judging by the easygoing style of American life, were judged to be incapable of matching Japan's prowess. Thus the Japanese were tempted into preparing their attack on the American outpost in the Pacific at Pearl Harbor; they struck on December 7, 1941. Four days later Hitler, thrusting aside all caution, declared war on the United States, thereby fusing the hitherto separate wars and making his war truly global.

III

Now the war had become World War II, and far more intensely global than World War I. It raged over most of Europe and Eurasia; it mobilized the fighting power of the United States for battle across the Pacific as well as the Atlantic. It covered all the islands, big and small, of the Pacific, all of East and Southeast Asia, reaching from Burma into India. Farther west, Iran was under divided occupation, Soviet in the north and British in the south, to keep the country out of pro-German hands. Nearby Iraq and Syria, though not occupied, were firmly pushed into an anti-German course. Egypt was a frontline country while the battles raged up the North African coast to Tunisia; the lands farther west soon became the staging area for the Anglo-American invasion of Italy. East Africa too witnessed war, as the Italians briefly struck out from Ethiopia before being ignominiously defeated with the help of Africans from British colonies in East and West Africa; these troops also served on the Burma front, defending India against the Japanese. All Africa was drawn into the worldwide net of military communciations, while suffering from the disruption of shipping services and of local economies.

The war also raged over large bodies of water, the Pacific Ocean as the American forces moved toward Japan, and the Atlantic, where the Germans in 1941 and 1942 pushed a naval offensive, largely through submarines, attacking British and American ships carrying supplies to England and the Soviet Union. Only the countries of the Western Hemisphere from Mexico to Argentina stayed out of the war, yet agitated by the disarray in the world economy and by divided loyalties. In order to counter pro-German (or fascist) sympathies, the United States launched a program of economic assistance to Latin America, promoting a pro-American neutrality. The impact of modernity helped to stimulate local initiative and develop local resources.

Directly or indirectly, the war mightily advanced the world revolution of Westernization. It made itself felt through the fast-moving military technology and communications as well as through the necessity of creating large-scale organizations serving soldiers, their weapons, and the civilians engaged in war production. It provided unprecedented mobility to military personnel, mingling people of different cultures and bringing new perspectives and expectations into hitherto out-of-the-way places. Everywhere the war news made the biggest headlines, teaching lessons in geography as well as politics. Victory or defeat in local battles affected all military operations around the world, and the future of the belligerents as well. Never before had the common fate of a world embattled been so intensely demonstrated, as politics, economics, and life in general became geared to the war. The world seemed to shrink into a manageable single entity, certainly as seen from the United States, which straddled the war in the Pacific and in Europe. Never before had a common event—a global war—injected such drastic changes into so many peoples' lives. For what ends and by what means?

IV

To take the ends first. Hitler's aims (no need to consider an ever more ineffectual Mussolini) have already been outlined: a land-based living space for an enduring German empire outshining its British model. At the height of military success his fantasies carried him into India or even to outright world domination. The beneficiaries of the current bloodshed were to be the future generations of racially homogenized Germans lording it over the inferior peoples of the East. As he put it in one of his table talks in October 1941:

> The new territories in the East seem to us like a desert. . . . We'll take away its character of an Asiatic steppe, we'll Europeanize it. . . . As for the natives, we'll have to screen them carefully. The Jew, that destroyer, we shall drive out. As far as population is concerned . . . we shan't settle in the Russian towns, and we'll let them fall to pieces without intervening. And above all, no remorse on this subject! We are not going to play at children's nurses; we

are absolutely without obligation as far as these people are concerned. To struggle against the hovels, chase away the fleas, provide German teachers, bring out newspapers—very little of that for us! We'll confine ourselves perhaps to setting up a radio transmitter. . . . For the rest, let them know just enough to understand our highway signs. . . . There is only one duty: to Germanize this country by the immigration of Germans, and to look upon the natives as Redskins. . . . [1]

As for western Europe, Hitler's cultural arrogance was more restrained. While never holding out hope for a cooperative European order under German leadership, he at least decreed as a minor concession that Germans henceforth use the Latin script as a common bond with other western Europeans. Though not quite as barbaric as in the East, German rule in western Europe was decidedly oppressive. It won no popular backing; fascist internationalism made few converts. Hitler's vision for the future remained vague. What counted during the war was brute force, intensified by the lack of popular support in the conquered lands.

For the Japanese generals, as for Hitler, the war was waged for the building of a majestic Japanese empire. The war stretched from Manchuria southward through China into Indochina—Saigon was occupied in July 1941—and after Pearl Harbor into the Philippines, Dutch Indonesia, British Malaya, and Burma, into the islands deep into the Pacific, with an expedition directed even toward the Australian coast. The goal was a "New Order in East Asia," first proclaimed in 1938 and, after Pearl Harbor, enlarged as the "Co-Prosperity Sphere." Civilian leaders hoped to gain goodwill among the many peoples in that area by proclaiming Japanese rule as liberation from Western colonialism. The generals, however, proceeded with the ruthlessness already demonstrated in China, while the terrorism of the military police matched the brutality of the army. Skillful Japanese diplomacy might have recruited an Indian national army clearing the way into India, but the generals treated its potential leader, Subha Chandra Bose, with contempt. And the "Co-Prosperity Sphere" produced neither prosperity nor goodwill. While destroying the mystique of Western superiority, Japanese occupation fomented a deep anti-Japanese hatred lasting for decades, unaffected by the sham independence for occupied lands proclaimed during the last days of the war.

On the other side, among the victims of aggression, the foremost and immediate aim was self-preservation. In the Soviet Union the war was waged not for any global aims, but for stark survival. Soon known as the Great Fatherland War, it drew on the deepest instincts of religion and patriotism; the virtual suspension of Marxism-Leninism in the face of catastrophe showed how shallow the ideological indoctrination had remained. Disloyalty and desertion were particularly frequent during the precipitous retreat of the first year. Morale improved when the barbarism of the German troops became public knowledge. But Stalin's security forces never relaxed their grip on the country, even at the front. In the second year the battles between German and Soviet forces reached a pitch of ferocity matched only in the last stages of Japanese resistance to the advancing Americans. The merciless destruction wrought by the retreating Germans left the country and its cities in shambles.

But whatever the damage to the country, a seachange had taken place in Russian history. By the end of the war the country had victoriously withstood the biggest onslaught in history; henceforth it enjoyed a measure of territorial security denied to it in the past. While the Communist Party took credit for that enormous feat and the population remembered—grimly, sadly, and proudly—the sacrifices that had made it possible, Stalin, like any Russian statesman under similar circumstances, was eager to enhance the future security of his state. State security in the Eurasian vastnesses had always hinged on the command of territory. During the war he insisted on regaining the territories lost to the Poles in 1920, granted to him by Hitler in 1939 and lost again in 1941; he also eyed control of the exit from the Black Sea to the Mediterranean. If suitable opportunities arose, Stalin would grab them. As the Nazi empire collapsed, Stalin's armies advanced through eastern and central Europe up to the Elbe River in the heart of Germany, where they met their English and American allies.

His partners in the war, Britain and the United States, conditioned so differently by their traditional territorial security, viewed their war aims in a different light. The Western allies set their sights not only on the defeat of Germany and Japan but also on the reassertion of their universalist mission. Soon after Hitler's attack on the Soviet Union, Franklin Roosevelt and Winston Churchill, as leaders of their countries, met in the North Atlantic and issued a joint declaration, known as the Atlantic Charter, outlining their war aims. In that document they repudiated any claim to "aggrandizement, territorial or otherwise," stipulating that in the coming peace settlement all boundaries of governments be subject to popular approval. They also held out hope for the fullest cooperation in the future between all nations for the sake of a common prosperity, with the help of a wide and permanent system of general security enforced by worldwide disarmament. Expanding Woodrow Wilson's vision, Churchill and Roosevelt also reaffirmed its universalism, practicing quite unaware a form of cultural expansion that again hastened the pace of the world revolution of Westernization.

Under the cover of their universalist ideals they too were engaged in power politics, as was obvious from the heated discussion at that meeting about whether the Atlantic Charter also applied to the British colonies. Roosevelt, undercutting communist propaganda, put the United States in the lead of anti-colonialism, strongly advocating independence for India, over Churchill's angry protest. The issue was left unresolved; but Clement Attlee, speaking for the Labour Party, soon afterwards interpreted the Atlantic Charter to a group of West African students as a promise for future self-government (thereby cementing African support for the Allied war effort).

All the while, the Americans, drawing on their own superior strength, projected their economic influence into areas of the world hitherto under French or British control. Secure at home from the ravages of war, they also took the initiative in planning the postwar "system of general security," sensing the opportunities growing out of their commanding position linking the European with the Pacific war. The conviction grew that the war had raised the curtain on "the American century." The aggrandizement was economic, cultural, political, and psychological. Under the umbrella of a gigantic military mobilization and at a very small cost in American lives or disruption of American tradition, America, "the most powerful and vital nation" of all, was "to assume the leadership of the world," as Henry Luce, a prominent publisher, put it.[2] From the left of the political spectrum Vice-President Henry Wallace proclaimed that "the century on which we are entering—the century which will come out of this war—can and must be the century of the common man . . . the people's revolution is on the march . . . . "[3] Appropriating the charisma of the British Empire and echoing Lenin's thunder, the Americans began to claim leadership in the world revolution of Westernization. In the pursuit of such ambitious ends they made a major contribution to the means of warfare.

V

And now the means. In the face of such extravagant goals of wartime statecraft in involving the fate of nations, the character of the global state system, and the course of humanity's future, what did the individual count? The more exalted the political vision, the bigger the ends, the greater the intensity of war and the ferocity of means. On the battlefields as in conquered enemy territory, the brutality of the soldiers on all sides—not excluding the Americans—increased with the length of the war. But the incentives for savagery rather came from the political leaders safe from combat. For Hitler, the product of the hells of Vienna and World War I, the war was the ultimate—the metaphysical—solution for all ills he had known. Thus he proceeded on a double yet functionally integrated course: the open war for a territorially secure world empire and the hidden war for the purification of the German race, the final solution of the Jewish question.

Both wars were fought with all the technological and organizational refinements available; in both, the inhumanity escalated to the insensate slaughter of millions, mixing patriotism, duty, and callousness in curious combinations, the culmination of a collective brutalization started even before World War I and rising ever since. The victims of the war for empire, soldiers and civilians, were Russians, the peoples of eastern Europe, and to a lesser extent Italians, Frenchmen, Belgians, and the Dutch, not to mention the Germans themselves. The other war, the extermination of the Jews—six million of them—stood out for the repulsive methodical refinements in the mass slaughter of helpless men, women, and children, the war's most barbarous inhumanity.

The escalation of slaughter was obvious also in the growing intensity of air raids on open cities, placing civilians in the frontline. The Germans claimed that British planes had bombed one of their cities; soon fleets of German planes bombed London. Eventually British and even more effective

American bombers practiced saturation bombing on German cities in even larger scale. Some of Churchill's advisors pointed to the military ineffectiveness of such attacks, urging concentration on strategic targets instead. To no avail: Churchill too was willing to repay terror with terror, further raising the human costs of war (maybe the motive force of modern war is not Clausewitzian martial rationality but the frenzy of irrationality). Consider Winston Churchill's state of mind in the summer of 1944, when he addressed his staff:

> It may be several weeks or even months before I shall ask you to drench Germany with poison gas, and if we do it, let us do it one hundred percent. In the meanwhile, I want the matter studied in cold blood by sensible people and not by that particular set of psalm-singing uniformed defeatists which one runs across now here and there.[4]

Privileged Englishman that he was, he too suffered at least a touch of that war hysteria which in the first world war had gripped Hitler and shaped his political program thereafter.

The irrationality of air war reached its peak in Europe in February 1945 when in the fire raid on Dresden over twelve hundred bombers dropped their explosives on a defenseless city of no military significance. In the Far East the firebombing of the more inflammable Japanese cities was even more devastating, the individual human tragedies left unrecorded in the fiery hell. Aerial bombing, of course, was utterly impersonal in its unseen indiscriminate effects when compared with the ground-floor inhumanity of Auschwitz or the S. S. *Einsatzkommandos* with their mass execution of Jews or Poles. But the elaborate organizational preparations, the technological refinements, and the ultimate aim of destroying the enemy were not so different in each case.

The most monstrous and abstract of the inhumanities occurred at the very end of the war in the dropping of two atomic bombs on unsuspecting Japanese cities, Hiroshima and Nagasaki. The development of these two bombs revealed the bizarre and yet inescapable fusion of a total-war mentality with the most sophisticated scientific ingenuity. Out of that ominous mixture rose a new epoch, the most dangerous in all human destiny, in which uncomprehending human minds are endowed with the power to destroy the human race and much of nonhuman life on earth as well.

Before the war scientists in England, France, and Germany, following Albert Einstein's work, had begun to probe into the structure of the atom, sensing in it a source of unlimited energy. On the eve of the war a German chemist, Otto Hahn, had traced the fission of uranium atoms, suddenly opening up the possibility of human control over atomic fission (or even fusion) and over the huge energies thus released. Equally suddenly, as the war started in Europe, the thought occurred to German refugee scientists dreading Hitler's ambition that these energies could be used for war. In their panic they assumed that Hitler would develop an atomic bomb; England and the United States must follow suit if only for deterrence. Out of these considerations emerged the Manhattan Project, the secret American research effort to develop an American atom bomb. By 1943 the British intelligence service knew that Hitler planned no such weapon. But the American preparations continued unchecked and in utter secrecy, with no discussion of ultimate purpose allowed among the participating scientists. In July 1945, after the end of the war in Europe, a prototype bomb was exploded in the New Mexico desert, watched by awestruck scientists and matter-of-fact military men. Should this superweapon, tapping for the first time in human existence the hidden energies of inert matter, now be used against Japan?

The decision to drop the bombs (only two were available) fell within the prerogatives of the political and military leaders, to President Truman and his close advisors. The director of the Manhattan Project, Robert Oppenheimer, considered his agency merely an instrument of government, although other informed scientists, aware of the long-run implications suddenly revealed, expressed their doubts with increasing alarm. In the inner circles of power, however, short-term considerations prevailed, with some reason.

All people were impatient to end the long and cruel war by any means available; as the fighting approached the Japanese home islands, the battles became more ferocious. What did a final atrocity matter if it ultimately saved lives? Admittedly, the Japanese government had given some signs of surrender, but the Japanese hard-liners were resisting to the bitter end, uncertain whether the Americans would respect the person and office of the emperor in their demand for unconditional surrender. Equally important, American leaders felt the need to impress American superiority upon Stalin, whose troops, after advancing to the center of Germany, dominated all of eastern Europe. Now that Hitler and National Socialism were dead, Stalin and Soviet communism rose in the minds of many American policymakers as the next grave danger. And, in any case, should a scientific-military enterprise that had cost billions of dollars be wasted, especially if it could save American lives?

On August 6, 1945, the first atom bomb, affectionately dubbed "Fat Man" (in contrast to "Little Boy," the second bomb), was dropped from a plane blessed by a Catholic priest and named after the pilot's mother-in-law. The pilot and his crew, their plane rocked by the blast even at its high altitude, observed a gigantic mushroom cloud rising above the city of Hiroshima before returning to their base, feted as heroes, the pilot awarded the highest decoration within the jurisdiction of the air force. It was years before he— and the stunned world at large—learned what had happened at ground zero.

The decision to drop that bomb (and the second one on Nagasaki) has subsequently been bitterly criticized, as the full extent of the consequences became known and the issues leading to that decision were reviewed in the light of postwar rationality. But it is difficult to see how toward the end of the war any political leader responsive to the public mood could have decided otherwise. President Roosevelt, who at the outbreak of war in 1939 had implored the belligerents to spare the civilian population (as stipulated under international law), had subsequently no compunction about using that weapon. President Truman felt no remorse for having used it: had the bombs not brought instant peace? By the summer of 1945, if not before, the moral sensibility of people in the positions of power had been dulled by a succession of incredible horrors; they had readily discarded the humane provisons of international law. The scale of their considerations had shrunk down to the necessity of ending the war no matter by what brutality. In the age of the final solution the slaughter was not aimed at soldiers in battle but against people indiscriminately. Given that attitude among heads of state—it was widely shared by their constitutents—dropping the nuclear bombs in August 1945 was a foregone conclusion; the firebombing of Tokyo with its 140,000 casualties had morally prepared the way for Hiroshima and Nagasaki. Only a few scientists, safe and sane in their laboratories, sensed the horrendous dimension of that event, some of them putting their hope for the prevention of nuclear war in the future on the United Nations, launched in the spring of that fateful year.

And so Hiroshima and Nagasaki were imprinted in the human memory as symbols of a quantum jump in human control over nature and in human destiny too. What counted was not only the almost instant deaths in Hiroshima of nearly 130,000 human beings at the time of the explosion and shortly after, but also the radiation damage lasting for many decades and generations afterwards, all from a single weapon (and one, by present standards, highly ineffective); the casualties in Nagasaki were less appalling—and less dramatized. Even more fatal was the new knowledge of a vast source of energy, more readily used for war than for peace. Who among the participating scientists, engineers, and administrators, who among the decision-makers at any level, had been willing to trace the long line of linkages between their own work and the human victims, between the present application and the future role of atomic weapons? The more impersonal the process of destruction, the greater the inhumanity and the more incomprehensible the connection between cause and effect—another indication of how much their world has grown over people's heads. Maybe Dostoyevsky's underground man was right: "Civilization has made man, if not always more blood-thirsty, at least more viciously, more horribly blood-thirsty."

VI

The atomic bombs, a product of the West, were the culmination in the adjustment of means to the extreme ends of global war. They also bore witness to the superior resources available for military use among the Western allies. Throughout the war its outcome was determined less by the valor of the soldiers and the civilians caught in battle than by the resources of intelligence, organization, and productivity.

In this respect Nazi Germany suffered from Hitler's intellectual shortcomings. He was woefully unrealistic in his assessment of both the Soviet Union and the United States. The invasion of the Soviet Union was shoddily prepared, he had no inkling of the resources and the resolve of the Soviet peoples—nor of the Americans. As for the uses of science and technology, Speer has testified that he was "filled with a fundamental distrust of all innovations which . . . went beyond the technical experience of the First World War generation."[5] Jet engines and rockets, let alone atomic bombs, were beyond his grasp. Because of Hitler's hesitation the first two went into production too late to affect the outcome of the war; luckily, the last remained unexplored. As regards effective coordination in the war effort, Hitler spared the German people the necessary all-out mobilization until it was too late. His "totalitarianism" even at the height of the war covered a variety of overlapping and competing bureaucratic chieftaincies, all limiting the war effort.

Admittedly, Hitler possessed valuable assets, the traditional patriotic docility of the German people, which kept the army fighting to the bitter end. He also benefited from the technical proficiency of German workers and administrators, not to mention an effective propaganda apparatus and a ruthless secret police. But these assets were insufficient to offset superior Soviet manpower and American war production, or to prevail in a many-fronted war spreading to North Africa, covering most of Europe, and extending over the high seas.

What counted most heavily against him, however, was his lack of statesman-like competence. He never undertook a rational analysis of his chances, balancing his resources, realistically assessed, against the obstacles likely to be encountered. He remained an uprooted Austrian peasant carried beyond his depths by the extravagant expectations of world power aroused after the turn of the century. Although gifted with vital insights into the dynamics of mass politics in central Europe and into European diplomacy, he was hopelessly disqualified for coping with world politics. His bid for final solutions was an irrational, hopeless throw of the dice. When his all-or-nothing gamble had collapsed, he committed suicide, as he had long anticipated. Sticking to his Wagnerian style to the last, he endowed his personal final solution with a grand flourish: as he himself had failed, so Germany too had no right to survive. Fortunately, his orders to that effect were not carried out. But Germany's unconditional surrender crushed forever German dreams of world power.

In the case of Japan, the lack of rational preparation and forethought as well as of effective cooperation among the different branches of the government was even more obvious. Proudly relying on Japanese military tradition, the generals were ignorant or even contemptuous of the civilian skills of large-scale planning and all-inclusive coordination of the country's resources. In underestimating the United States, they also remained blind to the importance of technology and organization. In this respect the retreat into tradition spearheaded by the army since the late 1920s had also been a retreat into backwardness.

As a result, the economic and intellectual potential of the country, although considerable, was never fully harnessed to the war. Japan remained especially backward in the military application of science, unaware to the end, for instance, that the Americans had cracked the country's secret codes and knew of its military moves in advance. Insufficient forethought showed up also in the lack of shipping to gather the riches of the Co-Prosperity Sphere in support of the war effort. For want of tankers as well as of electronic equipment protecting them, oil, though plentiful in Indonesia, became desperately scarce as the Americans approached the Japanese homeland. Under the threat of invasion the bravery, endurance, and self-sacrifice of the Japanese people—soldiers and civilians alike—remained exemplary to the cruel end. Had the emperor not interceded, the hard-liners would have sacrificed, Hitler style, the future of their country, crowning their own incompetence by the irrationality of national suicide.

In the Soviet Union, by contrast, the rationality of war, although often carried in battle to the borders of irrationality, retained the upper hand. Unity and cooperation were enforced by extreme and inescapable adversity, reducing the hardened willfulness of ordinary life. The Stalin revolution had prepared the population for straining its endurance as well as its flexibility and intelligence; the war carried the strain to the limit. Stalin himself, unaccountably blind to the menace of a German invasion and personally responsible for the lack of military precaution, quickly reasserted his leadership. Sometimes blamed by his generals for making strategic mistakes, he yet infused his country with his own indomitable will while coordinating the country's resources for the war effort.

The huge losses of territory in 1941 and 1942 forced a vast eastward relocation of people and machines under constant improvisation hampered by extreme shortages of all necessities. Food production fell disastrously; multitudes of people not directly engaged in the war effort died of starvation. Yet, despite the staggering initial losses of industrial capacity, the war economy (with some assistance from the Western allies) soon delivered weapons superior in quality to those of the Germans, increasingly more numerous, and used with dogged determination.

The heroism and endurance under unspeakable adversity shown by the peoples of the Soviet Union was without parallel, as was their capacity, despite frightful losses of men and equipment—and despite the shocks of the preceding decades—to surpass German military might. The Great Fatherland War, claiming twenty million lives, was the culmination of a long crisis of survival for a sovereign Eurasian state. No other country in the 20th century has passed through such dark valleys of torment and death. After 1914 it was war, revolution, and civil war; after a brief breathing spell came Stalin's revolution with its cruel dislocations and terror; then the second war struck, exceeding all previous horrors. Threatened with extinction, the Soviet peoples, their system of government, Stalin's statesmanship, all steeled and toughened by adversity from the start, proved their mettle in the most remorseless war yet. Elevating the perspectives even of simple soldiers, the war provided a source of civic consolidation and collective pride to the present day. The peoples of the Soviet Union certainly contributed far more of their vital substances to defeating Hitler than the Western allies.

How easy, by comparison, was the waging of total war for the British and Americans! England had gone to war in support of Poland; soon it was defending itself, its imperial possessions, and the world order in which it had risen to prominence, against Germany and Japan as well. On these issues the country was united, able to draw on a law-abiding population which, despite its marked class divisions, was ready for cooperation in case of emergency. In Churchill it possessed a worthy wartime leader setting a personal example of dogged endurance and determination. In addition, the country was spared the fate of its continental neighbors. Although its traditional security was diminished by attacks from the air and the German challenge to the British domination of the surrounding seas, it was still invasion-proof. Safe from the devastation of battle, it became an effective center for the coordination of the worldwide British war effort and a staging area for the invasion of the continent, aided throughout by its traditions of empire.

Under the stimulus of war economic life, long stagnant during the depression, revived quickly. More food was produced than before the war; industrial output climbed (except in textiles). Although the wartime diet remained monotonous, for once there was enough food for all. While workers' incomes rose, life under universal rationing remained bleak, yet not without promise. The sacrifices of war fostered a new spirit of social solidarity. With the help of the Labour Party Lord Beveridge, in 1942, submitted an official report on social insurance. The following year a new Ministry of Reconstruction began preparations for the postwar extension of social services, especially in health and education, designed to transform Britain into a welfare state; even during the war extensive measures were taken to mitigate the human consequences of wartime dislocation. More significant for the war effort was the remarkable mobilization of science and technology, contributing not only lifesaving penicillin and a wide range of radar techniques crucial for minimizing the destructiveness of German air attacks, but also careful analysis of all military operations, improving their effectiveness.

As a result of its assets, geographic and human, Britain survived the war relatively well. It suffered, despite heavy German air raids, less damage overall to life and property than in World War I, its military casualties down to half their previous rate. More important, its political system had taken a turn toward social democracy, solidifying the ideal of government by consensus and of social equality in the body politic, both essential to the continued vitality of the Western model. Only in one respect—a crucial one—did the war diminish Britain's stature: the country became dependent on the United States, without whose help it could not have prevailed. It was the last war which Britain waged as a world power; in contributing to victory it passed its former preeminence to its transatlantic kin, the United States.

VII

World War II at last propelled the United States, its strong isolationist reservations mixed with a messianic outreach, into global preeminence. In the worldwide power contest its immense assets, both geographical and cultural, emerged in their true scale, determining the outcome and setting the course of global development for decades to come.

The war released the country's productive and creative capacities long thwarted by the Great Depression. In the military mobilization of manpower it lagged behind the Soviet Union and Germany, but in economic productivity it excelled them by far. It supplied the ships for war in the Pacific and the Atlantic, for transportation around the world and for amphibious operations; it mass-produced tanks and fleets of aircraft for moving troops, supporting soldiers on the ground, and—in the most massive scale—bombers for the destruction of the enemy's cities; it developed the atomic bomb. At the same time, it supplied more food than ever for its own people and its allies in need. In mobilizing industrial output for war it nearly doubled its GNP, even slightly increasing the total of goods and services available for civilians. Unemployment disappeared; more women than ever were recruited into industry; incomes rose. Drawing on the services of American Negroes, the war also contributed to agitation for racial equality, a process fostered by the war aims but causing considerable conflict; the army continued to exclude Negroes from combat. Other minorities too gained political recognition, thereby demonstrating the advantages of American democracy. Only Japanese-Americans, their loyalty suspected, suffered grievously from wartime hysteria.

There was no question, everything considered: the American political system worked miracles as industrial leaders moved into the government to administer the far-flung war effort. Despite innumerable tensions, the people and their government virtually fused under the leadership of Franklin D. Roosevelt. In the biggest organizaitonal task ever undertaken, they harnessed the resources of the largest industrial nation for victory in global war and for the assertion of their country's ideals in the postwar order. The United States emerged into global leadership because of its long-established capacity for large-scale human cooperation and its experience in worldwide business, which had been unaffected by its political isolationism. It was the mainstay of the alliance that defeated Germany and Japan, its elite more practically tuned to global perspectives than its peers elsewhere. It fostered reasonably harmonious collaborations with its allies, including Stalin—all at minimal human costs: American casualities amounted to roughly 2% of the Soviet total. And the gains were enormous. At the end of the war the United States was the only belligerent physically untouched by battle, its prosperity and system of government enhanced and its power in the world unprecedented. Whatever the country's postwar stance, it had earned its preeminence in the world thanks not only to its civic virtues—much advertised at the time—but also to the privileged historic conditioning and geographical advantages that had made possible its immense cultural resources (including its virtues).

And so, under American auspices, the war that had grown into a contest of total cultural resources came to an end. The American-led victory boosted the universalism of the Atlantic Charter and of Western aspirations generally. The Western model, now domiciled in North America, was restored to preeminence after the crisis of self-confidence in the inter-war years. The militaristic anti-universalism of fascism had been defeated. The victorious ideals were "freedom, justice, and peace in the world," the transcendent ideals embodied most prominently in the new United Nations, prepared during the war among the nations united against Germany and Japan, and established on American soil in April 1945, shortly before the war's end.

Admittedly, the effectiveness of the United Nations was limited from the start by the hardheaded realism guiding key policymakers in the leading states and by deep divisions among key members. Its high ideals were bound to remain paper promises; no people, no governments, were ready to conduct their affairs according to transnational perspectives. Yet the United Nation's widespread popularity at the time testified to the hopes released by the untold sorrow of thirty-one years of war and revolution in the world, as well as to the universalism of those hopes. Everywhere around the world a new global awareness had been created. Out of suffering and horror a better future could only rise from worldwide cooperation based on key ideals of the Western liberal-democratic experience. As the Charter of the United Nations announced: " . . . to save succeeding generations from the scourge of war, which twice in our lifetime has brought untold sorrow to mankind, and to reaffirm faith in fundamental human rights, in the dignity and worth of the human person, in the equal rights of men and women and of nations large and small . . . to promote social progress and better standards of life in larger freedom," the victors and their allies—the United Nations—promised "to practice tolerance . . . and to maintain international peace and security" as well as "to employ international machinery for the promotion of the economic and social advancement of all peoples. . . . "[6] In establishing the United Nations and endowing it with a sweeping vision addressed to all humanity, the victors celebrated the defeat of fascism in its many forms, grandly setting forth their own ideals and historic experience as the guidelines for the future, thereby raising worldwide expectations to the high level of their own accomplishments.

Under the cover of these hopes, however, new and more ominous tensions lurked. The former Great Powers of continental Europe lay in shambles; even Britain's power was reduced. With the decline of the old West, two geographically non-European superpowers emerged, foremost the United States and secondly the Soviet Union, the latter more preeminent by territorial sway and Leninist ambition than by actual strength. Moreover, following the decline of the old European empires, the emergence of independent non-Western states patterned after the European nation-state was merely a question of time.

In this manner the second world war, ending the deadly rivalries resulting from the first world war, prepared new power struggles within a full-blown global state system even more fiercely competitive than the old European state system, and with still more jarring discrepancies between ends and means, between ideals and reality, between the requirements for peaceful cooperation (or even survival) and the human capacity for understanding them.

IV

The United States:
The Foremost Superpower

The Culmination of the World Revolution of Westernization

After 1945 the peoples of the world, gradually forgetting the horrors of World War II, entered a new era of their existence. Now they lived in a single, though sharply splintered, entity, in an anarchic global community, in which homogenizing and divisive forces held each other in precarious balance. Human life now proceeded down a broad avenue flanked by the threat of nuclear war on one side and the promise of the United Nations on the other. It veered between the unilateral exercise of power armed with the deadliest weapons yet invented and the vision of an ever more profitable interaction in a global state system and an interdependent economy, all outgrowths of Western experience.

The Western-led unifying integration of the world was accompanied by a massive overall rise in material security. As a result populations, especially in non-Western societies, increased rapidly. With the help of the new prosperity and the new technologies of communication the past and present creativity of all the world's peoples became mutually accessible on an unprecedented scale. The interaction established, under Western auspices, a single set of universals, a single perception of modernity. Yet the Great Confluence also deepened the cultural disorientation, especially in societies too suddenly propelled into the global network. Among the latter, the established ways of life were subverted more massively than before. Perspectives became stretched beyond past landmarks; people unable to forget their roots were driven to search for more inclusive perceptions, authorities, and institutions, for more universal truths conforming to Western precedents. Yet eventually that search also overburdened people in the metropolitan centers by the magnitude of the tasks undertaken and the complexity of the detail to be mastered. Under the surface of modernity the overload nourished a deep-seated aver-

149

sion against the new globalism, blocking with traditional ways the uncertain passage into a strange and overly demanding future.

The fast-paced quarter-century after World War II was dominated by the emergence of two rival superpowers, the United States and the Soviet Union (treated in Parts IV and V), and by the rise of independent states among peoples hitherto living under Western domination or outright colonial control (Part VI). At the end of that momentous change the world found itself organized into a system of states fashioned after the model of the Western nation-state. The militantly competitive interaction of the European state system was now reconstituted on a global scale amidst even greater cultural incompatibility compacted more tightly on a shrunken planet endowed with finite physical resources. In that densely packed world the United States occupied the center of the stage.

15

The United States to 1945: A Most Privileged Nation

Seen from a high altitude a nation's presence in the world reflects that nation's historical substance, the sum total of its experience through time. Viewed in this manner the United States was after 1945 propelled into global leadership, channeling its own distinct accomplishments into the global mainstream. It contributed its qualities and skills developed in the unceasing interaction between its physical environment and the human resources carried over from Europe and enlarged as indigenous opportunities provided. What was the United States like at that historic turning point in its evolution?

A summary of the American collective experience up to 1945, necessarily impressionistic and sketchy, like a portrait, requires a double focus. Drawn by an American it is, in some respects, a self-portrait representing the national identity as seen from within (from one of many possible angles). Yet the United States also needs to be viewed multidimensionally, from the outside, in comparison with the rest of the world. The comparative perspective, admittedly, carries the risk of viewing the country in too rosy a light and thereby disconcerting the critical insiders who, in the best American tradition, try to mend the flaws in their society. Yet can they escape the urgency of enlarging their awareness beyond traditional bounds—without losing their reforming zeal in their local communities?

There is no denying the fact that, starting with their geopolitical setting, Americans have enjoyed exceptional advantages; they have grown up historically and geographically under a special dispensation which blessed their institutions, collective values, and ways of thinking. Their New World habitat provided a bountiful environment not too different, despite greater extremes of climate, from that of northwestern Europe and decidedly richer in natural endowment. Geography and chance also offered protection against hostile invasion, guarding the country east and west by oceans, north and south by weak neighbors. The United States was part of the cultural competition of Western civilization, yet safe—even in its early quarrels with the seafaring states of western Europe—from the relentless military pressures of the European state system; armed force, apart from the civil war (an internal conflict) played virtually no part in its domestic politics.

While the rest of humanity was caught amidst ceaseless wrangles with fairly equally matched rivals and beset by greater adversities of climate, soils, and geographical location, Americans enjoyed, as the human condition goes, a uniquely privileged existence, living in an island of free security. The struggle for collective or individual survival was never as extreme or politically surcharged as among the peoples living in the interlocking landmass of Eurasia or Africa. As Goethe had enviously observed: *"Amerika, Du hast es leichter"* ("America, you have an easier time"). The American national identity inescapably reflected that exceptionality. It reproduced on a larger scale the beneficial security of the British Isles, the largest single source, through language and political culture, of the traditions channeled into the American mainstream.

II

And now the human story, forever dramatized for self-assurance in the uncertainties of a new world, much admired especially in its early stages, and always closely watched. The American identity was shaped by self-assertive, enterprising, and practically minded people predominantly originating in northwestern Europe, the center of Western civilization. The most vigorous representatives of European expansionism, aware of their power over the indigenous peoples, the American settlers were determined to make the most of their advantage. Some of them were solaced in their pioneer hardship by the promise of a purer life, trying to set up a shining light in the wilderness as a guide to universal perfection; others sought to better their material fortune, but were ready too, in idealistic sincerity, to ennoble their struggles in the wilderness with visions of superiority drawn from the highest European aspirations. Prompted by the ideals and the discipline of the Judeo-Christian tradition as well as by the rationalist ambitions of the Enlightenment, the colonial settlers developed a collective ego morally justifying expansion territorially, politically, and culturally, over other human beings and over nature too. Living on the frontiers of European culture, they were even determined to prove that in their simple practicality they were better than the cultivated Europeans, who, in the inevitable comparison, continued to humiliate them (little though the Americans cared to admit the fact).

Their guiding sentiment was well expressed by John Adams in 1765 when he said: "I always consider the settlement of America with Reverence and Wonder—as the Opening of a grand scene and Design in Providence for the

Illumination of the Ignorant and Emancipation of the slavish Part of Mankind *all over the Earth*" (italics added).[1] A few years later the Declaration of Independence proclaimed the ideal of life, liberty, and the pursuit of happiness as the guiding motto for the new republic, which, as its Great Seal announced, created no less than the "New Order of the Age." In Thomas Jefferson's more modest estimate, it was destined to "be a standing monument and example for the aim and imitation of the peoples of other countries."[2]

Unlike other nations proud of their past, the United States was considered by its founders to be a new start endowed with a universally applicable vision of a superior polity, free from the corrupt practices of the Old World or lesser civilizations, and preordained to spread its blessings around the globe. Most significantly, its promise lay in the future, untarnished by the imperfections of the European past.

These justifications for American independence set forth a powerful claim to national superiority, the essence of which was expressed (for Russia's benefit) by Dostoyevsky: "If a great people does not believe that the truth is only to be found in itself alone; . . . if it does not believe that it alone is fit and destined to raise up and save all the rest by its truth, it would at once sink into being ethnographical material, and not a great people. A really great people can never accept a secondary part in the history of Humanity, nor even one of the first, but will have the first part. A nation which loses this belief ceases to be a nation."[3] That faith in American superiority soon became the accepted American civic religion, a national ideology with a profound hold over popular imagination and individual self-esteem. Eventually institutionalized in a thousand rituals sustaining civic and individual life (and covering up many shortcomings), it was forever reinforced by spontaneous agreement, a perennial stimulant for a self-serving, ego-boosting, outward-oriented, forward-looking, and eminently practical idealism. As Woodrow Wilson observed at a time of a new American self-confidence in comparison even with Europe: "Sometimes people call me an idealist. Well, that is the way I know I am an American. America is the only idealistic Nation in the world."[4]

That practically oriented idealist belief in national superiority, no doubt, was anchored in personal self-interest. What better stimulus for overcoming self-doubt and for individual opportunity and growth? Whatever the present discrepancies between the ideals and the realities—and they were a source of unending public agitation—a better future lay ahead; with proper effort it could be achieved. The long-run record of achievement indeed encouraged the rooted optimism. The achievements took many beneficial forms. The early settlers, for instance, had brought from the homelands of the industrial revolution the popular culture of mechanical tinkering. It was put to good use in an empty continent encouraging the development of labor-saving devices. Mechanical aptitude fortified by moral zeal was soon turned to business advantage; Americans became leaders in the applied mechanical arts, in industrialization. Mass production and the ever expanding domestic market, protected by tariffs and aided by ample foreign investments, made economic prosperity into a proverbial American attribute.

The capacity for transcending social and cultural barriers was proved in action. Quite early in the 19th century universal suffrage for white males became a fact. By 1865 slavery was abolished. All along, the cultural source of immigration was broadened, admitting southern and eastern Europeans and eventually even Orientals and Hispanics—all of them integrated, in the ceaseless wrangles of ethnic politics, into the American creed. To cite Woodrow Wilson once more, addressing a group of newly naturalized citizens: "You have taken an oath of allegiance to a great ideal, to a great body of principles, to a great hope of the human race."[5] And as late as 1963 Martin Luther King Jr., pleading for racial equality, invoked "the American dream" in proclaiming his own "dream that one day this nation will rise up and live out the true meaning of its creed: 'we hold these truths to be self-evident that all men are created equal.'"[6] Everything considered, Americans developed a remarkable tolerance for the racial and cultural diversity in their midst. Thanks to the latter, they escaped settling into a constrictive, rigid class structure.

The American civic creed with its "bounding pulse of youth,"[7] as it appeared to the English observer James Bryce in the 1880s, also promoted a remarkable capacity for coping with change. Social, economic, and political change came faster than even in western Europe; it was a permanent feature of American life, enhancing individual self-reliance and collective adaptability. From the same source sprang an infectious optimistic capacity for surmounting all social and cultural obstructions to unity. That optimism helped to make the United States *e pluribus unum*, transforming a mini-world of human diversity into a model universe, at least in comparison with other polities. Examined by critical insiders, the ideal was forever tarnished; much effort was needed to give it reality. Yet the unifying magnetic idealism kept beckoning, symbolized by the Statue of Liberty inviting the hungry and oppressed of the world to the American shores. Americanization may not have lived up to expectations—it produced its own insecurities—but disillusionment has never yet led to an appreciable repudiation of the guiding civic religion or of the political institutions upholding it; the pervasive optimism in the sheltered environment of American exceptionality was a psychological bulwark against the doubt and disorientation threatening all modern life. All told, the vast orchestra of American life, with its immense variety of instruments and many brassy notes, produced a surprising overall harmony.

Yet how was it possible that the emphasis on freedom, equality, and happiness for all did not create anarchy? The prevailing ideological accent certainly rested on self-assertion, individualism, and competition. These terms were part of the civic religion, stimulated for a long time by the opportunities of the open frontier and subsequently of a remarkably open society and economy. But why would a society of such competitive individualists not collapse like a sheet of sand despite all rituals of national unity? How did a society plagued by xenophobia, by extreme religious visions, by sharp contrasts between rich and poor, by incessant rapid change, by vanity, arrogance, and violence hold together with a minimum of civic strife?

For an answer we need to probe into a still unexplored layer of collectivized individual motivation, into the culturally structured superego submerged beneath the surface of individual awareness. Foremost among the invisible factors providing the individual discipline that made freedom and individualism into a socially constructive force simmered the Judeo-Christian tradition embedded also in the rationalism of the Enlightenment. The Christian religion in its many variants played a significant role in American life, in part because the separation of church and state encouraged individual initiative in religious affairs. Regardless of their differences—and they were considerable—Christians lived under the symbol of the cross, exhorted to follow a celibate, martyred Christ and, for the most part quite unawares, to practice an ascetic self-discipline for respectful service to others and for bracing themselves against adversity. As James Bryce noted: "Christianity influences conduct, not indeed half as much as in theory it ought, but probably more than it does in any other modern country."[8] Yet the seeming selflessness of American churchgoers also enhanced their selfhood constructively; it elicited encouragement and support from fellow citizens, thus working for their own advantage. Altruism is but a socialized form of selfishness. In the American setting the extent of selfish socialization (or socialized selfishness) was carried further than elsewhere.

More visible than religion were the formative effects of the frontier, as observed, for instance, by de Tocqueville. Here extensive social cooperation was a spontaneous guarantee of self-protection and self-advancement; it remained so in the open society after the end of westward expansion. In this manner the enterprising sociableness of English society fused with the communalism of the independent churches and grew into a collective American trait. The infectious chumminess of "a nation of joiners" has always impressed visitors and newcomers from the Old Country, as have the expansive friendliness of individual citizens and their marked skills in handling human relations. In the judgment of James Bryce, Americans were a "good-natured people, kindly, helpful to one another, disposed to take a charitable view of wrongdoing. . . . Nowhere is cruelty more abhorred."[9]

Throughout American history these qualities have produced an extraordinary social adhesiveness. They allowed the creation of large-scale organizations in which individuals still could thrive and affirm their personal identity by acting in an outgoing, horizontally oriented socializing manner, reaching as many people as possible. Thus a distinct American ego emerged, sharply contrasting with that of Europeans, especially from central or eastern Europe. There people were forced by external restrictions to develop a strong-willed, vertically oriented inwardness at the expense of their social outreach. To them Americans always appeared shallow and superficial, lacking in individuality and depth, or even in intellectual creativity. But the American fluidity of individual wills and the malleability of individual opinion—the penchant for expansive social interaction that created the seeming uniformity overarching the rich diversity of American life—constituted a major human achievement in its own right.

The artistry of social discipline was collective rather than individual and therefore less visible by traditional standards. But it proved its merits in managing modern technology and a multitudinous, complex society. It made individuals more adaptable in dealing with demanding machines; it also helped to tie them willingly into ever larger networks of organizations and communities stretching across the continent and beyond. The association of immigrants from cultures incompatible elsewhere was not only transformed into a powerful polity endowed with a common purpose and eager for outreach into the world; it was also raised to leadership in industrial productivity and social management. Americans voluntarily, even unthinkingly, accepted the tighter controls needed for extensive social cooperation because such cooperation enhanced personal opportunities. Miraculously, the socialization of selfishness progressed as part of American life. Put differently, the disciplined integration of individuals into the collective will—a process attempted under totalitarian regimes only in special "struggle sessions" and under compulsion—had always been practiced subconsciously in this "land of opportunity" without impairing the individual's sense of personal identity or freedom. To a remarkable extent the polity and the individual "soul" operated on the same wavelengths. Whatever its shortcomings, American society patently furnished the needed assurances of community.

Admittedly, in its exceptional external security American society remained remarkably loosely meshed; it contained protective niches for almost any variety of lifestyle or conviction, unlike the cramped polities of Europe compelled under external pressure to achieve a far greater degree of national cohesion. It even tolerated a far larger volume of corruption and incidental violence; but violence, "as American as cherry pie," remained almost entirely outside the political process (certainly as regards the federal government). The Civil War may be considered an exception. But it did not threaten the continuity of the American creed entrenched in the North; indeed it broke the South's resistance to essential aspects of that creed. Nor was it aggravated by foreign intervention. Even so, the psychological damage to the South reflected, in milder form, the adverse effects of military defeat so common in European history. As a regional debility it did not alter, however, the direction of American political and social evolution, though it may have slowed its pace. Whatever the obstacles, the course remained set for a reasonably open democratic society.

A key trait in the evolution of the United States, of course, was democracy under a federal republic. It was a system of governance rising in ascending tiers from local politics through the states to the federal government, operating through consensus and conformity at every level, all loosely linked, for the most part, through two political parties. The art of democratic politics consisted in acting within the bounds of that consensus, to sense it, express it, shape and reshape it, and finally enact it by majority decision in legislation. American politics too was a creative achievement in its own right. Coarse and even corrupt, down to earth and therefore humble, it contributed its share to the prevailing good feelings. And more: it miraculously reduced, for the most part, the political, social, and cultural differences of a large country to competition between only two parties. Both parties, porous and flexible in interacting with public opinion, made possible government by consensus, the goal toward which, according to Bryce, "the Americans have marched with steady steps, unconsciously as well as consciously. No other people now stands so near it."[10] Public opinion had become a political power ever since universal (male) suffrage had been introduced; the introduction of female suffrage in 1918 even strengthened that trend.

Long before the beginning of the 20th century American politics had turned into mass politics in an orderly and constructive fashion, an integral part of popular culture with its capacity for transcending social and cultural differences. It was "democratic" politics, practiced by "democrats" endowed with a special moral character. As Bryce observed, "the adjective [democratic] is used to describe a person of simple and friendly spirit and genial manners, 'a good mixer,' and one who, whatever his wealth or status, makes no assumption of superiority and carefully keeps himself on the level of his poorer and less eminent neighbors." Democratic consensus came easy to people given to "a sort of kindliness, a sense of human fellowship, a recognition of the duty of mutual help owed by man to man," all qualities "stronger than anywhere in the Old World. . . . "[11]

The physical hardships of American life were sometimes considerable, especially in the early stages of settlement and expansion; yet the collective setting of state and society proved exceptionally favorable: Americans indeed had an easy time. Government sat lightly on a society keenly responsive to individual enterprise and private initiative for private as well as public welfare (for that reason even in the late 20th century the American welfare state never matched its European equivalents). Throughout American history social and economic activity offered better outlets than politics for ambition and discontent. The idealism of the American tradition, moreover, served as a constant prod not to discredit but to improve the system of government. The endless and often sharp criticism by social reformers or ambitious politicians reaffirmed the guiding ideology which, in addition, mercifully discouraged political violence.

An even bigger boon for the American way derived from the persistent triumphs in all invidious comparison with the rest of the world. Where else were people so free and well off? Even Bryce, a British peer, admitted that "there are elements in the life of the United States which may well make a European of any class prefer to dwell there rather than in the land of his birth."[12] What other governments, what other countries, could provide more persuasive justification for loyalty? That loyalty in turn reinforced the capacity for consensus, that pervasive good nature which replenished or even increased the lubricants of compromise in the body politic. Americans generally were spared the destructive cynicism toward government (or any authority) so prevalent in some European countries.

And best of all, victory in invidious comparison also provided Americans with a unique collective freedom. Liberated, in their own estimate, from their Old World past and leading in human progress, they felt free to conduct their own experiments spontaneously, never burdened with the necessity of following an alien model. They readily took advantage of European intellectual accomplishments, always on their own terms and trying to give them practical application. What greater incentives for creative experiments in all aspects of collective life could one find in the world? All told, one might facetiously argue, Americans faintly approached the blessings of communism as ideally envisaged by Marx.

These traits, of course, are part of the American self-portrait as seen and felt from the inside, tinged with the optimistic benevolence evolved by the American experience and shared by sympathetic outsiders like James Bryce. Hostile observers, judging from the cynical perspectives of less favored polities, have always drawn a grimmer image of American life, calling attention to the high price of competitive "capitalism," to its unemployment, widespread violence, discrimination, and to its ruthlessness to American Indians or ex-slaves. In their judgment the American creed has been a myth covering up the shortcomings of American society.

Yet how would the hostile outsiders account for the absence in the American mainstream of any revolutionary movement or any repudiation of the existing system? How are they to explain the persistence of the American dream, even as a myth, with persuasive appeal to the most oppressed members of the community? What reasons can they give for the fact that the tight social, economic, and political order of American society, which a very German Max Weber dreaded as an "iron cage," appeared to Americans as an invitation to individualism and free enterprise? As outsiders, these critics were blind to the invisible supports, to the hidden hand of cultural conditioning accounting for the marked resilience of the population amidst the not inconsiderable adversities of an open society. The substructures supporting the social discipline may have suffered somewhat in the giddy 1920s, but they were still much in evidence during the Great Depression; they proved their worth in World War II.

The complacent insiders, however, also were blind when it came to exporting their ideals to less favored peoples around the world. Take the crucial ideal of freedom. Viewed properly in the contexts of American life, it constitutes a complex cultural molecule replete with the invisible restraints of individual and collective discipline; taken out of context, it easily decays into license. Likewise, the concept of democratic pluralism, so common in the analyses of the American system, denotes merely the surface of a political practice deeply rooted in an underlying unity; separated from its roots it promotes disunity and anarchy. How, then, was it possible to export American democracy and laissez-faire enterprise without also exporting the conditions that had created these boons? Could democracy American-style work—at least among countries competing for power around the world—without the benefit of unprecedented external security and the ingrained conviction of worldwide superiority?

III

Americans—to put now their most characteristic qualities into this thumbnail sketch—certainly were the most successful expansionists in the world. From the start, expansion was an elemental necessity of self-preservation and self-affirmation, self-righteously viewed in moral terms. As the Rev. Increase Mather wrote in 1676: the "Lord God of our Fathers has given us for a rightful Possession" the land of "the heathen People amongst whom we live," to which he added his thanks for " . . . the wonderful Providence of God, who did . . . lay the fear of the English and the dread of them upon all the Indians. The terror of God was upon them round about."[13] In this spirit the white man claimed dominion over all lands from the eastern seaboard to the Rocky Mountains and beyond. Resenting the encroachment of arrogant strangers, the Indians fought back, sometimes in utter desperation, to no avail; in the ceaseless Indian wars the best Indian, as the saying went, was a dead Indian. Between the Indians and the frontiersmen backed up by the U.S. army, the contests of power were so unequal that by the end of the 19th century the Indians had ceased to be a political factor deserving recognition or serious attention; the Indian wars, while nurturing romantic fantasies, left no scares on the civic creed. In the process of fighting the Indians the seaboard settlers grew into proud Americans adding new states to the United States from the Atlantic to the Pacific.

In the westward thrust through the continent lateral security posed no problem. To the north the British in Canada offered few incentives for territorial expansion, although in one instance, over a border dispute in the Northwest, Americans threatened to fight if they did not get their way; through the 19th century Canada was eyed as a potential enemy. The temptations were greater to the south, towards the countries of Latin America. Here Americans, some decades after declaring independence, took a sweeping view, claiming to be the protectors of the entire hemisphere against European intervention. In the famous statement by President Monroe (1823), the American government declared that it would regard any attempt by European nations "to extend their system to any portion of this hemisphere as dangerous to [its] peace and safety." Much depended, of course, on the interpretation of "peace and safety," but the Monroe Doctrine projected a bold pretension to an American hegemony over the Western Hemisphere. How bold was shown not long afterwards, when trouble brewed over American settlers in Mexican Texas. For their sake Americans went to war with their southern neighbor, conquering Mexico City in 1847 and annexing California as well as Texas; all told, the Americans took two-fifths of Mexico's territory.

The Mexican war marked the culmination of a feverish expansionism in the name of an American "manifest destiny to overspread the continent allotted by Providence for the free development of our yearly multiplying millions." Its exultation was perhaps best expressed by a senator from New York in January 1848:

> Whoever will look back upon the past and forward to the present, must see that, allured by the justice of our institutions, before the close of the present century, this continent will teem with a free population of a hundred million souls. Nor have we yet fulfilled the destiny allotted to us. New territory is spread out for us to subdue and fertilize; new races are presented for us to civilize, educate, and absorb; new triumphs for us to achieve in the cause of freedom.[14]

Subsequently, however, Americans respected Mexico's territorial integrity (with minor infringements under President Wilson). Why fight a country when it posed no threat militarily, economically, or politically? By 1900 American capital dominated Mexican finance; Americans owned 78% of the mines, 72% of the smelters, 58% of the oil, 68% of the rubber plantations, and about two-thirds of the railroads. Almost as attractive as Canada, it offered opportunities for the peaceful expansion of American business.

Tempting opportunities also beckoned further afield. After the Mexican war Americans began looking across the Pacific. "Westward will the course of empire take its way," observed Commodore Perry after forcibly opening Japan to Western commerce in 1853, adding that "to me it seems that the people of America will, in some form or other, extend their dominion and their power until they shall have brought within their mighty embrace multitudes of the Islands of the great Pacific, and placed the Saxon race upon the eastern shores of Asia."[15]

After the issue of slavery had been resolved and the West been settled, "Manifest Destiny" reemerged, encouraged by the final thrust of European colonial expansion at the end of the 19th century. "Whether they will or no, Americans must now begin to look outward,"[16] wrote Captain Mahan, the prophet of American seapower. And outwardly they looked, especially on their own continent and in the Pacific. As for the Americas, Secretary of State Olney in 1896 announced that "the United States is practically sovereign on this continent, and its fiat is law upon the subjects to which it confines its interpretations." It soon showed its strength by annexing Puerto Rico, exercising protectorates over Cuba, Panama, and Nicaragua, as well as by making the British government toe the American line in a South American border dispute. In the Pacific Admiral Perry's vision came closer to reality with the annexation of the islands of Hawaii, Midway, Guam, and Wake; after the war with Spain the Philippine islands were added. In West Africa Liberia, settled by American blacks, constituted a more inconspicuous (and often overlooked) outpost of American influence.

By 1900 the United States—now come of age, as it was said—had acquired colonial dependencies and become a world power. Reflecting a common sentiment and setting an American model for the great anti-Western challengers of the 20th century, the influential Senator Henry Cabot Lodge wrote: "Small states are of the past, and have no future. The great nations are rapidly absorbing for their future expansion and their present defense all the waste places of the earth. It is a movement which makes for civilization and the advancement of the race. As one of the great nations of the world the United States must not fall out of the line of march."[17] The "advancement of the race" sometimes took rather uncivilized forms, as in an incident during the conquest of the Philippines when the people of a rebellious village, men, women, and children, were herded into an extinct crater and cold-bloodedly gunned to death from the rim. As the *Washington Post* in 1898 summed up the new American mood: "A new consciousness seems to have come upon us—the consciousness of strength, and with it a new appetite, a yearning to show our strength. . . . Ambition, interest, land-hunger, pride, the mere joy of fighting, whatever it may be, we are animated by a new sensation. . . . The taste of empire is in the mouth of the people, even as the taste of blood is in the jungle."[18]

That bloody "taste of empire" in all encounters with alien peoples and creeds had been an ingredient in the American character from the start; nourished by the Indian wars, it was now channeled, especially under Theodore Roosevelt's New Nationalism, into global politics. It remained, however, subordinate to the idealism of the American civic creed, held in check by isolationist sentiments or even morally deplored. As compared with the urgency of foreign policy in European states, American foreign relations remained a peripheral issue in public awareness. In any case, territorially secure, the country could expand its sway more effectively in nonterritorial ways and more in line with its ideals.

In reaction against the blatant imperialism of the previous decade President Taft argued that dollars were better than bullets. In an increasingly interdependent world, one of his spokesmen asserted that "international commerce conduces powerfully to international sympathy."[19] Aware of the rapidly growing strength of the American economy, American businessmen had long looked for opportunities abroad, in Europe, Latin America, the Far East, and the Middle East too. Now they worked hand in glove with the government as agents of American policy, eager to spread American influence and promote prosperity. Thus in the Progressive Era American expansionism took an economic turn, demanding an Open Door wherever it saw new opportunities. Thus began the age of the ubiquitous American multinational corporations, led by Standard Oil. Superior technology of mass production aided the process, introducing mechanical novelties like the telephone, phonograph, typewriter, sewing machine, electric streetcars, and, of course, the automobile.

Promoting American business in an interdependent world could easily be fitted into the door-opening American civic creed. American business would draw together the whole world in mutually advantageous relationships, as happened within the American polity. This vision contained even a pacifist strain. Traditional power politics impeded the peaceful flow of goods and people. For that reason President Taft advocated the settlement of international disputes by arbitration.

It was President Wilson, however, who completed the idealist reinterpretation of American expansionism, drawing together the best in the American experience as a model for global cooperation. Under the banner of making the world safe for democracy and guaranteeing lasting peace under the League of Nations (as told in an earlier chapter), he presided over the peace conference which ended World War I. Prompted by the current globalization of power politics, Wilson, the most high-minded among American presidents, projected the American ideal over the world at large. He annexed it to the American dream, holding out the vision of an Americanized humanity.

His initiative, as we have seen, impressed another idealist, Lenin, whose vision by contrast was based on theory rather than on an exceptional historic fate. But he raised a basic question: did the American vision, abstracted from its moorings, really possess universal validity or was its export merely another form of imperialist expansionism? Admittedly, Americans carried with them the benefits of their historical exceptionality and the fruits of their prosperity, in sharp contrast with Lenin's backward Russians. Yet, however attractive their gifts, did Americans infuse into international relations the equality so stressed at home? In their pride of global mission they let their ego grow large in 1917, ready to face the sacrifices of war. Yet, giving victory to the countries upholding the Western liberal-democratic tradition, were they aware at their moment of glory that they inspired others, who were outsiders to their traditions, to attempt the same preeminence? What they felt to be idealism might appear to the others merely as raw power disguised as moral righteousness. Who, for that matter, was morally entitled to reshape the world in their own image? Did there exist a universally accepted political morality?

IV

After the war even American opinion, always inclined toward isolation, became disillusioned with Wilsonian idealism in foreign policy. Economic expansionism continued in the Far East and in Europe until curtailed by the Great Depression. In their cultural outreach, likewise, Americans spread their influence—think of jazz or Hollywood. But as for politics, the leap from the American experience into the world at large was a leap into the dark, threatening to impose intolerable obligations not soon (if ever) repaid in goodwill and profit. Democratic internationalism required an international awareness and a worldwide response in kind, which at the end of World War I did not exist; the benefits of self-serving altruism obviously stopped at the water's edge. Tying the United States to the League of Nations overstrained the prevailing sense of American self-interest, which allowed only limited support for the emerging agencies of international cooperation.

More generally, tying the Untied States to world politics opened up chilling uncertainties. It exposed the country to the sense of external insecurity that had been the fate of continental Europeans (the more so the farther into

Eurasia one looked). Inevitably, the increased sense of insecurity also threatened disunity at home. As domestic division crept into foreign policy, foreign policy aggravated domestic division. National security, in any case, is a complex phenomenon defying objective definition; its essence is psychological, even spiritual. Individuals endowed with an inward sense of security can cope with world affairs in a more rational manner, while those inwardly troubled tend toward a more alarmist and violence-prone attitude toward foreign policy.

Confronted with the unsettling dynamics of immersion in international affairs, was it surprising after World War I that Americans wondered what was to be gained from leaving the comfortable womb of geographic isolation? Shrinking from the brink of involvement in world politics, they inclined toward transforming their traditional geographic isolation into a state of mind known as "isolationism." Some militant isolationists, influenced by fascism and willing to betray their traditional idealism, even talked of creating a "Fortress America" defying a hostile world.

In one ominous respect, however, even isolationist Americans could not escape being hooked into global affairs. The universalization of the American experience, implicit in the American creed and recently spelled out in Wilson's crusade to make the world safe for democracy, could not be scaled down. Now an equally universal counterideal had been launched by communist Russia. Could Americans ignore the challenge? Anti-Russian sentiment had been an integral part of the liberal tradition in Europe; it had crossed the Atlantic in the late 19th century; it was inflamed to a new intensity by the Bolshevik revolution. Touched at the core of their political identity, Americans—and particularly the official guardians of national security at the Department of State—took a hostile stance from the start, interpreting the Bolshevik revolution (in the words of one of them) as an effort "to make the ignorant and incapable mass of humanity dominant in the earth."[20] The Bolshevik threat was to be contained, even by American participation in the anti-Bolshevik intervention in the Russian civil war. Longer than any other major country the United States subsequently delayed recognition of the Soviet government.

All along, alarmed American diplomats, like security-minded conservatives in western Europe, sensed communist inspirations behind every rebellion against the existing order; they took communist propaganda at face value, viewing revolutions anywhere as part of the Soviet bid for world domination. On the other hand, a small minority of Americans dissatisfied with the American dream were impressed by the even more inclusive communist vision and founded the American Communist Party, with sympathy from American socialists. Meanwhile, some American businessmen concluded profitable deals with Moscow. While introducing a new source of division into American opinion, Soviet power, however, was too weak to cause major controversy. In the 1930s President Franklin D. Roosevelt took a pragmatic attitude toward Stalin's Russia, according it diplomatic recognition as a potential ally against the Japanese and later against Hitler; he even distrusted the anti-

Soviet attitude of the American experts assembled at the State Department. Obviously, Soviet power and aspirations did not loom as a major challenge to the run of Americans who had decided to stick to their own hemisphere. Besides, was not American industrial efficiency as preached by Frederick W. Taylor an ideal among many communists; was not "Detroit" held out as a much admired symbol of industrial efficiency during Stalin's first Five-Year Plan?

During the 1930s isolation-minded Americans, uncertain about the drift of events and sobered by the Great Depression, reduced their outreach, reversing past intervention in the Caribbean area and promising to be good neighbors in Latin America. They made little effort to stop Japanese aggression in the Far East; they kept out of European affairs as well, thereby indirectly aiding Hitler and Mussolini. Yet they could not escape being part of the world. Alarmed after the fall of France by Hitler's threat to Britain and the invasion of the Soviet Union, President Roosevelt cautiously extended American aid to these countries while trying not to offend isolationist opinion confirmed by the effects of the Great Depression. Yet after Pearl Harbor and Hitler's declaration of war, the fatal die was cast; the United States could not stay out of the war. By the same logic it could not stay out of the global power struggle.

Compelled by necessity to lead the war in Europe and the Far East and to inspire public opinion, President Roosevelt revived Woodrow Wilson's idealist vision, embodying it in the Atlantic Charter and the directives for the United Nations. Inevitably, the war also rekindled past ambitions for power, for spreading the American model around the world, mobilizing the American experience with all its diversity for global leadership. Without equals in the world and willing now to accept the responsibility of victory, Americans, whether they wanted or not, had crossed a historical threshold, moving from the exceptionality of their geographic isolation into the center of the world's stage, joining the human race, as it were, yet still carrying with them the boons of their past.

Never before in human history—to put the crowning touch on this summary of the American experience to 1945—had one country risen so high above all others, so towered over the world. In the course of World War II, Americans had moved into that exceptional preeminence not by their own lust for conquest or power, but by uniquely favorable circumstances: their own privileged evolution in a sheltered hemisphere and the power vacuum created by the near-universal exhaustion and widespread destruction at the end of that war—a war for which they bore only a faint responsibility by not having actively opposed aggression in its early stages.

At that moment of glory, Americans also could claim that they were uniquely prepared for their global preeminence by having transformed within their own country a great variety of races and cultures into a viable polity, imperfect yet all-inclusive, accustomed to transact their business in an unusually flexible and magnanimous manner conducive to building consensus. What other people had been better trained for dealing with the world's human variety? Admittedly, they had just acquired possession of the most devastating weapon yet invented. That fact, however, had barely left its mark at this historic turning point in the country's destiny.

How did that privileged American sanctuary fare now in an ever more tightly interdependent world in which it stood out as the universal model, the heir of the great western European empires?

16

The United States after 1945:
Exceptionality Eroded

In 1945 the United States stood at a peak of power, physically unharmed, its standard of living improved despite the war's sacrifices, its political system in good order, and its traditions buoyed by military triumph, the sole possessor of the atomic bomb. The Great Powers dominating the world in the past had either been crushed, like Japan or Germany, or diminished, like France or Britain. And the Soviet Union, suddenly propelled into a threatening prominence as heir to Hitler's domination over eastern Europe, suffered from historic handicaps aggravated by the postwar disarray of ruling over the biggest and most brutalized battlefields of the war. By common sense, one might argue, the United States was utterly secure. As President Truman, flushed by the victory just completed, told his Americans: they possessed "the greatest strength and the greatest power which man has ever reached."[1] Stalin no doubt agreed with him, like other observers around the world.

Two years later, however, the mood had profoundly changed. In trying to come to terms with their new fullness of power, Americans in high positions saw their condition in a sharply different light. Look into the corridors of power in Washington and listen to George Kennan, the foremost American Soviet expert and perhaps the most influential figure among American leaders trying to define their country's new role in the world affairs. He concluded a series of lectures at the National War College in the summer of 1947 on a gloomy note:

> Today we Americans stand as a lonely, threatened power on the field of world history. Our friends have worn themselves out and have sacrificed their substance in the common cause. Beyond them—beyond the circle of those who share our tongue and our traditions—we face a world which is at the worst hostile and at best resentful. A part of that world is subjugated and bent to the service of a great political force intent on our destruction. The remainder

is by nature merely jealous of our material abundance, ignorant or careless of the values of our national life, skeptical as to our mastery of our own fate and our ability to cope with the responsibilities of national greatness.[2]

In expressing his apprehension, Kennan by no means advocated a retreat into a Fortress America. On the contrary, like his sponsors in President Truman's cabinet, he was eager to have the United States assume the responsibilities of "national greatness." But how? How in the troubled postwar years of domestic readjustment could the country in its lonely preeminence be geared to global leaderhip? How could it mobilize its power to meet the adversities Kennan had outlined? And assuming that American resolve could be mobilized, what guarantees existed that American leadership would be able to generate the goodwill needed to mitigate the rising tensions in the global community?

I

Consider first the negative and positive conditions under which the American people, now numbering almost 150 million and rapidly multiplying, moved across the massive barrier between traditional isolation and global leadership.

That transition called for a drastic reorientation of a country propelled into wartime globalism only in reaction to aggression. What was required now was a permanent peacetime commitment to active participation in an uncongenial or even hostile world without abandoning traditional values and ideals premised on external security. Popular resistance to immersion in world affairs was bound to be deep. Why should Americans change? They had come to their new world to escape from the evils of the old; except for a small minority, they had little contact with the outside—foreign trade accounted for but a very small percentage of the country's GNP. In addition, the consensus-building give-and-take of domestic politics, as of social interaction generally, absorbed much vital energy; little time and effort was available for foreign affairs. In any case, ever more foreigners began to arrive in the United States, the center of the world, to learn from Americans: why should Americans venture out beyond their customary truths?

And how, moreover, could American tradition and values be related to that un-American world? Among themselves Americans, on the whole, were a peaceable and even generous people, shying away from the human costs of military conflict. Yet they also retained traits of that frontier violence revealed, say, at the time of the Spanish-American War (and cherished by the National Rifle Association). And how could they keep their own tensions from spilling out into the world at large? Who could foretell how the little-known psychodynamics of public sentiment under stress would affect the country's foreign relations?

The biggest and most crucial question in this momentous transition was whether Americans could escape looking at the world from the perspectives of their privileged tradition. Could they avoid practicing "cognitive imperialism," viewing the world from the busy ground floors of American life and assuming that all humanity should behave like Americans? Put differently: having so successfully overcome in their privileged isolation the differences among immigrants from many cultures, would they now be able, amidst the grimmer realities of the world, to see the others, at least to some extent, as the others saw themselves? Could they promote global consensus by adding a touch of compassionate tolerance toward other cultures to their own convictions?

The answer was predetermined—one of the given facts of the post-1945 world: a population of such size and so preoccupied with its own welfare could not but judge all the others by its own limited insights. That fact introduced a major source of conflict into American politics and the conduct of American foreign policy, with profound consequences also for the world as a whole. In the old days foreign policy was left to the experts; now that it had moved closer to public attention, it became more dependent on majority opinion among an ill-informed public. That public craved collective reassurance; in every presidential election it backed the candidate promising greater fortitude in guarding the country's pride as the best and the strongest in the world. How then could the government follow the advice of a minority better informed and more humbly open to other peoples' visions of human progress?

Considering the power of public opinion in the United States, one might well argue that, despite being better prepared for global responsibility than other peoples (as pointed out below), Americans still added to the anarchy in the global community; American democracy with all its boons adversely shaped the American presence in the world. Foreign policy was determined by the parochialism of the voters, to the dismay of more globally oriented citizens, of American allies abroad, and certainly of governments in conflict with the United States. The transition from traditional hemispheric security—and innocence—to global involvement was a profoundly disturbing challenge to American culture and identity, and an additional strain on the world as a whole.

Yet there was another side, a mixture of goodwill and selfishness. The United States, despite Kennan's apprehension, did not come with empty hands to that world. It brought with it all the boons of its exceptionality, beginning with the future-oriented grandeur of advancing the "the illumination of the ignorant and emancipation of the slavish part of mankind all over the earth." Suiting the scale of the new globalism, that mission was embodied in the Atlantic Charter and the Charter of the United Nations, a universal inspiration reverberating around the world no matter how difficult to follow in practice. Next came American generosity, expressed by help to others for rebuilding their countries. That motive had been at work even before the end of the war, through the United Nations Relief and Rehabilitation Agency.

Admittedly, American self-interest played a large role throughout. Americans craved to expand their collective and individual ego to global proportions. American generosity likewise was self-serving. All though the war, spokesmen for American business had worried about the country's economic future. Full employment required foreign customers and a world freed from the economic restraints imposed by the Great Depression. Encouraged by the new globalism, American interests were advanced in the Bretton Woods agreements of 1944, which envisaged worldwide free trade and a common currency based on the American dollar. Yet that potent self-interest, often taunted by foreigners, represented merely a globalized version of the traditional self-serving socialization of American life; in helping themselves, Americans intended to help others as well. They granted independence (with strings attached) to the Philippines in 1946; they pressed for decolonization throughout the non-Western world. They revitalized and expanded world trade, exporting their technology and managerial skills, their habit of doing things in a big way. American business particularly preached the material idealism of a global economy promoted by free enterprise, American-style, to illuminate all humanity.

Unfortunately, there were limits to these assets. Having worked reasonably well within the cultural confines of the United States, American business encountered obstacles abroad. Its social effectiveness extended as far as Western culture prevailed, to western Europe and beyond, to Western dependents in the non-Western world. But among the latter—as in Latin America—the cultural differences inescapably came into play. Americanization through international trade became an issue of the power politics of cultures: which ways of doing business were to prevail, foreign or indigenous? Who was to reap the advantages, the immensely favored Americans or their hard-pressed local partners? It was the familiar set of questions: who imposed change? Who was made to change—how and to whose benefit? In the postwar world American goodwill raised awkward issues of political power. Other people around the world did not feel or act like Americans, least of all the leaders of Soviet Russia.

Patently, there was some truth in Kennan's reference to a resentful or even hostile world. Under these circumstances, safeguarding the country and—as the ultimate security—making the whole world conform to the American way required a double effort. On the one hand, Americans could work through the inherent persuasiveness of their achievements—an easy task. On the other hand—and ever more prominently—they felt it necessary to proceed by deliberate acts of power, and at times even brute power.

II

The exercise of raw power, of course, had become commonplace in the war recently concluded. In its aftermath a major new threat emerged, perpetuating wartime habits of thought and feeling into the peace that followed, all too soon transforming it into an indefinite Cold War. In the wake of Hitler's defeat the Soviet Union had moved into central Europe, eliminating pro-Western political parties and harassing their constituencies; the countries under Soviet occupation were becoming Sovietized. In western Europe, barely liberated from Hitler's terror, fear rose of a further advance of Stalin's communism. Did history not testify to Russia's unlimited appetite for expansion? American government officials responded to the anti-Soviet agitation in Europe and to Churchill, its most notable mouthpiece, noting also the challenge to American ideals of freedom and self-determination. Recalling the prewar alarms about communist subversion, they also remembered the brutalities of Stalinism. Only the United States, they concluded, could stand up to the Soviet threat, which, when viewed through the blinders of wartime emotion, loomed as huge as the well-remembered Nazi horrors. Whether the Western reading of Soviet motives bore any relation to Stalin's assessment of postwar conditions or Soviet reality was a question never explored. Even Kennan, who had some insight into the dynamics of Russian history, read the evidence through hostile American eyes.

In a famous report of 1946[3] Kennan called the Soviet regime a "negative," "conspiratorial," "destructive," "insidious," and "irrational" adversary, a "malignant parasite" preying on the potentially peaceful relationship between the peoples of the United States and the Soviet Union. The Soviet system stood beyond the pale of civilized intercourse, he argued, to be treated as best a physician can treat "unruly and unreasonable individuals." He had no trouble converting more influential Americans, including President Truman, to his views. His prescription[4] for dealing with the Soviets' "expansive tendencies" set the course for subsequent American foreign policy.

He recommended "intelligent, long-range policies no less steady in their purpose, and no less variegated and resourceful in application, than those of the Soviet Union itself." In these words he outlined a political stance matching Soviet expansionism (as he saw it) with a proportionate American outreach. Thus started the "Cold War" and with it the symmetry of Soviet-American relations, known on the American side as "containment." The gist of that policy lay in "the vigilant application of counter-force at a series of constantly shifting geographical and political points" wherever the Soviets showed "signs of encroaching upon the interests of a peaceful and stable world." In other words, the potential reach of American power was extended to the borders of the Soviet Union from Europe through Asia to the Far East. Containment implied the biggest spurt of military expansionism in American history, usually with the consent of countries bordering on the Soviet Union.

There was yet more to containment. With its help, Kennan argued, the United States could undermine the Soviet system itself by projecting "among the peoples of the world the impression of a country which knows what it wants, which is coping successfully with the problems of its internal life and with the responsibilities of a World Power, and which has a spiritual vitality capable of holding its own among the major ideological currents of the time." By playing the trump card of American exceptionality and by counting on the internal strains confronting the Soviet system, Kennan speculated that containment might reduce Soviet Russia to chaos "in forms beyond description" within ten or fifteen years. There seemed to be a chance for an ultimate solution to the communist threat (though by the dialectics of U.S.-Soviet relations the American counterthreat was bound to perpetuate the very conditions which had shaped the Soviet system).

With these hopeful but ill-founded arguments Kennan made "the issue of Soviet-American relations . . . in essence a test of the overall worth of the United States as a nation among nations." Rather than despair over the Soviet menace, he confessed, in terms reminiscent of British imperialist rhetoric, to "a certain gratitude to a Providence which, by providing the American people with this implacable challenge, has made their entire security as a nation dependent on their pulling themselves together and accepting the responsibilities of moral and political leadership that history has plainly intended them to bear." These words touched the church bells of American patriotism, which set off a more boisterous alarm than Kennan had intended (and which he subsequently came to deplore).

Circumstances meanwhile provided an opportunity for a dramatic demonstration of containment. In 1946–47 civil war raged in Greece, pitting Greek communists against the Western-oriented government. On the mistaken assumption that the communists were Stalin's agents, President Truman in March 1947, with congressional approval, dispatched military and economic aid to Greece (and to Turkey as well), affirming that "it must be the policy of the United States to support free peoples who are resisting attempted subjugation by armed minorities."[5] In his mind—and that of many Americans—the complex, shifting world was neatly divided henceforth between free and totalitarian ways of life, and the United States was the providential defender of freedom. These sentiments, elevated as the Truman Doctrine, cleared the way for the peacetime placement of American troops abroad under regional agreements with American allies lacking military stength of their own. Thus began the global buildup of American military power, in western Europe, around Africa, in the Indian Ocean, and in the Far East. In this manner the United States became the eager defender of the "free world" as it defined that term. Its closest allies were the Western Europeans plus Canada, soon united, together with Greece and Turkey, in the North Atlantic Treaty Organization (NATO).

The year of the Truman Doctrine witnessed another major breakout from American isolationism, the Marshall Plan. In the postwar years American allies in Western Europe required financial assistance for repaying their wartime debts and restarting their economies. For the longer run, American firms needed partners and customers, especially among Western Europeans, in order to expand their productivity and strengthening into the bargain America's allies against communist influence. Rebuilding Western Europe, the Americans also expected to rebuild the world economy according to their specifications. Under the cover of economic aid the Untied States insured that communists were eliminated from positions of influence in Western Europe and free enterprise was promoted over socialist planning. American economic aid, political pressure, and military power under NATO combined in shaping postwar state and society in Western Europe, most obviously so in the new German Federal Republic. At the same time, it blandly made friends with fascist dictatorships in Spain and Portugal which allowed American military bases on their territories. Meanwhile, the justification for burdening American taxpayers with extraordinary outlays for distant lands and uncertain or delayed returns to themselves had to be persuasive—good reason for dramatizing the communist threat.

III

As Americans crossed over the threshold to globalism, traditional isolationism with its ignorance and fear of foreign cultures was subtly refashioned into anti-communism. The need for American involvement in world affairs could not be denied, but Americans were not prepared to accomplish the necessary mental reorientation as well. Anti-communism affirmed American power in the world and at the same time perpetuated or even justified the country's self-centeredness. Oversimplifying the American role in world affairs and responding to the postwar mood of unrest, it hardened American convictions. Who was to be blamed for the widespread uncertainties? The segment of the population intellectually, materially, and psychologically most vulnerable to change easily targeted Soviet communists and their American sympathizers as the root cause.

Even their more liberal and sophisticated fellow citizens could not escape a deep feeling of injured innocence in the face of Soviet hostility. Convinced of the superior rectitude of the American presence in the world as well as blind to the power contained in their proud posture of innocence, they charged the escalation of tension to power-hungry Soviet leaders and to the wickedness of communist ideology. How guiltily innocent Americans could be in their sense of mission was shown by an eminent Harvard historian with close ties to the intelligence community. "The United States," William Langer wrote as late as 1960, "should, at all times, exert its influence and power on behalf of a world order congenial to American ideals, interests and security. *It can do this without egotism because of its deep conviction that such a world order will best fulfill the hopes of mankind*" (italics added).[6] Was trying to make all the peoples of the world conform to one's own ideals an act of selflessness? A positive answer was possible only if the others already shared congenial ideals and values. Did they really?

Under these conditions there was little hope for introducing into American opinion a note of humility, allowing accommodation to the pride of peoples less favored by history. More patently than before, American universalism in these years tended to become merely another version of the expansionist nationalism common to European history.

The new quest for ideological mobilization made rapid progress in the government. Already in 1946 Kennan, perhaps subconsciously inspired by the Soviet model, had urged that the American public be deliberately educated about the realities of the Soviet challenge. In addition, he called for "courageous and incisive measures to solve the internal problems of our society" plus "self-confidence, discipline, morale, and community spirit." And more: the government should project to other nations "a much more positive and constructive picture of the sort of world we would like to see. . . . "[7] The drift of his argument was clear: greater government leadership, consolidation of American consensus by official encouragement, and more active propaganda abroad. Since Americans did not automatically take up the Soviet challenge, their government had to substitute official encouragement for the lacking spontaneity. Inevitably, leadership in world affairs massively increased the weight and authority of the federal government in the conduct of American democracy.

The Truman administration soon responded, in the spirit of Kennan's advice, with its own grass-roots-oriented militancy. Frightened by the evidence of wartime espionage on behalf of the Soviets, the government conducted a thorough security clearance in its own ranks with the help of secret agents and secret testimony; tested loyalty now was a precondition of government service and even of citizenship or residence in the United States. While American intelligence agents protected Nazi war criminals knowledgeable in Soviet affairs, the attorney general deported aliens suspected of radical views. Meanwhile, the public was reminded of the American civic creed. A Freedom Train displaying relevant memorabilia toured the country, stirring up patriotism. Soon the Voice of America began to broadcast the world's news and the image of the United States in accordance with official American convictions.

These measures were topped by a drastic escalation of the country's warmaking capacity. The armed forces were put under the unified command of the Department of Defense while all political-military intelligence activity was concentrated and expanded in the Central Intelligence Agency (CIA). In his most central responsibility, the president, furthermore, was henceforth advised by the new National Security Council. Although President Truman's call for compulsory military service went unheeded, the United States achieved an imposing military posture undergirding its global ascendancy with raw power, proud of its monopoly of atomic bombs (at least until 1949).

Part spontaneous, part induced and intermingling with contrary traditional impulses, the new anti-communism quickly took hold of American society. It flourished at the grass roots, a populist trend, yet was represented alsoin government and established society. A powerful Committee on the Present Danger spread the message among bankers, government officials, politicians, and academics. In Congress the House Un-American Activities Committee (HUAC) conducted investigations against individuals and organizations suspected of pro-communist leanings with little regard for their constitutional rights; from the Senate Joseph McCarthy experimented with anti-communist demagoguery by wild charges of disloyalty directed especially against the State Department. As he exploited the irrational dread of uncomprehended change, his anti-communism began to resemble some aspects of European anti-Semitism.

These changes introduced and consolidated in domestic politics the vested interests of a self-confident and ambitious military establishment with powerful allies among industrialists and scientists—a novel phenomenon in American political culture. Now the United States, measuring itself against the Soviet Union and ready, if necessary, to enhance its influence by military pressure, had become a world power of the Old-World variety, its exceptionality eroding by leaps and bounds.

IV

In 1949 the United States had finally crossed over into the old-fashioned ruts of power politics. Three years into the Cold War, after the Berlin crisis of 1948–49, the Soviet Union exploded its first atomic bomb and China turned communist under Mao Zedong. Mao's victory raised the intensity of American anti-communism, but communist China never menaced American ambition as did its Soviet neighbor. China was a newcomer to world politics; it had been an object of cultural curiosity, economic opportunity, and political sympathy in the past, but never a political threat. Its new communism seemed hardly capable of revolutionary appeal in the world; the new regime faced too many problems at home. By 1950, however, the red flag over China signalled a further extension of Soviet power, the main source of American fears. How were responsible American officials to guard their country in this treacherous world?

Their attitude was revealingly summed up in 1950 by a report of the National Security Council (NSC-68).[8] Written by Paul Nitze, the founder of the Committee on the Present Danger, it consolidated and consecrated the conservative American view of the world as an enduring cultural force; his definition of national security has guided the watchdogs of American power ever since.

Outdoing Kennan in pessimism, Nitze painted the consequences of shifting from geographic security into global leadership in apocalyptic terms. "The integrity and vitality of our system is in greater jeopardy than ever before in our history. Even if there were no Soviet Union we would face the great problems of the free society" in the contemporary world. The presence of the Soviet Union escalated the perils to an extreme.

> Conflict . . . has become endemic and is waged, on the part of the Soviet Union, by violent or non-violent methods in accordance with the dictates of expediency. With the development of increasingly terrifying weapons of mass destruction, every individual faces the ever-present possibility of annihilation. . . . This Republic and its citizens in the ascendancy of their strength stand in their greatest peril. . . . The issues that face us . . . [involve] the fulfillment or the destruction not only of this Republic but of civilization itself.

In the grim world of the Cold War two irreconcilable political systems competed. Drawing an even more Americanized image of Soviet intentions than Kennan, Nitze wrote that "the Kremlin's design for world domination . . . calls for the complete subversion or forcible destruction of the machinery of government and structure of society in the countries of the non-Soviet world and their replacement by an apparatus and structure subservient to and controlled from the Kremlin." By contrast, "the fundamental purpose of the United States" is "to assure the integrity and vitality of our free society which is founded upon the dignity and worth of the individual"; liberty is "the most contagious idea in history."

How, then, was the United States to meet the threat? Nitze's answer was clear and simple: affirm the basic American values at home and develop economic and military power abroad. "As we ourselves demonstrate power, confidence and a sense of moral and political direction, so those same qualities will be evoked" not only in Western Europe but also throughout the world, at the same time undermining the loyalty of the Soviet masses to their regime. American leadership in the world required that the government "strengthen the orientation toward the United States of the non-Soviet nations, and help such . . . nations . . . to make an important contribution to U.S. security . . . in a system based on freedom and justice, as contemplated in the Charter of the United Nations."

Yet, as for the specifics of American policy toward the Soviet Union, Nitze's single-mindedness was slightly dented. On the one hand he argued that the United States and its allies "should always be ready to negotiate with the Soviet Union" in the hope of winning its consent to the "equitable terms" offered by them. On the other hand he recommended the ultimate solution: "our policy and actions must be such as to foster a fundamental change in the nature of the Soviet system." He expected the United States to accomplish that goal aggressively, by turning "the current Soviet cold war technique" against the Soviet Union itself.

Was the United States, then, to become as vicious as it considered the Soviet Union to be in its conduct of foreign relations? Nitze honestly wrestled with the awkward problem of ends and means in power politics, asking what else in the face of force the United States could do but use counterforce. "Our free society confronted by a threat to its basic values naturally will take such action, including the use of military force, as may be required to protect those values." For patriotic support of this Machiavellian contention he quoted from the *Federalist Papers* that "the means to be employed must be proportionate to the extent of the mischief." In other words, in the conflict between ends and means the United States would always match the tricks of Soviet expediency.

But for Nitze there was a moral escape from the dilemma that threatened American moral superiority. "The essential tolerance of our world outlook, our generous and constructive impulses, and the absence of covetousness in our international relations are assets of potentially enormous influence." Or even more to the point: "the integrity of our system will not be jeopardized by any measures, covert or overt, violent or non-violent, which serve the purposes of frustrating the Kremlin designs." In the last analysis, the known sensibilities of American society, Nitze implied, would exercise a moral restraint over the means to be used—provided, one might argue, that the Congress exercised effective oversight. In this fashion the exceptionality of the American tradition contrasted with the brutishness of Russian history preserved the moral superiority of American foreign policy; a polity so favored by history could morally do no wrong. The logic of that argument was most succinctly expressed some years later by Senator Barry Goldwater: "extremism in the defense of liberty is no vice."[9]

In this fashion the traditional belief in the providential mission of the United States was carried over into global power politics, with no holds barred. "Our aim in applying force," Nitze advised, "must be to compel the acceptance of terms consistent with our objectives, and our capabilities for the application of force should, therefore, within the limits of what we can sustain over the long pull, be congruent to the range of tasks which we may encounter." In short, maximize American military and political force to the limit of national capacity.

As for national capacity, Nitze affirmed that "the United States now possesses the greatest military potential of any single nation in the world." But the potential had to be made real. Urging a threefold increase in American military might, he called for a "reduction of Federal expenditures for purposes other than defense and foreign assistance, if necessary by the deferment of certain desirable programs." It was guns before butter, armaments before the legislative implementation of the American dream of economic equality and social justice, yet not without protest about the underlying assumptions. Prominent American experts on the Soviet Union and patriots like Kennan or Charles Bohlen, a former ambassador to the Soviet Union, protested the crisis-promoting oversimplification in Nitze's assessment of Soviet intentions, to no avail. Nitze successfully pleaded that if his argument would begin with a convincing statement which maximized the nature of the Soviet threat, the burden of proof for the rest of his case would be reduced.[10] Had not such exaggeration ever since 1947 been the most effective goad to prod the American government and its people from isolationism into world politics?

Yet consider for a moment how one-sided Nitze's assessment of national security was. It left out the major assets of American power: the country's secure borders, its vast and efficient industrial capacity, the unquestioned loyalty of its citizens, the consensus-based conduct of government, and the innate appeal of the American model around the world. Stressing military power above all else, Nitze's assessment ran counter to the idealist humanitarianism of American life; it diminished or even demeaned the best qualities in the American tradition. Who could tell, however how those qualities could be applied to the harsh contests of global power politics?

There was indeed cause for apprehension. The country's exceptional freedom from external pressures was gone for good. Now it was deeply immersed in world affairs; the alien realities outside demanded attention, a disturbing force requiring endless internal adjustments. Were those who for so long had determined their destinies by themselves now to allow outsiders to subject them to external necessities? The natural inclination was to resist the outside pressures to the utmost. Let the others change instead—especially as, by American estimates at least, they had so much less to offer for the progress of humanity. In this manner Americans were drawn to take their stand in the central issue of modern power: who forces whom to change?

In this power game the majority of Americans, like Paul Nitze, had their minds made up; they were determined to resist all externally imposed change, making sure that they could hold their own, if necessary by brute force. A minority, however, extended the idealist tradition, asking whether the American capacity for transcending cultural differences could not after all be universalized to encompass *all* humanity. Their voice too could be heard in all foreign policy discussions, but for the most part only in pleas for moderation.

Inescapably, the run of Americans, unable to contribute from their protected past sufficient rationality or stamina for disarming hostility in a divided world, let themselves be willingly persuaded by the alarmists among their leaders to become almost as obsessed with national security as the men in the Kremlin (and on the latters' terms). Soon after Nitze submitted his report, the Korean War seemed to lend support to his contentions, while the nuclear arms race proceeded to the development of hydrogen bombs and thereafter inter-continental ballistic missiles ready to deliver them. The ICBMs constituted a deadly threat, subjecting the United States to massive or even fatal destruction, ending for good America's external security, the source of its exceptionality. For its protection, according to the mind-set shaped by Nitze's arguments, the United States had to stay ahead in the arms race, the champion of the free world guided by a coherent policy covering the entire globe. As President Eisenhower observed in his first State-of-the-Union Address, "the freedom we cherish and defend in Europe and in the Americas is no different from the freedom that is imperilled in Asia."

Yet the United States by itself, Eisenhower pointed out, could not "alone defend the liberty of all nations threatened by Communist aggression," no matter how powerful. Under his presidency, therefore, two more imposing-sounding alliance systems were added to NATO: CENTO (the Central Treaty Organization, consisting of the United States, Britain, Turkey, Iraq, Iran, and Pakistan) and SEATO (the Southeast Asia Treaty Organization, consisting of the United States, Britain, France, Pakistan, Thailand, the Philippines, Australia, and New Zealand). Now the Soviet Union was surrounded from Scandinavia to East Asia with anti-communist bulwarks backed by American weapons.

After South Korea had been secured and the global arc of alliances put into place, special attention was given to the countries around the Persian Gulf from which the all-essential oil flowed into the economies of Western Europe and the United States. When in 1953 these interests were threatened in Iran, the United States together with Britain engineered the overthrow of Prime Minister Mossadegh, who had planned to nationalize his country's oil industry. Subsequently, the United States built up the Shah of Iran, no democrat, as its most loyal ally on the very borders of the Soviet Union. But countries nearer home were also closely watched, following the traditional American urge to control the course of events in Central America. In these operations the CIA played the leading role carefully concealed from public view. All the while the effectiveness of nuclear weapons were further refined, again largely in secret.

The drift of events illustrated how the American government tried to recreate the country's past exceptional external security through a deliberate policy of establishing an unbeatable military and political superiority whatever the costs to domestic politics or traditional identity. By 1956 it had concluded entangling alliances with forty-two other states in strategic locations around the globe, involving itself in their domestic politics. It was interfering in the internal affairs of many other states as well, operating a worldwide network of intelligence, espionage, and covert action while freely dispensing military aid and thereby subtly militarizing American domestic politics and the world at large. By the logic of their rivalry the two superpowers became more alike in their ruthless search for security. How far at this time the approximation had advanced in shaping American attitudes could be read in the official reasons for barring J. Robert Oppenheimer (of the first atomic bomb) from further security clearance on the dubious grounds of disloyalty: "There can be no tampering with the national security, which in time of peril must be absolute, and without concession for reasons of admiration, gratitude, regard, sympathy, or charity."[11] National security in defense of freedom was apparently no respecter of traditional American freedoms.

V

By comparison with the Soviet Union, however, the United States still retained its massive assets of good will and tolerance carried over from the past and reinforced, to a large extent, by its new global preeminence. What if the loyalty now in demand required, in the words of an eminent contemporary historian, "the uncritical and unquestioned acceptance of America as it is"? Did contemporary American reality as measured by traditional standards call for repudiation? In the overall perspective—and certainly in the ground-floor experience of the average citizen—American life, bolstered (but hardly shaped yet) by the government's concern for national security, continued to flow in its traditional, immensely capacious, upward-bound channels, despite the Cold War, the arms race, confrontations with the Soviet Union, the Vietnam War, and the civil rights movement. Admittedly, in 1957 American pride was jolted by *sputnik,* the first artificial terrestrial satellite heralding the space age, launched by the Soviet Union. Yet the American response only released additional civic energies into a wide range of activities, including science and education. In the 1950s and 1960s, amidst the contrary crosscurrents and the mixed motives in this era of transition, the American spirit of enterprise rose to a high pitch; it became the driving force behind the world revolution of Westernization. It was "the locomotive at the head of mankind, and the rest of the world the caboose" (as President Kennedy's national security advisor, McGeorge Bundy, called it).[12]

That locomotive gave an impressive spurt to the country's prosperity, sustained by a rapidly growing population and an expanding domestic market. By 1960 Americans enjoyed a higher standard of living than ever before; in the next decade the country's GNP nearly doubled. Suburbia flourished with its new homes and shopping centers, its cars, and its easy access to downtown offices located in shiny steel-and-glass skyscrapers, while a new network of interstate highways speeded commerce and travel for recreation. Promoting and intensifying the new material wealth, the media (and especially television) radiated the glory of American consumer culture far and wide. Constituting only 5.5% of the world's population, Americans consumed more than half of the world's consumable goods, raising in their new visibility the expectations of all the others, whatever the consequences.

In the United States itself popular expectations advanced, even though the gap between rich and poor widened. The biggest gainers in the upward thrust were the well-to-do and well-trained with access to the great business corporations most suited to take advantage of the postwar globalism. Yet the charm of the American civic creed still worked, absorbing all social tensions in a remarkable drift to voluntary cooperation and conformity; even the poor were swept along in the prevailing optimism. Unemployment remained relatively low and public concern for the underprivileged was keen, as shown for instance in President Johnson's "War on Poverty" under the promise of his Great Society program. The volume of charitable, humanitarian, and cultural services by private agencies likewise grew. As before, none of these activities warranted complacency; reality always lagged behind the civic ideals. But patriotic goodwill was never endangered by comparison with other countries.

In addition, the country made big strides in realizing Martin Luther King's dream of racial equality for the black minority. The American civic creed with its call for equality in life, liberty, and pursuit of happiness, much advertized in the postwar assertion of American national identity, had accelerated the agitation for decolonization around the world; the emergence of independent states in Asia and Africa, in turn, boosted the civil rights movement in the United States. At times the struggle waxed bitter, reflecting the escalation of tension that came with the globalization of vital issues. Yet, starting in the 1950s and culminating in the next decade, a new measure of racial equality was enacted into law and into human attitudes as well. The results affirmed the vitality of American tradition. The rising black middle class, though still small, was determined to succeed by adjusting to prevailing norms; it proved far more influential than the militant minority demanding a separate black state. In public affairs racial confrontation ceased, while the interminable private accommodations ran their troubled course. More generally, the influx of respected aliens from around the world—students, scholars, businessmen, diplomats, delegates to the United Nations headquarters in New York—introduced a new tolerance toward human diversity which in time radiated abroad as part of the new globalism.

In characteristic ambivalence, the visitors to the United States were often critical. Yet they could not escape being impressed, in deeper layers of their awareness, by American life with its affluence-affirmed middle-class confidence, its fluent ways of getting things done, its well-established two-party system, its government by consensus. However troubled on the surface, the American political system stood as a model of constitutional government for the newcomers to statehood in Asia and Africa. American policymakers, especially in the Kennedy administration, were admired by Western European leaders for the enlightened globalism of their prespectives. And even against the backdrop of the escalating unrest over the Vietnam War, the landing of the American astronauts on the moon in July 1969, televised around the world, left an enduring impression of unrivalled superiority. That feat out-trumped *sputnik* and all subsequent Soviet space exploits.

VI

The outward radiation of the boons of American exceptionality was perhaps most powerful in the fields of finance and industrial productivity catering to the worldwide hunger for material prosperity. The government made generous contributions to the agencies—the World Bank, the International Monetary Fund, as well as the United Nations—guiding the expansion of the global economy, thereby encouraging the outward flow of private investment. The United States was the biggest foreign investor, its investments abroad reaching nearly $100 billion by 1972, most of them pouring down the road of least cultural resistance, into Canada, Britain, West Germany, and into other Western European countries; smaller sums went into Japan and even less into the developing countries just emerging from colonialism (where 70% of the world's population lived). American investments stimulated indigenous enterprise, mostly among firms capable of managing business in a big way— the times were made for bigness. The money was put to good use in transferring and increasing industrial productivity.

Setting a model for dynamic industrial innovation came easy to Americans under the special conditions prevailing in their country. Ample research and development facilities existed, with close ties, geographically and institutionally, to industry and government; useful technological spin-offs flowed from the ever expanding weapons industry. Moreover, the high work morale fostered by the Puritan ethic still persisted, certainly in the high-tech industries or legal and financial services. "The real secret of American productivity," the London *Economist* enviously observed in 1953, "is that American society is imbued through and through with the desirability, the rightness, the morality of production. Men serve God in America, in all seriousness and sincerity, through striving for economic efficiency."[13] Freely sharing the relevant skills, Americans now helped others to help themselves, proud givers and gainers in the global exchange of goods and services, yet also stimulating, through the virtually threefold increase in world trade, a massive rise in material welfare for the rest of the world.

The most visible vehicle of the American outreach, of course, were the new multinational corporations engaged no longer merely in exploiting raw materials but in manufacturing durable consumer goods, including automobiles, refrigerators, and the ever growing variety of electronic equipment. The secret of their success lay in peculiarly American capacities. Technological expertise counted heavily, but even more important were superior management skills, market analysis, salesmanship, and personnel policy—the arts of large-scale cooperation. Soon Europeans, and later the Japanese, began to copy the American giants with major ventures of their own—the opportunities were open regardless of national boundaries for those capable of seizing them. The world was becoming a single market, a worldwide shopping center, a global factory. The planning of operations, the production of components, the assembly of the final product and its sale could be arranged over the entire world, wherever the costs were low and the profits high. Worldwide transportation and communication hardly posed a problem thanks to fast ships, airplanes, and telephones (like computers) linked instantly by space satellites.

The multinational giants, American or European, included GM, IBM, Pepsi-Cola, GE, Pfizer, Nestlé, Shell, VW, or Exxon—to mention only the most prominent. They ran virtual empires, shaping the lives of peoples in many lands with budgets exceeding those of many small states. Was it surprising that they showed little respect for petty nationalist rivalries? As the president of the IBM World Trade Corporation boasted:

> For business purposes the boundaries that separate one nation from another are no more real than the equator. They are merely convenient demarcations of ethnic, linguistic, and cultural entities. They do not define business requirements or consumer trends. Once management understands and accepts this world economy, its view of the marketplace—and its planning—necessarily expand. The world outside the home country is no longer viewed as a series of disconnected customers and prospects for its products, but as an extension of a single market."[14]

At another time he went even further, observing that the "critical issue of our time" is the "conceptual conflict between the search for global optimization of resources and the independence of nation-states."[15] Big business was a pioneer for thinking globally in terms far more inclusive than those of the political globalists dominating the previous generation. It made excellent use of the idealist universalism of the American tradition—"commerce conduces powerfully to international sympathy"; "dollars are better than bullets." The multinationals further broadened the new globalism. Through their advertisements they set common expectations, common habits of consumption ranging from automobiles to toothbrushes, toilet paper, and Coca-Cola. In addition, the ceaseless coming and going of people (not to mention the flow of political refugees) spread the message of globalism, popularizing the attractions of the great Western metropoles, and New York foremost. The emerging capitals of the new nations were designed to match the Western prototypes despite all cultural and political differences. And their elites, even for their local prestige, conformed to the latest American—or Western—fashion in dress and lifestyles, especially after having lived abroad (without, however, having become aware of the cultural underpinnings of Western society). From the worldwide traffic emerged a common transcultural sense of modernity, reaching into all parts of the world, persuasive even where it ran into anti-Americanism or was officially rejected (as in the Soviet Union). American tastes too became more cosmopolitan, in food, in cars, and—marginally—in religious creeds.

Not all of the new globalism was superficial. Alongside the production and sale of material goods we observe a deeper and often unconscious process of sharing, training, inspiring, and cultivating a novel sense of enlarged perspectives. Far more persuasively than after World War I, a good-natured and creative challenge was issued to the entire world to catch up to the boons of American exceptionality, to look forward, away from the past, and rise to an all-inclusive perspective, "illuminating the ignorant and emancipating the slavish part of mankind."

Carried aloft by the new globalism, that impressive extension of American universalism called the United Nations became the focus of the new transnational idealism. Its message, as expressed in its key documents of 1945 and 1948, was "to promote social progress and better standards of life in larger freedom," and thereby protect "the inherent dignity" and "the equal and inalienable rights of all members of the human family" as "the foundation of freedom, justice, and peace in the world." In these years the United Nations Secretariat and its many agencies worked valiantly toward that goal as far as their limited means and capabilities allowed, trying to improve world health, advance food production enlarge its members' educational and cultural resources, enhance their technological resources, and keep an eye on the world's life support system.

Associated with the Untied Nations were a large number of newly constituted or newly enlarged agencies ranging from the International Chamber of Commerce (a most influential body), to the World Federation of Trade Unions, the International Air Transport Association, the Salvation Army, Rotary International, and, less conspicously, the International Federation of Margarine Associations and the International Shrimp Council[16]—who did not go international in the heydays of the new globalism? All of them, large or small, copied in their own settings the gatherings of heads of states or of their counsellors in the great international economic agencies like the World Bank or the International Monetary Fund.

The original thrust of the new globalism had come from the United States, but it soon released a mighty response, perhaps more disinterested and public-spirited, from Western Europe and Scandinavia, particularly Sweden. With the emergence of new states in Asia and Africa new supporters arrived, pleading their own cause through the globalism of the United Nations. In the early 1970s a long debate began over the creation of a worldwide New International Economic Order.

Soon the new globalism began to follow its own dynamics prompted by its successes. The Western achievements currently modelled by the United States were becoming universal property: the nation-state (preferably under a democratic constitution), science and technology, the ambition to achieve a respectable standard of living and to be respected in the global state system by whatever means were needed to manifest power. These properties were now constituted as universals regardless of their origins. What had been Western—or American—in the past was now merely modern, accessible to all regardless of previous cultural orientation.

Yet wherever modernism brought its promised results, it was blessed by a touch of that American exceptionality and idealism that reached out into the world at large, spontaneously, even unthinkingly, with little or no policy intentions—elementally. The disinterested humanitarian zeal was perhaps best demonstrated in the American Peace Corps, established by President John F. Kennedy "to represent the United States and its values in the most positive way possible." As most Peace Corps volunteers interpreted that mandate, they served best by selflessly dedicating their American skills to the improvement of local folk, regardless of official political and economic goals. In addition, the American public contributed generously to organizations dedicated to mitigate human suffering around the globe.

The American touch was perhaps a bit abstract and impersonal, but "deep beneath the anonymous American smile," so the Canadian Catholic philosopher Jacques Maritain sensed, "there is a feeling that is evangelical in origin—a compassion for man and a desire to make life tolerable."[17] That smile was well-meant and sincere, but winning friends only where people had been culturally tuned to appreciate facial messages of that nature. Interpreted by the sign language of different cultures, that smile might turn into a condescending smirk or even a threat.

VII

Power, the capacity to shape the conditions of work and life according to their preference, was never far from the minds of the promoters of the new globalism, whatever their good intentions. The multinationals, despite their professions of loyalty to a transnational global economy, remained tied to the nation-state and to an international order of which the United States was the ultimate guardian and they the principal beneficiaries. The American firms profited from intimate relations with the Washington establishment, through defense contracts, personal contact and service in the highest (and sometimes most secretive) government posts. All of them stood for free enterprise (as they understood the term), opposed to Marxist policies of nationalization and central planning—they were "capitalists." Some of them admittedly expanded their operations into the Soviet Union, yet without surrendering their freedom of action in their home offices. Naturally, they preferred a world that let them maximize their opportunities in the service of the common good as defined by their cultural values. When challenged, they turned to their government for support; American business and government had long been allied.

As it happened, the multinationals made trouble for themselves by their own actions (though they would hardly concede the point). More intensely than ever before, the new globalism not only introduced novel consumer goods into alien countries and cultures but also broadcast consumer ideals like freedom, equality, social justice, and self-determination. Popular expectations rose and traditional authorities were discredited; the revolution of Westernization—now transformed into the revolution of rising expectations—exacted its customary toll, thrusting disoriented societies into political turmoil. In addition, Stalin's successors, attracted after their master's death by their opportunities in a rapidly changing world, broadcast their Leninist claims with renewed vigor. The agitation of decolonization strengthened their case.

How under these conditions could the multinationals conduct their business and spread their wares in Third World countries? In general, they relied on the culturally most congenial and economically most cooperative elements in the population, essentially the elites. Thus they were allied to forces determined, despite the inroads of the new ideals of liberation emanating from the United States and the United Nations, to maintain the stability of the past that guarded their privilege. Inevitably, the elites' embattled resistance to change radicalized the democratic opposition. In the ensuing contests, the liberal ideals of the universalized American civic creed proved inapplicable or ineffective; reformers were turned into revolutionaries, liberals into Marxists. Soon the polarization of local society was reinforced by the polarization of global politics. Social revolution caused by local conditions provided opportunities for spreading Soviet ideas on the one side and bolstering American anti-communism on the other—pity the nonpolitical local folk caught in the middle.

Confronted with majority-backed reformist governments trying to curtail their power and to achieve a more equitable distribution of indigenous resources, the American multinationals—often pursuing enlightened social policies at home—turned for assistance to sympathetic American government agencies engaged in covert action and to the armed forces. In the eyes of security-minded American government officials, economic and strategic interests coincided; "communists" threatened both. Circumstances differed from country to country, from continent to continent; but the basic problem remained: how could American business interests be reconciled with local democratic self-determination? As President Nixon once formulated the issue with diplomatic discretion: "There is no more delicate task than finding new modes which permit the flow of needed investment capital without a challenge to national pride and prerogative."[18] How delicate would American policy be?

In the case of the small Central American republics little restraint was needed. In Guatemala 2% of the population owned 70% of the land. When President Arbenz, elected with a strong mandate, talked of taking land from the United Fruit Company without meeting its terms for compensation, the CIA, in 1954, sent a small invasion force assisted by American fighter planes to overthrow the president and replace him with an American-trained colonel who was well rewarded with American aid after he had cancelled his predecessor's social legislation.

In the next decade, under President Johnson, the Dominican Republic suffered a similar fate. After the CIA-engineered assassination, in 1961, of Rafael Trujillo, an unsavory dictator who had embarrassed his American sponsors, his democratically elected successor, Juan Bosch, clashed with American corporations operating in his country; he was ousted after two years by a military coup led by Reid Cabral. Threatened in 1965 by a popular revolt, Cabral called for American help. At once President Johnson dispatched some 20,000 U.S. Marines to restore order on the ground that "American nations cannot, must not, and will not permit the establishment of another communist government in the Western hemisphere."[19] Did it matter that no other American nations had been consulted nor the facts of "communist" influence been proven? As Bosch lamented: " . . . a democratic revolution [had been] smashed by the leading democracy in the world."[20] One may doubt, of course, whether any democratic experiment in countries lacking the prerequisite cultural underpinnings could have worked. In any case, if it defied the United States, it would most likely never run its course.

In one embarrassing case, however, the United States failed to have its own way: in Cuba. In 1959 the Batista regime, characterized by the then Senator John F. Kennedy as "one of the most bloody and repressive dictatorships in the long history of Latin America,"[21] was overthrown by Fidel Castro. Courting American help for his land reforms and eager to establish good relations (on his own terms), he was denounced as a "communist." Hard-pressed for economic support, he then turned to the Soviet Union, taking advantage of the polarization of global politics. Provoked in turn, President Eisenhower authorized the CIA to train an invasion force of Cuban exiles for overthrowing Castro.

The invasion—the ill-fated Bay of Pigs operation—took place early in the Kennedy administration. Escalating tension, the invasion attempt persuaded Castro to seek salvation in yet closer ties with the Soviet Union. Concerned about the security of his regime, he even permitted the Soviet Union to station short-range nuclear missiles on his island, offsetting similar American weapons placed for some time in Turkey near the Soviet border. In the course of the Cuban missile crisis, the first nuclear confrontation between the superpowers, President Kennedy forced Khrushchev to change his plans; but continued American hostility—between 1960 to 1965 the CIA, according to the Senate investigation of 1975,[22] made at least eight attempts on his life—deepened Castro's dependence on his Soviet ally without, however, reducing him to a docile pawn.

In any case, the Monroe Doctrine notwithstanding, the Soviet-American rivalry had been spilled into the Western Hemisphere, promoting further polarization wherever American and local interests clashed. What was at issue, however, was essentially not Soviet power, but local pride and the pre-rogative of self-determination in the face of Yanqui imperialism. On this psy-chological battlefield there was little room for compromise as long as Amer-ican corporate righteousness allied with the CIA set the pace, pushing ardent revolutionaries like Che Guevara to the opposite extreme. It was easy, within limits, for individual Peace Corps volunteers to adjust to alien cultures. Unwieldy agencies like the multinational corporations, let alone government, both closely tied to the cultural roots that upheld their institutional fabric, enjoyed no such freedom of accommodation. At the cultural frontier their Americanism was hardened—and their idealism eroded.

Admittedly, President Kennedy, more sensitive to the American idealistic tradition than his predecessor, tried to tone down American military brash-ness through his Alliance for Progress, which promised to "transform the American continent into a vast crucible of revolution in ideas and efforts" as proof that "liberty and progress walk hand in hand."[23] While helping to improve health and education in Latin America, these well-meant endeavors could not subsequently prevent a rash of military coups against democrati-cally elected "revolutionary" regimes. The most signal case of American intervention in Latin America for the sake of American corporate interests occurred under President Nixon: the overthrow, with CIA prompting, of the Marxist regime of Salvador Allende in Chile by a military junta under General Pinochet. In the eyes of American alarmists Marxism was tantamount to communism.

The communist threat required the assertion of American power around the world. By 1967 seven hundred thousand Americans were stationed in strategically located countries. The United States was the chief agent in five regional defense alliances; it was additionally tied by mutual defense agree-ments to forty-two states. It was an active participant in fifty-three interna-tional organizations and provided military and economic assistance to almost one hundred countries in all continents.

VIII

All along, and rising to a major crisis in American politics and conscience under President Nixon, the most militant containment of communism had been practiced on the other side of the world, in Vietnam. Starting as a small operation after the Korean War, American intervention had steadily grown, deliberately kept out of the limelight in order not to distract public support from more essential issues of domestic policy. Yet in the late 1960s it had built up to major proportions. Lack of military success in Vietnam gave visible proof of a condition which the country could not handle by its traditional resources of mind and power. The key issue was Vietnamese nationalism, the desire of an Asian people for emancipation from colonial dependence and foreign domination. Its hero was Ho Chi Minh.

Ho Chi Minh's political career had begun among the young Asians gath-ered in Paris at the end of World War I. Like many Chinese, he had been inspired by Wilsonian idealism and then turned communist as a result of his disillusionment with the Paris Peace Conference; in addition, the French government and the French socialists had denied his plea for Vietnamese independence. Later he had worked as an agitator and organizer in the Soviet Union, China, and Southeast Asia, spending the war years under Japanese occupation. When the Japanese withdrew, he declared Vietnam's independence from France, citing in justification also the American Declaration of Independence. The French again denied his plea and fought his liberation army, which was supported by the Soviet Union and communist China. After the French defeat at Dienbienphu in 1954, the country was nominally divided between a communist North and a pro-Western South, caught up in the polarization of global politics, a target of American containment. If South Vietnam turned communist, it was argued under the impact of the Korean War, all of Southeast Asia would fall domino-style to the communists. To avoid that calamity, the Eisenhower administration decided to build up South Vietnam as a separate nation holding its own in the war against North Vietnam and also against the Vietcong, the communist guerilla force in South Vietnam. Yet how could that be done without falling into the pattern of domination familiar from French rule?

The task of nation-building was entrusted to Ngo Dinh Diem, a Catholic anti-communist patriot with long residence in the United States. He was saddled with the impossibility of reconciling his patriotic desire for self-determination with incessant American interference. Denounced as a tool of American imperialism by the Vietcong and Ho Chi Minh in the North, he found it difficult to command sufficient loyalty among his people for managing a responsible government. In his methods of governance he was no democrat, proving himself a liability for his American supporters keen on establishing democracy. If he showed his independence toward the Americans, he weakened their support and even risked being overthrown and assassinated (as happened, with the help of the CIA, in 1963). But how could consensus-based government make progress in the clash between indigenous communist nationalism and American intervention feeding alien weapons, goods, manpower, and ideals into a country internally divided by religion, politics, social status, and forever wracked by disorder and civil war?

South Vietnam's weakness and the American determination to prevent communist rule called for ever greater infusion of American resources; inescapably, the results were further disorientation and persistent peasant sympathy for the communists. Trapped by their resolve not to appear weak, American presidents from Eisenhower to Johnson let themselves be tempted into ever deeper involvement. A first climax came in 1964, when President Johnson decided to force a decision by escalating the conflict. Fabricating a provocation to American power out of confused reports of a North Vietnamese attack on American gunboats in the Gulf of Tonkin, he rammed through Congress a resolution giving him a free hand to take "all necessary measures to repel any armed attacks against the forces of the United States and to prevent further aggression." In the aftermath American armed forces entered the conflict on a large scale; American intervention became the Vietnam War.

The war flowed into the mainstream of American politics brimming over at the time with crosscurrents of significant portents, soon polluting its visible surface with unprecedented bitterness. It was an important turning point, warranting a closer look.

IX

Early in the war American self-confidence, although touched by the Cuban missile crisis and President Kennedy's assassination, stood high. Some social scientists believed that there existed no contingency which human reason could not resolve; the future seemed finally under control. That confidence was reflected in the slogans characterizing presidential aspirations, John F. Kennedy's New Frontier or Lyndon Johnson's Great Society. Yet by the dynamics of the American civic creed, these programs were soon tested against the raised expectations, giving rise to criticism which the new programs were only partially able to meet. At the time of the great American outreach public attention typically focused on unfinished domestic business, favored by a slight relaxation in superpower rivalry after the Cuban missile crisis.

Among the boiling domestic issues in the Vietnam era the agitation for racial equality ranked top attention (as already mentioned). It reached a high point in the brutal attack on a column of civil rights marchers by state troopers and local police in the city of Selma, Alabama, a stronghold of Southern racism. Soon afterwards, triumphant demonstrators from far and wide in the United States marched safely from Selma to the state capitol at Montgomery, this time under federal protection. Both marches and the public outrage against the murders of civil rights advocates speeded the passage in Congress of President Johnson's Civil Rights Bill. They also fed the incipient opposition to the Vietnam War; civil rights and war did not harmonize.

The struggle for civil rights was associated with another and far more complex phenomenon, the counterculture among American youngsters, particularly those from affluent families, yet with links to urban ghettoes. The counterculture was partially a product of the perennial conflict between parents and adolescent children, between the strict social discipline needed for responsibility and success in society, on the one hand, and, on the other, the natural resentment of growing youngsters against enforced libido-restructuring socialization. Yet in the 1960s that conflict assumed an additional dimension, raising it to especially strident proportions.

The children were the first generation to experience the inroads of the new affluence as well as of the new globalism, the first to respond to the world just opened by their parents' labors and still unknown to them. The outside world appealed to them through its lifestyles, its values, even its religions, and most temptingly through its denial of the tight, socially oriented self-discipline built into the America tradition. There had been inroads before, from European romantics, from Marxists, from American blacks; they now recruited additional followers. But the new cultural subversives came directly from Asia, Latin America, or Africa, from peoples unwittingly revenging their own cultural disorientation. Thus arrived drugs, accentuated sensuality and sexuality, self-indulgence in its many forms, a rebellious cultural counter-revolution, a "barbarian intrusion."[24] In part it carried a constructive message: traditional ways of living and thinking—even of feeling—were too narrow, too set for the new globalism. Yet its main thrust was negative. Responsible globalism requires a new intensity of selfless and ascetic discipline, a more potent universal puritanism; the globalism of the counterculture was anarchic and destructive.

The superficial unity of humanity which it paraded soon foundered in an aimlessly messy, or even quarrelsome, degeneracy, while its moderate followers were absorbed into the new hedonistic consumerism or more immediately into anti-war protest. The subversive pressure from without, however, remained. In the age of globalism the socially oriented puritanism of American society—like that of Western society generally—was put under a permanent if subtle siege maintained by non-Western cultures (except perhaps Japan). The siege was abetted from within by the ever-present human temptations held in check now more by impersonal cultural conditioning than by individual effort—temptations additionally enlarged by affluence. All along, the subversive anti-puritanism threatened to advance from the counterculture into the strongholds of the establishment.

Viewed in this light, the rise of the counterculture was symptomatic of the changed conditions of American life in the years of the Vietnam War. Despite—or because of—its preeminence, the United States was irrevocably set into the cultural pluralism of the world, subject to the influx of alien ideas and values. At some level of their awareness, Americans were now compelled to cope with cultural diversity, a diversity patently not subject to the integrating universals of their civic creed. Those universals were losing their potency as American contact with alien creeds and values (which claimed a universality of their own) grew more intense. As a result, there rose over the horizon a profound metaphysical threat never faced before in the United States: an intolerable relativism akin to the disorientation imposed on non-Westerners by Westernization. What, in the Great Confluence of cultures, religions, ideologies, and ways of life—what stood out as the Truth? Did the world, more Westernized than ever, constitute a meaningful universe or was it a Tower of Babel?

More specifically at the time of the Vietnam War, what was ultimately right: American power or an elusive transnational consensus built on a common human desire for survival? That question, posed by the anti-war protestors, also divided American policymakers in their disputes over policy in Vietnam. Should they unleash the fullness of American might, possibly with the help of atomic bombs, on the ground that "there is no substitute for victory"? Or should they extricate their country from a costly failure by redefining American interests for use in a more cooperative and more decentralized world? Yet what transcendent premises existed for even a limited understanding between Americans and North Vietnamese communists or, ultimately, between the superpowers? Was it surprising that in the face of these questions many Americans were tempted to retreat into their past, their roots, and shore up their tottering verities by brute force? What convincing contrary vision was available to people on the American ground floor?

X

Undeterred by the domestic turmoil, President Johnson had meanwhile expanded the Vietnam War into a major effort. Under cover of extensive secrecy and deliberate deception of the public, nearly half a million men in the army and air force had been shipped to South Vietnam with the latest equipment in chemical and electronic warfare. Yet, despite the massive infusion of American military power, there was no prospect of victory. Worse: the tide of opposition rose throughout the United States, among young men to be drafted for service in a hopeless war in distant jungles, among TV viewers looking at battle casualties in their living rooms, among church groups voicing their moral distress, and among all the potential beneficiaries of the Great Society. Where was the money going? As Senator Fulbright pointed out, between 1946 and 1967 $904 billion had been spent on military power and only $96 billion on education, health, and welfare, the core of the War on Poverty. The military-industrial-intelligence complex had become the outstanding feature in the country's economic and political landscape. What, then, about social justice in the United States?

All along, the distant war grew fiercer. In January 1968 the anti-war agitation was boosted by the enemy's Tet offensive, proving his continued vitality; the protest assumed massive proportions during the presidential election campaign of that year. After President Johnson withdrew from the race, disillusioned about his prospects for reelection, the Democratic National convention, held in Chicago, became the scene of raucous anti-war demonstrations. As the Chicago policemen charged into demonstrators, journalists, and bystanders with unnecessary ferocity, they provided the American public with another taste, after Selma, of politicized violence American-style.

In the November election the law-and-order-minded majority of voters typically backed the Republican candidate, Richard Nixon, who had taken a tough stand on the war. Yet, after becoming president, Nixon and his chief advisor, Henry Kissinger, realized—as Johnson had at the end of his presidency—that the United States had to withdraw from Vietnam, achieving "peace with honor." While preparations were made for reducing American involvement and for negotiations with North Vietnam, honor was pursued by a further extension of the war, into Cambodia and Laos, and, with increasing reliance on air power, into North Vietnam. The cruellest phase of the war was reached in 1972 when more bombs were dropped on enemy targets, including women and children, in North and South Vietnam than in all previous years together (the total tonnage of bombs dropped was well above that of World War II).

At the same time, however, President Nixon, with Kissinger's help, prepared a new perspective for the exercise of American power in a slightly less polarized world. The first Strategic Arms Limitation Treaty was signed in 1971 amidst signs of a broader détente, leading eventually to the Helsinki Agreements of 1975. Relations with communist China were established for the purpose of letting that country take its place as a counterweight to the Soviet Union. As for the American role, Nixon arranged a timely retreat from the earlier worldwide commitment to "containment." As he stated in 1971, "the United States will participate in the defense and development of its allies and friends," but no longer "undertake all the defense."[25] The burden had to be more equally distributed, which meant most immediately that South Vietnam had to fend for itself. American military aid thereafter was more generously allocated to friends and allies, but the Nixon doctrine (as it was called) did not save South Vietnam. In 1973, by agreement with North Vietnam, the United States withdrew its remaining military forces, without much damage to the overall balance of power in Southeast Asia; American influence remained safely entrenched in the friendly dictatorships of South Korea, the Philippines, and Indonesia; even communist China had turned against the Soviet Union. After Nixon's resignation, American aid was curtailed and South Vietnam quickly fell to the communists in March 1975.

Thus ended a futile war which had cost the United States in sixteen years nearly sixty thousand soldiers killed, three hundred thousand wounded, and untold others suffering from drug addiction, the effects of chemical warfare, or psychological distress; in money the price tag was a cool $150 billion. North and South Vietnam, of course, had suffered infinitely worse damage, some of it long-lasting. But after staggering sacrifices—a million and a half people killed, not counting the soldiers—Vietnamese nationalism had prevailed against the mightiest nation in the world.

The victory of the Vietnamese communists was a victory for indigenous anti-Western, anti-American nationalism. It took a communist form because only the communists, following the Soviet model, possessed the skills and techniques of organization needed for self-determination in a polarized world. What suitable resources of political experience could Americans have offered without offending native susceptibilities? The limits of American power resulted not from insufficient application of brute force, but from the unrecognized cultural difference between the two countries. Using unlimited military power, the United States could save Vietnamese democracy only by destroying it (to adapt the words of an American officer beholding a community his men had levelled while rescuing it from the Vietcong). The gains for the Soviet Union were incidental and minor; the real winners were the hard-line Vietnamese patriots like Ho Chi Minh and his heirs. Their goal had been set in 1919, inspired by Woodrow Wilson. How the world had changed in the intervening half-century!

XI

How the United States itself had changed since the end of World War II was starkly revealed as the Vietnam War ground to its end. Paul Nitze's recommendation that Americans employ Soviet Cold War methods against the Soviets had been applied with a vengeance. The CIA and other national security agencies had grown into a veritable national security state within the government, fenced off by tight secrecy and operating under insufficient congressional oversight. The public never knew what was done until disillusionment with the Vietnam War forced a closer look. The results of the subsequent congressional investigations were shocking indeed. The CIA had been involved in several assassination plots against unwanted foreign leaders, not only against Trujillo, Castro, or Diem, but also Patrice Lumumba of the Congo and a Chilean general standing in the way of Allende's overthrow.[26] Covert action as part of containment had assumed far more sinister proportions than had been suspected; there was a dark underside to the American defense of the free world.

Covert action had even reached into the United States itself. Enemies were suspected wherever opposition to government policies or even the status quo cropped up. The CIA and the FBI had felt it necessary to keep troublemakers under surveillance, using spies or even provocateurs goading suspects into actions designed to discredit them. The domestic security operations covered a wide spectrum of the American public: civil rights agitators (Martin Luther King was kept under FBI scrutiny until his assassination in 1968), anti-war groups, religious organizations with pacifist convictions, or academics judged soft on communism. FBI spying on political leaders for use by the president, it was now revealed, dated back to the days of Franklin D. Roosevelt; but never had domestic undercover operations against dissenters reached such proportions as at the height of the Vietnam War. The CIA had been designed for protecting the United States abroad; now it blatantly operated within the country as well. In the polarization of global politics the alarmists were tempted to divide even Americans into friends and enemies, the latter to be thwarted by methods learned in the Cold War.

Was it surprising, then, that in 1972 some CIA-trained agents of the Committee to Reelect the President (appropriately code-named CREEP) should be encouraged to break into the national headquarters of the Democratic Party located in the capital's Watergate complex? The first time they succeeded in planting listening devices and copying office documents; the second time they were caught in the act. Their capture set off an even more intricate conspiracy, involving the White House and the attorney general, for covering up the misdeed and protecting the president's chances for relection. But thanks to two reporters of the *Washington Post* braving the incredulous indifference of the media, the awful truth won out. Soviet opinion, incidentally, favored the president's side in the revelations that followed.

In the upshot, however, the Constitution prevailed. Forestalling impeachment, President Nixon resigned and Congress probed into the lawlessness of the CIA and FBI, overriding all attempts to thwart the investigations in the name of national security. As Senator Franck Church, one of the chief investigators, vindicated the America civic tradition: "We concluded that despite any temporary injury to our national reputation, foreign peoples will, upon sober reflection, respect the United States more for keeping faith with its democratic ideals than they will condemn us for the misconduct revealed. We doubt that any other country would have the courage to make such a disclosure, and I personally believe this to be the unique strength of the American Republic."[27] Had Nitze, back in 1950, been right in assuming that the moral fibre of the American system would prevent its covert actions from descending to the Soviet level? If that was the case, it had been a close call.

The novel interaction between the methods of covert action in foreign policy and in domestic politics as well as the dispute over democratic ideals and national security reflected a deeper problem: how to conduct an effective foreign policy in the competitive global state system with institutions and attitudes inherited from times of geographic isolation and political security. How could the spectacular latitude of American democratic politics with its constitutional system of checks and balances be geared to a united and forceful presence in the world? Any administration in power was tempted to use—or abuse—its prerogatives in order to achieve its aims in foreign policy. To prevent abuse by the executive branch, Congress in 1973 passed, over Nixon's veto, the War Powers Act stipulating congressional approval for committing American troops to prolonged service abroad. But that legislation merely compelled a determined president to shape public opinion more deliberately, seeking unity in terms of the lowest common denominators available: the fear of communism and pride in the country. Inevitably, global responsibility was reshaping the American political system as well as American identity.

Meanwile, other changes had occurred in American fortunes. To start with, the combined costs of the Vietnam War and the social programs of the Kennedy and Johnson administrations had considerably exceeded government revenues. The resulting inflation affected also the international monetary system, reducing American influence. When in 1971 Nixon revalued the dollar and ended its free convertibility into gold, the Bretton Woods agreement of 1944 came to an end; the dollar had ceased to be the common international currency. Not long after, the Organization of Petroleum Exporting Countries (OPEC), in existence since 1960, decided to show its strength in the world market, quadrupling the price of oil, thereby causing an energy crisis and demonstrating American dependence on outside producers, the strongest of whom were Muslim countries around the Persian Gulf hostile to Israel. At the same time, the foremost beneficiaries of the great outpouring of American skills, Japan and West Germany, were catching up to their American models in some crucial branches of industry. American firms and American labor resented the impact of foreign competition, while alarmed patriots began to fear for their country's industrial leadership. And what about the demands of the emerging countries in Asia and Africa for a New International Economic Order?

In these years the country experienced the paradox of preeminent power. Preeminent power meant continuous involvement in world affairs. Yet such involvement inevitably also limited the exercise of power; it required accommodation to alien interests and conditions. American ascendancy after World War II had led to an unrivalled economic and political expansion around the world. Expansion had led to complex relationships with, as well as dependence on, allies, partners, suppliers, customers, collaborators, and competitors. Such dependence imposed a huge burden of negotiation, manipulation, and change reaching into the very fabric of American life, a source of weakness as well as of strength. The country and its citizens were caught in the silent power game: who forces whom to change most? The temptation came naturally: preserve the past advantages by using compulsion based on superior strength. By contrast, constructive adaptation required accommodation to alien perspectives even within the relatively congenial confines of the free world. Were Americans willing or able to make such adjustments?

The crosscurrents intruding from without ran into others issuing at home. There rose a new concern over the quality of life and the deterioration of the natural environment by pollution and other consequences of affluence and high technology. Nuclear power was pitted against conservation and the use of natural sources of energy. Out of the energy crisis emerged the spectre of a finite ecological system, of limits to growth, for Americans and all of the rapidly multiplying humanity; the global consequences of the great outpouring of Western civilization in terms of population growth and pressure on natural resources were just beginning to show—the future, after all, was not under control. On the other hand, one could also observe the rise of a self-indulgent consumerism and psychological withdrawal from the mounting pressures of a world growing over peoples' heads, a tendency toward abdication of critical judgment on the big and abstract issues projected by the new globalism. Let the media, above all television, or the president deal with those issues.

Yet, whatever the changes, the country was still the strongest on earth. After the heated confrontations of the Vietnam War and the Watergate crisis, the political scene calmed down. The tensions of the 1960s and early 1970s had roughed up the surface of life; their deeper impact was buried in the collective subconscious. By all appearances, the reservoir of optimistic goodwill at the center of collective life still held; the continuities of the civic tradition, though battered, persisted. The country's unity and the loyalty of its citizens were never in doubt; nor were its industrial potential, its wealth, its impressive presence in the world. Idealists like Senator Church could draw reassurance even from the recent setbacks. Could any other country have withstood failures like the Vietnam War, scandals like Watergate, or the subsequent revelations of misconduct in the security agencies with such complacency?

Yet did the senator's complacency itself not prove how the country's new power had corrupted its civic integrity? Viewed from a high altitude, the United States had imperceptibly moved into a new era, changing its character.

XII

By 1976, two hundred years after declaring its independence from the Old World in order to "be a standing monument and example for the aim and imitation of the peoples of other countries," the United States had left the protective shell of its exceptionality, adapting itself to the anarchic and competitive global community of which it now was an integral part. Remembering the expansive idealism of its tradition, it had stepped forward as the driving force of the world revolution of Westernization, serving, by its preeminence and visibility alone, as the model for good government and the good life, the admired and envied leader in human accomplishment, the patent winner in worldwide invidious comparison. That role had endowed it with the hidden but very real power of compelling other polities, other peoples, to follow its lead. With their help (wittingly or unwittingly) it had poured into the world its managerial skills for large-scale organization, its technology and material accomplishments, and even its ideals. It had immensely enriched humanity in material resources and creative potential.

At the same time, however, it had been prompted, amidst the uncertainties of global interdependence, to take the counteridealist precautions of military power common to traditinal European statecraft. In its new dependence on forces beyond its control, it had increasingly buttressed—and limited— its position by political and economic pressure and military presence, if not outright combat. Reassured by its new might, the United States had carried its cultural expansionism to new heights. Spreading its affluence, advancing the material security of life, and raising expectations throughout the world, it had thereby not only highlighted global inequality, intensified cultural protest, and escalated the competitiveness of global politics; it had also been forced to deny its own ideals when espoused by others for their own self-interest. It had propelled the ambiguities of the world revolution of Westernization to their historic culmination, at a price to itself.

In their global leadership Americans had grown immensely rich, sustaining the promise of their civic creed despite all discontent. Yet their government's redoubled emphasis on raw power also had begun to affect the quality of their collective life. Compared with the heyday of American exceptionalism, the government had waxed bigger as well as more secretive and authoritarian; social and economic inequality had become more accentuated, public opinion more polarized under stress, consensus more brittle, and social awareness dulled, while individual citizens had found their liberties restricted. More broadly, the traditional idealism of the civic creed had been tarnished by hypocrisy or reduced to myth by the cynical practice of power politics creeping into domestic affairs. The expansive good nature bred into the American past could not be easily reconciled with the world's harsh realities intruding into daily life; it tended to shrink into a defensive excuse for overlooking the shortfall from the American promise. And the informed goodwill and heightened awareness needed for matching the American preeminence with a constructive sense of global responsibility had proved to be beyond the capacity of Americans preoccupied as before with their own well-being.

In addition, Americans did not escape the subtler consequences of being open to the world and its alien cultures. They too now were subject to subversive influence entering from without and undermining the traditional spiritual discipline so crucial in the country's historic evolution. There seeped into American life a silent counterrevolution, fed from more easygoing cultures, against the demanding social discipline inherited from the past and weakening it from within as a result of success and affluence. Thus the new globalism burdened even Americans with a trace of the metaphysical uncertainty so common among the victims of Westernization. Simultaneously, Americans began to suffer from the mental and psychological exhaustion produced by the overload of detail to be managed in their affluent daily lives—lives tied, in one way or another, to an ever more actively interdependent world. Under these pressures, Americans tended to become divided, many inclined to harden and contract their traditional convictions, a few others experimenting with novel lifestyles.

Slowly but relentlessly, the adversities resulting from immersion in global affairs began to affect the American role in the world. The example Americans set for the rest of the world became more uncertain and commonplace, their idealism corrupted by the Machiavellian realism of power politics. Integrated into the global state system and the world economy, taking the lead in raising the overall standard of living and intensifying global interaction, the United States had conformed to the world's anarchic competitiveness, too self-centered, too ignorant, to apply its exceptional endowment of culture-transcending peaceful cooperation to the anarchic global community. While universalizing Westernization in the form of compulsory modernization advanced by non-Westerners as well, it also had universalized the militant competitiveness of the European (or Western) state system and its traditional arms race. Still immensely privileged, it had become absorbed, for better or worse, into the human condition.

It had also become more like its archrival, the Soviet Union; and the more alike, the further apart from it in ideology and political stance wherever their interests collided.

V

The Other Superpower: Victim, Rival, and Counterrevolutionary

17

Building a State
in Backward Eurasia

I

A portrait of the Soviet Union summing up the Russian experience in a more comprehensive manner than has been possible in previous chapters shows, in striking contrast with its American counterpart, a ancient body politic with features marked by hardship and adversity. In the background stretches the vast openness of the Russian heartland set into the western reaches of Eurasia, the biggest landmass on earth. From the North Atlantic to the Pacific, and, except for the Artic north, open on all sides down to the high mountain ranges in the south, the geographic setting offered no natural shelter against enemies. Here secure political boundaries had to be man-made, placing an unending strain upon human energy and introducing a tragic source of violence for the sake of creating the nonviolent continuity and stability needed for cultural growth and humaneness.

Where on this vast continent would the limits of the country's power or the guarantees of its security lie? Expansion was the only alternative to being conquered and deprived of self-determination. Yet expansion added its own insecurities. The conquered neighbors, unlike the Indians of North America, retained some power and remained rebellious. Worse: there always lurked further dangers beyond. In the ethnic quilt of Eurasia the same peoples lived on both sides of the frontiers. There existed no natural borders.

Greater threats brewed further afield. To the west and southwest the tsars of Russia encountered the ambitious rulers of central and western Europe, all of them expansionist in their own way; in the Far East Japan rose as a dangerous opponent. Was it surprising that under these conditions war and military power were built into the vary marrow of the Russian state? Its symbolic center of authority resided within the awesome fortress walls of the

Moscow Kremlin, constructed after two centuries of Mongol domination to ward off raiders from the steppes to the south and southeast.

And yet the state was always weak, poorly endowed in its western regions with natural resources, subject to a harsh climate, and hampered by unreliable communications, especially along its main West-East axis. Together with the country's size, these obstacles inhibited the development of its potential; they impeded the human circulation which promoted cultural homogeneity and political consensus. Even more debilitating weaknesses stemmed from the fact that Russia was always a cultural hinterland to more advanced metropolitan centers, first of ancient Byzantium, later of Europe, "the West," as paradigmatically it came to be called in the 19th century.

Throughout its history Russia was cut off by geography, religion, and hostile neighbors from active participation in the fast-paced development of western Europe. It received no credit for having once buffered Europe against the Mongols; it remained apart, heir to Eastern Christianity and viewed in the West as a heretic and barbarian at the edge of civilization. All the same, it was drawn into Europe's power struggles, a full member of the European state system by the 18th century. In that competition it stood out for its size and was therefore feared—yet defeated in all invidious comparison concerning cultural refinement or statecraft, if not in war as well.

A self-conscious all-too-frequent loser, it was determined all the more to seek security by expansion—expansion frequently in response to attack and invasion or by diplomacy—southward against the declining Ottoman Empire, and to the west by annexing the Baltic peoples, parts of Poland, and later Finland. While the major states of western Europe built overseas empires, the Russian tsars followed the lead of geography into Central Asia, as much concerned with frontier security as with colonial gain. In the distant northeast Russian traders crossed the Bering Strait into the Western Hemisphere. While abandoning that outpost, Russia established a base on the Pacific Ocean by seizing the northernmost coast lands from a China weakened by Western encroachment; in 1860 it founded there the ice-free harbor of Vladivostok. More than the Asian and African subjects of the Western colonial empires, however, Russia's "colonial" peoples gathered in the process of expansion were joined to their colonizers by their common geographic destiny, whatever their cultural roots; none could escape the antagonistically interactive landlocked boundlessness of Eurasia.

In this vastness of territory the rulers of Russia could find little external security. They bragged about the size of their country: one-sixth of the world's land surface—yet much of it useless, its very magnitude a source of weakness rather than strength. And the greater their sway, the greater the fear and hostility around their country's borders, the greater their insecurity. In Europe a paranoid fear of Russia's expansionist barbarism became entrenched among intellectuals (including Marx) and in popular culture; along its lengthy Asian borders Russian power was likewise apprehensively watched. Meanwhile, the seafaring states of western Europe, far more secure within their boundaries and the envied leaders in human progress, had encircled the earth with their empires. It was as if geography had cursed the peoples of Russia (and their less powerful immediate neighbors as well).

II

That unpropitious setting shaped the human features of the Russian empire. The eastern Slavs, commonly called Russians, were a widely scattered peasant people of diverse traits not found in the West. The majority of them were serfs until the 1860s; restless, ready to escape from the burdens laid upon them; meek, long-suffering, passive, yet at times rebelling in merciless fury, always defeated yet never surrendering hope for liberation; inexperienced in collective action, without capacity for constructive outreach beyond peasant hut or village, taught at the price of life not to take risks; illiterate, superstitious under the veneer of piety as preached by the Orthodox Church or sectarian priests; hostile to outsiders, cunning and resourceful in terms of their hard lives, capable of immense heroic endurance; callous, even brutishly violent, yet also generous and selfless; generally without public outlets for their talents—a "dark people" and unpredictable, held in contempt by a minority of more privileged Russians. Yet were the latter better suited for citizenship? As late as 1958 the young poet Yevtushenko characterized himself in these words: "I am thus and not thus, I am industrious and lazy, determined and shiftless. I am . . . shy and impudent, wicked and good; in me is a mixture of everything from the west to the east, from enthusiam to envy. . . ."[1] It has been Russia's fortune to produce remarkably powerful, though temperamental, individuals—ill-suited, unfortunately, for the levelheaded conformity of good citizenship.

The merchants, constituting a separate and hardly prestigious social category, were almost as powerless as the serfs, but with wider perspectives trained in the Eurasian caravan trade rather than in commercial contacts with western Europe. They were largely illiterate, more pious yet also ostentatiously displaying their wealth, their lives patterned by the goverment under which they lived and which they had to serve; there never rose in Russia a resourceful and independent middle class. The landed nobility likewise lacked the freedoms granted to Western feudal lords who served their royal master under a common law. Unlimited rulers of their serfs, they were dependent on the tsar, ever more tightly integrated into the bureaucracy and compelled to conform to its requirements. Nowhere in Russia do we observe the active corporate life and the keen competition of guilds, professions, municipalities, or estates that stimulated individual and collective creativity in western Europe.

The Russian Orthodox Church too was different from Western Christianity; it lacked the dynamic drive of the competing creeds and churches. Closely allied with the state and otherworldy in its orientation, it imposed its own rigidity of dogma and liturgy, provoking and suppressing dissent and sectarianism, stifling rather than enhancing the spread of Christian discipline into the underpinnings of social life. Admittedly, Russian Christianity contributed

its share of saints; at its best it brought solace to the bitterness of life—
nowhere in the West was Christ's suffering so keenly matched in daily routine
or the cry for God's mercy so heartfelt. The Orthodox Church was spiritually
most powerful when it preached humble submission to adversity, to God's
will—and less so when it enjoined obedience to political authority. In any
case, it could never escape the obstacles hampering all cultural development
in the Russian lands. At a crucial moment, in 1917, it was found useless as a
source of social progress and political adjustment.

III

Limited indeed were the human resources with which the rulers of Russia
had to work in their efforts to provide external security and a minimum of
law and order. They faced a difficult task. Since time immemorial the Russian
state had been built around a military leader unchecked in his prerogatives
by his companions and subjects, or by any human law. That autocratic tradition was reinforced by the Byzantine heritage adopted in the 15th century;
the imperial Roman "caesar" became the Russian "tsar," head also of the
Russian Orthodox Church and bearer of the imperial ambition; the splendors
of Constantinople were reflected in the Kremlin's golden-domed cathedrals.
Yet that ambition was soon frustrated by comparison with the European
states to the west; the West was always painfully ahead. Since the troubled
times of the early 17th century the Russian state, trying to catch up with its
western neighbors, was built with increasing urgency from the top down, over
the heads of its ignorant and suspicious subjects.

The basic pattern, evolving since the 16th century and enduring to the
present, was laid down by Peter the Great at the beginning of the 18th century. In his "service state" all subjects had to serve the tsar unconditionally,
though in different capacities and with different rewards, in order to make
the country strong. The most important "servitors" were the nobility, compelled to conform to Western models of efficiency. The tsar himself—he now
called himself emperor—adopted Western ways, moving his capital from
Moscow to Saint Petersburg, his "window to the West." He has been glorified—and denounced—as a Westernizer, though he was moved by a patriotic
desire to preserve Russian institutions and identity. Realizing the need for
drastic innovation, he smoothed over the break from tradition by a deliberate
escalation of national pride; Russia henceforth was an empire, determined—
like the other major states of Europe whom it tried to copy—to leave its mark
on the world. But Russian power rose not from the innovating collaboration
between rulers and ruled, but from the will of the monarch mobilizing his
people by command as much as the conditions of his country would permit—
inciting him and his successors to claim ever greater authority in order to
overcome the obstinacy of geography and the sluggishness of the people.

Westernization, however, did not overcome the country's weakness. It
foisted alien practices and perspectives on unreceptive people, confusing and
dividing their loyalty. By proceeding through command and compulsion, it

discouraged creative initiative or drove it into opposition; by trying to enhance unity, it promoted rebellion or even revolution; by imposing modernization, it perpetuated backwardness. And by covering up the country's weakness through magnifying its ambition in the world, it augmented external insecurity.

What, then, in the face of these contraditions should Russia's rulers have done? Passivity and meekness in the lands between Europe and Asia meant surrender and enslavement. Did rulers and peoples in western Europe demonstrate these qualities in their relations with each other? In the Western case, however, self-affirmation did not require drastic alientation from tradition. Russia, unfortunately, was different. It was the first, the most dramatic, and the most powerful victim of the revolution of Westernization, the longest-running experiment of state-building under conditions of backwardness in closest proximity to the West. Its destiny was shaped not by the dynamics of internal adjustments (as was the case most prominently in England and the United States), but by the disorienting interaction of external influences with domestic realities, with the former always holding the initiative. And never willing to admit their cultural dependence, its elites lacked an honest understanding of their country's tragic condition.

The country's difficulties increased the faster western Europe forged ahead and the closer Russia was drawn into contact with it. When Russian soldiers and officers returned home from France after the defeat of Napoleon, they could not help remarking "how good it is in foreign lands";[2] Russia could not compare with the West. More directly, the principles of the French Revolution challenged the very system of Russian government: the ideals of popular sovereignty and representative government were ruinous for autocratic power. Under Russian conditions freedom spelled anarchy; it threatened the multinational Russian empire with collapse. At the same time, that empire could not do without Western science, technology, or cultural inspiration. Western schools and universities were as necessary as railroads, banks, or breach-loading rifles.

Promoting education proved a particularly troublesome challenge. Schools and universities were to bring into the country Western intellectual fire without burning down Russian government or tradition. Instead they gave rise not only to the intelligentsia, with all its agonies and glories, but also to a flaming revolutionary movement determined to overthrow tsarist rule and match the West's superiority by surpassing it. Yet however patriotic their zeal, the revolutionaries were merely agents of circumstances beyond human control. It was Russia's backwardness which made the West in all its aspects an effortless and even unintentional revolutionary force, compelling the Russian government to strike back as best it could.

Self-defense for the government meant no mean effort. In the first place, the tsars (the modern ones by descent more German than Russian) had to rally their wavering people to their banner, trying to cement popular loyalty by spreading a conviction of national superiority (in that manner, after all, national consensus had prospered in Western polities). For that purpose

Nicholas I in the 1830s artificially created an ideology of countersuperiority, adding to an earlier mission of universal truth (embodied in the Byzantine heritage of Russian Orthodoxy) the superior qualities of autocracy and Slavic culture; the West was made to appear decadent and Russia the hope of the future.

Obviously, that ideology had to be propped up by a variety of administrative innovations. They included the cultivation of an official literature backing up its extravagant claims, indoctrination in loyalty through the bureaucracy and the army, and strict censorship. Invidious comparison with the West revealing the vacuity of the official ideology, or hostile criticism at home, could not be tolerated. On all fronts, vigorous countermeasures were called for against the subversive effects of the Western impact. These measures required strengthening autocratic authority, making it even more repressive and repulsive by Western—or "civilized"—standards.

Again, what else could the government have done? Relaxing its controls meant anarchy at home and weakness abroad, conditions offensive to patriotic pride; they profited only the country's enemies. And so the tsars continued, with improvements copied from western neighbors (above all Prussia), to hold the country together by force and to suppress all opposition by whatever means came to hand, all in the name of a spurious but indispensable superiority. Armed force, violence, and cynicism in statecraft were in the marrow of Russian government. In the fight against revolution force was carried to callous refinement, training a cynically ruthless secret police, which in turn generated a core of ruthless revolutionaries, both with international connections. In order to counter hostile opinion in the West, the tsars paid for favorable press coverage, especially in countries on whose financial support they depended. Russia's troubles as well as its state power spilled out over its borders.

What chance was there under these conditions for the development in Russia of the good-natured, self-confident, and disciplined tolerance that built popular consensus in the Western liberal-democratic states, or of the more regimented industrial efficiency of Germany? The temperament of individual Russians was autocratic and self-righteous; psychic energy was invested in introversion rather than social outreach—and the depths reflected the confusing diversity of conditioning factors and cultural influence as well as a conviction of superior insight and strength. Overly assertive or submissive on the surface, most Russians cultivated a self-serving practical expediency while thirsting in their souls for purity and integrity. In its political system the country resembled an involuntary community of negatively charged particles; in human relations the individual particles were capable of infinite energy turned inward or to intimacy in a very small social circle, or else turned outward in rebellious self-assertion. The aversion to conformity also applied to human relations with machines; there had been little cultural apprenticeship to the mechanical arts, let alone to the large-scale organizations required for modern industry.

It was no wonder, then, that the tsarist regime approached a fatal crisis when Western constitutional democracy, industrialism, and the rising intensity of power politics put the Russian empire at an even more threatening disadvantage, despite hesitant effort at reforming the service state inherited from Peter the Great. Its backwardness was demonstrated most patently in war. In the mid-19th century the French and English defeated Russia in the Crimea; in 1905 Japan triumphed in the Far East; in both cases defeat set off dangerous tremors in the body politic. In World War I German armies tested the political, economic, social, and moral fabric of the tsarist regime as never before—and destroyed it. More vulnerable than its rivals (except for the Austro-Hungarian Empire, which disappeared for good), the Russian state faced not only revolution but also extinction.

IV

At that point, out of Russia's revolutionary tradition rose another spurt of Westernization, magnified now to American dimensions. Taking in their ideology a leaf from the American Founding Fathers, the Leninist revolutionaries mobilized their energies for a better future. In the name of the superior vision of communism, they promised liberty, equality, and peace to ground-floor peoples awakened by Western domination to global perspectives and stirred to action by the miseries of the first world war. Their Soviet republic, though poor and backward, was held up as the new City on a Hill, transforming the world according to its own vision, making it safer for human life and happiness than "capitalist" democracy ever could.

Denounced by conservatives as traitors to the Russian tradition, the Leninists, however, were doing no more than recasting their Russian past into a new ideological mold, covering up the structural disruption by a mighty escalation of political ambition; they repeated, from a far weaker base, the feat of Peter the Great in terms suitable for the age of competitive globalism. And with what resolve! Westernized anti-Westerners, they turned the tsars' defense against Western subversion into a bold offensive against the West, trying to match its comparatively effortless revolutionary outreach by a deliberate counterthrust of world revolution: all victims of Westernization were encouraged to unite in a global alliance under Soviet leadership. With this fervent gospel they returned from Saint Petersburg to the Kremlin, saved from German domination by a fortuitous constellation of world politics rather than by their own strength. All along, they staked their future, in contradiction to Marxist dialectics and common sense, on the combined capacity of willpower, intellect, and injured national pride to overcome all obstacles. Yet, offering their country, transformed by revolution, as a model for humanity, they did not escape their country's traditional adversities; indeed, they piled up new ones.

As anti-Western Westerners they had to reculture a people stirred up by the collapse of the tsarist regime as never before against all enforced modernization, tranforming them, on the surface, into a model of social cooperation according to Western specifications embellished with Marxist ideals.

For that purpose they evolved a new service state with a greatly intensified collective discipline for all subjects drastically recasting state, society, and the economy. They challenged head-on Russia's backwardness—Russia's marginality at the edge of the West—in a monstrous experiment in state-building for which the West offered no guidelines. No government in modern history had ever attempted such convulsive mobilization.

And more: by threatening the world order built up through Western ascendancy, the Leninists mobilized the West against themselves, with ambiguous results. By repudiating the tsars' debts, they lost badly needed economic aid (previously available to the tsars); by isolating their people from opportunities for invidious comparision with the West—an advantage—they also barred them from learning from it, except through officially approved channels. Pretending to pursue a policy of peaceful coexistence with the "capitalist" states, they not only preached class struggle but also promoted it around the world as far as their limited resources permitted, ingeniously clutching at revolutionary outreach as their only instrument of power in international relations. And in their relations with the outside world they practiced the cunning, secretiveness, and deviousness—the weapons of weakness—learned in the revolutionary underground; they were wily and tough opponents in diplomatic negotiations, thereby doubly antagonizing their partners. Promising peace, they made their own contribution to the intensification of global competition and to the bitter anti-communism erupting in fascism. And for what gain? The amorphous "proletarians of the world" possessed little capacity for unified action; they preferred national self-determination to following the Soviet lead.

Finally, the Leninists' assests—ruthless willpower, appropriate organizational skills, political astuteness, and experience with Russian mass politics—were frustrated by their visionary ideology. Useful as a bid for countersuperiority, Lenin's russified Marxism was a questionable tool for rescuing the country from backwardness; it created a false consciousness. It applied the lessons of a genuine social revolution (as interpreted by a German romantic speaking the language of Enlightenment) to a country in the throes of reculturation under external pressure; revolution in Russia—or in any backward country—had preciously little in common with the French Revolution. In the Russian setting, moreover, the future-oriented optimism of the revolutionary creed was belied by the pessimism implicit in the extensive reliance on compulsion and terror. Expectations were raised to unrealistic heights; the gulf between theory and practice proved a source of discouragement and disloyalty, which in turn necessitated still greater reliance on force. What unending and discouraging problems!

Yet could any politically alert Russian rulers determined to hold their own in the mounting competition for global strength and visibility have done otherwise? Should they have scaled down their ambition and pride, their sense of urgency, as many Western observers (and some of their own intellectuals) advised, just when their Western model—and the United States foremost—had triumphantly raised their ascendancy in the intensified global competition? Were they to submit to discouragement and concede the country's inferiority, admitting that they were imitating the West, and add to its cultural arrogance by kowtowing to it? Given their patriotic zeal, their effort to cover up their weaknesses under an artificially hardened facade of countersuperiority and to catch up at any price was only natural; it was part of Russian tradition—as well as of the human condition at the perilous edge of an admired and rejected superiority.

There was no escape from the tragic fact: backwardness could not be overcome by the will of a political elite which itself was the product of backwardness. Who, anyway, had a clear understanding of the nature or magnitude of the challenge? The hostile stance of the Soviet regime toward the "capitalist" world would hardly encourage a more perceptive comprehension of its problems among western observers, especially after the Russian revolution moved from Leninism to Stalinism. The ideological hostility on one side engendered and deepened, in a vicious cycle, a corresponding political hostility on the other side. Like its Western rivals, Soviet Russia was no messenger of peace nor a dispassionate analyst of political dynamics in an interdependent world. Admittedly, the hostility was muted during the 1930s and World War II.

V

At the end of that war the Soviet Union, aided as in the previous war by a second front diverting German strength to the West, emerged victorious, having overcome the most powerful onslaught in modern history, at a human price beyond measurement by Western experience. Victory gave to the Stalinist regime an elemental legitimacy and to the country not only an increment of external security but also added prestige without, however, diminishing its disabilities. A bastion of isolation and ignorance at a time of intensifying globalism, it still had reason to be afraid of comparison with the outside, while yet more than ever determined to present its idealized image to the world.

Viewed realistically, however, that image, in its harsh, even brutal features ennobled by a streak of heroism and lightened by a touch of pride over having survived unspeakable hardship, summed up centuries of adversity. Its oddly fitting Marxist-Leninist garb bedecked with medals of martial glory barely covered up the scars from wounds suffered in war and civil commotion. The portrait as a whole showed a self-conscious victim of aggression, both military and cultural, eyeing all outsiders with deep suspicion yet ready, for the sake of self-respect, to impress on them the superior merits of the all-embracing Soviet motherland.

With a stern countenance expressing the sum total of its past, the Soviet Union after 1945 looked forward to the future in a transformed world. In that world it was The Other Superpower pursuing a counterpolicy of containment, trying to stop the West's—the first superpower's—revolutionary outreach and yet in its own way promoting that outreach by having made it the measure for its own endeavors. Its goals were formulated in the West. By the psychodynamics of invidious comparison, Soviet state ambition was inseparably hitched to the American model, the Russian mirror image of American assertiveness.

Under these conditions, the Soviet Union faced an uphill struggle in the years after the end of World War II, when the world revolution of Westernization peaked. The West, now embodied in the United States, set the pace, keeping the Soviet Union, despite its new significance in world affairs, in the tragic pattern of rebellious submission to externally enforced reculturation. The Soviet Union continued the traditional Russian ambiguity, a proud counterrevolutionary rival and yet victim, still defeated in all invidious comparison.

18

The Strains of Catching Up

I

After the end of the war the Soviet Union found itself, superficially viewed, in a new historic condition of unprecedented external security. In the Far East the Japanese threat had been eliminated. Closer to home, the traditional Great Powers of Europe, its nearest and most dangerous adversaries in the past, were either destroyed or significantly reduced. The decline of Europe, the old West, made possible a drastic westward extension of Soviet power; the dreaded backward colossus of the East advanced into the very heart of Germany.

Admittedly, such expansion was not the product of deliberate design (though it could be fitted into the stereotype of an innate Russian expansionism); nor was it the result of local social revolution, nor even of unilateral military conquest—the United States and its Western allies surely had contributed their share to the collapse of Hitler's Germany. Essentially, Soviet power moved into a political vacuum left by Germany's defeat. Yet once established there, it was to endure, given the Russian craving for external security. Now, for the first time in history, Moscow controlled the western access to the Russian heartland as far as central Germany. The only problem was how to cope with the peoples of the occupied regions tragically placed by geography on the western approaches to Russia, all of them wedded to their national traditions, pro-Western and therefore strongly anti-Russian and anti-Soviet, and generally backed by Western European and American sympathies.

In the postwar years, the power of the Red Army loomed large over all of Europe, reinforced by patriotic exultation over a mighty and well-deserved victory which boosted the prestige of the Soviet system. Soviet power indeed stood out over the entire world, again on the rebound rather than on merit. It was Hitler who had reduced the French to helplessness and the British to dependence on the United States; the Red Army had completed the change in the global balance of power by driving the Germans from the Soviet father-land all the way to Berlin. Whatever the causes, the Soviet Union had risen to global prominence, almost in the same category as the United States. Not surprisingly, it probed for added influence even in areas not under Soviet control.

Yet even at this moment of triumph Stalin was acutely aware of his coun-try's weaknesses. Vital regions of the country had been ravaged by the war; during their retreat the Germans had ruthlessly destroyed what assets were left. The population, utterly spent, yearned for a quiet and less oppressed life. Stalin also had cause for suspecting disloyalty. Returning soldiers and prisoners of war had seen how good life—even wartime life—was in the West; their tales might embarrass the regime. In addition, he had to worry about his new subjects, the peoples of Eastern Europe, who always looked West, not East. As for the world at large, Stalin was impressed by the sharp contrast between his exhausted country and the United States, the latter at the apex of its military, industrial, and political might. It was armed with the atom bomb, even brandishing it as an instrument of power politics before Stalin's eyes.

Guided by his life's experience, Stalin knew that victory permitted no relaxation of his past policies; insecurity still haunted the country. Tightening the party's controls after the dislocations caused by the war, he doggedly resumed his previous course. By the fall of 1946 a new Five-Year Plan was announced for rebuilding the country (which was accomplished in remarka-bly short time) and then catching up to the Western model; there was to be no letup in the forced pace of "building socialism." At the same time, a fierce propaganda drive was launched against all Western influences seeping into the country through wartime contacts and the rise of American prestige around the world. In the chauvinism bred by victory Russians were dubiously credited with many key inventions of modern industry; they were leaders in science as well. Inevitably, this wave of propaganda was backed up by renewed terror against people suspected of "bourgeois" attitudes or other forms of disloyalty; anti-Semitism fit only too well into this security-oriented frame of mind.

At the same time, in escalating conflict with the United States, Stalin was consolidating Soviet control over eastern and central Europe. In 1948 he took drastic action, suddenly imposing a Stalinist regime on Czechoslovakia, blockading overland access to Berlin, and tightening communist rule every-where, except in Yugoslavia, where Marshal Tito defied him. Thus collectivi-zation and forced industrialization Stalin-style, together with rule by the Communist Party—all evolved in response to Russian conditions—came to the more advanced countries of eastern and central Europe, now renamed "Peoples Democracies." In response to the establishment of NATO and the preparation for economic cooperation in Western Europe, these countries were eventually integrated militarily, politically, and economically into the Soviet bloc under Kremlin leadership. While Western Europe was revived with American help, the eastern half of Europe was sovietized. As Stalin observed, each victor in the war imposed its own system of government on the territories overrun by its armies (he overlooked the vital difference: the United States shared its blessings; the Soviet Union expanded its Eurasian grimness). Even more important in the long run, Stalin mobilized his country's resources for the development of Soviet atomic weapons and of rocket systems for military use. He was not going to be intimidated by American nuclear bombs.

Victory thus brought more Stalinism, more demands for discipline. In Stalin's words at the party's 19th Congress in 1952: "The task is to put a firm stop to violations of Party and state discipline, instances of irresponsibility and laxity, and a formal [rigid and unthinking] attitude toward decisions of the Party and government; to work tirelessly to heighten the sense of duty to Party and state in all our officials, to root out untruthfulness and unconscientiousness mercilessly. . . ."[1] Victory also affirmed the traditional Soviet admiration for American efficiency. According to Stalin, duty to party and state meant more. "American business ability is that unconquerable force, . . . without which any serious constructive work is unthinkable. The union of Russian revolutionary drive with American business ability . . . is the essence of Leninism."[2] Reculturation after the American model continued by stern command backed up by the final blast of Stalin's terror.

II

Stalin died the following March. There were casualties in the crush of people attending his funeral; people openly wept—obviously he had touched them at their depths. There was good reason for strong emotion. Together they had lived through extraordinarily heroic and trying times unlike any ever experienced in the West. He had been in charge of the mobilization that assured survival in the fiercest war ever mounted against the country. Whatever the human price for the effort—and it was appalling—it has to be measured against the costs of national collapse and German domination, German domination as described by Hitler in October 1941!

Attempting a fair assessment of Stalin's historic role, one also has to view the human price in the grim contexts of Russian history as a whole, but especially of the crisis years before defeat in World War I, with the revolution and the civil war to follow. Stalin, the product and symptom of the extreme tensions of the late tsarist age, carried his brutal reflexes into the postrevolutionary era, embodying in his person both the patriotic desire for escape from the humiliating threat of political extinction and the ruthlessness growing out of the stubborn and anarchic character of his people.

In his work he was above all concerned with domestic affairs, with mobilizing his country's potential. Of the outside he knew virtually nothing, following within Lenin's ideological guidelines the maxims of traditional power politics that gave him no insight into the dynamics of Western (or "capitalist") development. But he never deviated from his goal to make his country and its Marxist-Leninist vision a universal model; he was dedicated to the ambition of catching up to and overtaking the West, first of all in terms of hard power. In his last years the Soviet Union developed both fission and fusion nuclear weapons; the groundwork was laid for the launching (in 1957) of *sputnik,* the earth's first man-made satellite, which opened the way for space exploration and inter-continental ballistic missiles.

Yet Stalin's death was also a historic turning point for his country, more perhaps than victory in 1945. It ended the cruel spell cast by the crisis that had beset the country since the late 19th century. With him passed the justifications for the extreme methods of coping with that crisis. Thereafter Stalin would be viewed in his own country more critically, or even with horror, by more humane standards engendered by greater stability and Marxist-Leninist idealism, reinforced by the expanding Western influence as represented, for instance, in the United Nations' Declaration on Human Rights. Yet even as an embarrassment to his successors—as to communists everywhere—and as evil incarnate in most Western eyes, he remains one of the great figures of the 20th century, thrust into prominence by the world revolution of Westernization at flood tide. He can be fairly judged only in that context—and, in strict morality, only by people who have risen through similar ordeals to positions of high responsibility.

And thanks to Stalin, his successors could start from a sounder base in a fast-changing world that was slipping from his grasp. He and the brutal school of catastrophe—the only school for teaching, under proper leadership, recalcitrant peoples the necessity of tighter social discipline and expanded social awareness—had transformed the loose-knit and fractious Russian empire into the Soviet state. The old penchant for violence and anarchy—for extremeness—had been reduced by the excesses of his revolution and by the war. A solid core of vested interests in the regime and its efficiency, centered in the privileged ranks of the *nomenklatura,* guaranteed both greater stability and continuity in the far-flung operations of party and state. Stalin also had implanted a unifying ideology that furnished common goals steeled with determination and a touch of idealism; its goal of communism furnished a vision of ultimate perfection which in the long run could restrain the ingrained ruthlessness of governance in the Eurasian vastness.

All told, Stalin had supplied the country with reasonably solid foundations for effective statehood in the competitive globalism of the 20th century. What was needed now was building on that still uncertain foundation the splendors promised by Lenin—or at least a quality of life comparable to that of Western Europe or the United States.

III

Stalin's successors came to their job somewhat better prepared than he. They had grown up in the years of revolution, civil war, and the New Economic

Policy, finding demanding and constructive outlets for their revolutionary zeal in the Stalin revolution. Supporting Stalin, they had been at his mercy, good Stalinists all, yet without the cynical ferocity trained in the revolutionary underground. They were *apparatchiki*, Soviet-style organization men, cooperative yet fractious and of authoritarian temperament. One of them, Nikita Sergeyevich Khrushchev, was refreshingly human, at his risk. The experience of these men was practical in the large-scale, summary way of the raw days of the Stalin revolution. Would it be sufficient for advancing the country from socialism to communism? And what of the heritage of Stalinist terror and inhumanity?

Was their training sufficient, moreover, for helping their country keep up with the world? After Stalin's death, Western "capitalist" prosperity soared to unprecedented heights, spreading from the United States into Western Europe and the Far East, encircling the world and advancing the lot of the working classes everywhere. The emancipation of the colonial peoples in the wake of World War II did not follow Lenin's blueprint of world revolution. It was carried out essentially under Western auspices, transplanting Western constitutions and norms into the Third World. Only where a former colonial power resisted the drive for independence did communists attain a foothold (as in the case of French Indochina). Both the "proletariat" of the "capitalist" countries and the colonized countries rushed to take advantage of the new prosperity, refuting the logic of Marxist dialectics. Even communist China, like Tito's Yugoslavia, proved an unreliable partner.

Equally disturbing for Soviet prospects, the rapid upswing of the world economy promoted an infectious internationalism repugnant to the traditional protective isolationism of both tsarist and Soviet Russia. By its ideology the Soviet state should have been a leader in progressive globalism. In fact, however, the new global openness encouraged cosmopolitanism and invidious comparision, both posing a deadly threat to the loyalty of the population and the cohesion of the country. The party's substitute cosmopolitanism, called "proletarian internationalism," was a poor imitation, lacking the creative touch of spontaneity.

In the larger setting, in short, the Soviet Union still suffered from its traditional adversities. In all contacts with the outside world its own peoples were unreliable, while the eastern European satellites were kept in line only by military force. And the country's general progress was measured, even by the Politburo, in relation to "capitalist" achievement. World conditions, to be sure, allowed much opportunity to discredit the model; and in a few instances (as in early space exploration), the Soviet regime could score a resounding success. Yet, by essential indices like standards of living, let alone the formal guarantees of freedom, the comparison always went against the Soviet regime.

The key source of humiliation, aggressively stressing its advantages, was of course the United States. It furnished the incentives and the guidelines for Soviet policy in domestic and foreign policy, forever taunting the Soviet leaders with the challenge of "catching up and surpassing" its own glory.

IV

The first leader to come to grips with these problems was Khrushchev. A quick-minded, simple-mannered, and sometimes embarrassingly uncouth man, brimming over with vitality and patriotic zeal, he tried in his folksy ways and ready talk to bring the government closer to the people, to give it a human touch. An indefatigable traveler, he covered the length and breadth of the Soviet Union, mingling with the crowds, listening with sincere interest, yet always ready to preach proper communist thought and behavior. Eager to prove that in the new globalism the Soviet Union too could reach out into the world, he even ventured abroad, to the satellite countries, to China, India, and Indonesia, to Egypt and to France, and twice to the United States, the first time, in 1959, with considerable aplomb. By his inexhaustible and impatient enthusiasm he came to represent, in his own person, his country's ambitions—and its problems.

He came to power gradually, by peaceful means and on merit. None of Stalin's surviving associates could match his imaginative boldness in ridding state and society of Stalinist rigidity and facing the demands of the times. From the start he shifted government attention toward agriculture, neglected in Stalin's emphasis on heavy industry, opening up the virgin lands of Kazakhstan to farming—at considerable risk, considering the semi-arid climate of that region. He let collective farmers pay more attention to their private plots and put them in charge of mechanized equipment; he introduced the cultivation of corn (maize), following American precedent with but scant regard for local conditions. Throughout he hurried agricultural productivity, without paying sufficient attention to detail or to organizational effectiveness.

He also pleaded for more attention to consumer goods and consumer needs, trying to make Soviet standards of living more respectable; he took the Stalinist bombast out of Soviet architecture, providing more housing. He raised popular expectations for a better life, counting, overoptimistically, on public willingness to work more effectively. Liberating the country from Stalinist terror, he himself set an example of constructive criticism of shortcomings in the party and the government, hoping to involve ordinary citizens in raising efficiency and work morale by loosening the tight grip of bureaucracy. By his design the "dictatorship of the proletariat," as the Soviet state had been characterized since Lenin, was transformed into "the state of all the people." He tried to give the regime a broader base and make it appear more democratic. Would the people rise to the challenge?

At his boldest, he rocked state and society by probing into the crimes of the Stalin era. Soon after Stalin's death, the agencies of the secret police were reorganized, investigations started, and innocent victims rehabilitated. Thus began "the thaw." In 1956, in a famous, not-so-secret speech delivered in his semi-impromptu style toward the end of the 20th Party Congress, Khrushchev allowed a shocking glimpse into the depths of Stalin's terror. His revelation affirmed more civilized rules of governance, encouraging a new moral integrity in the body politic and a new phase in Soviet literature.

Yet Khrushchev's boldness also had adverse effects; it tended to discredit the Soviet system at home and abroad. At home many leading party figures were implicated in Stalin's crimes; lesser officials had participated in line of duty. Even Khrushchev was asked about his role in those days. The public at large was stunned. Suddenly the revered architect of Soviet achievement, the great leader and symbol of their survival, was discredited. Soviet intellectuals with moral convictions rejoiced, officialdom hesitated in indecision, the hard-liners frowned. And simple folk felt offended, especially if they had held jobs in the camps.

Consider the responses of the latter, of "semi-literate people and . . . of quite limited mentality, . . . extremely undeveloped in both the intellectual and moral sense," according to the dissenter Roy Medvedev.[3] Having been convinced by Stalin that they were doing something useful for the country, they felt betrayed. "Often they themselves found life in the camps not particularly pleasant. And, of course, they do not like to think now that it was all to no good end . . . and that they themselves were accomplices in crimes of extreme gravity. There is a certain tragic aspect to all this. . . ." Who was more important to the party, a few heroic moralist critics (like Solzhenitsyn) or the common people characterized by Medvedev?

Khrushchev found himelf tragically trapped in the middle, forced to limit the investigation and the rehabilitations, under rising pressure to consider the consequences. The French communist Thorez warned him about the damage to foreign comunist parties. Worse: the American government had quickly got hold of his speech and disseminated it widely, letting the world at large draw its own conclusions. Most threateningly, soon after Khrushchev's speech, Poland and Hungary stirred in protest against their own Stalinist regimes, leading in Hungary to a short-lived repudiation of Soviet control.

It was a measure of Khrushchev's ability that he survived that crisis, rising, after sharp confrontations with his rivals, to the peak of his power in 1958. And at the 22nd Party Congress, in 1962, buoyed by recent Soviet space feats, he resumed his attack on what by then was known as "the cult of personality," contrasting Stalin's self-glorification and abuse of power with Lenin's modesty. De-Stalinization had been necessary, he said with some justification, in order to bring the party and the people closer together and to strengthen the position of the country in world affairs. As a result, Stalin's body was ejected from Lenin's tomb on Red Square. But meanwhile, Khrushchev, ever eager to dramatize his goals, had come under suspicion of fomenting a personality cult of his own.

V

Khrushchev's most dramatic impact, lending support to that charge, lay in the field of foreign policy. He began cautiously, continuing trends visible in Stalin's last years. Concerned over the growing power of the United States, the arms race, and the rising tensions in international affairs, Khrushchev aimed at greater security through stressing the need for "peaceful coexistence." As the weaker country (and a recent victim of brutal aggression), the Soviet Union made the most of the growing fear of nuclear war, trying to enlist the support of the nascent peace movement as well as of nationalist stirrings in Western Europe or among the emerging new states in Asia and Africa. And, as before, Marxism-Leninism defined the ideological framework for Soviet policy. "Socialism," Khrushchev contended in 1956,[4] was on the way to become a "world system"; the "general crisis of capitalism" was deepening. The colonial empires were falling apart and new states, likewise opposed to nuclear war and imperialism, arose in their place, with the blessing of the Soviet Union, the champion of peace and freedom. Now, as "international relations are becoming genuinely worldwide relations," the Soviet Union had to broaden its perspectives.

In "correctly estimating the requirements" that had arisen in foreign policy since Stalin's death, Khrushchev, at the party congress of 1956, gave crucial attention to the "capitalist system." He argued that, despite its crisis, "we must study the capitalist economy attentively, . . . study the best that the capitalist countries' science and technology have to offer in order to use the achievements of world technological progress in the interests of socialism."[5] He even allowed foreign news and opinion to circulate more widely among the elite, cautiously opening a window to the West.

Within the "capitalist system" the United States, of course, offered the key challenge to Soviet socialism, to patriotic pride, and to Khrushchev's personal ambition. When in 1957 the Soviet Union launched *sputnik*, the first probe into space (and a shock to the American sense of superiority), Khrushchev's confidence knew no bounds. As a result, U.S.-Soviet relations became his obsession, a psychological duel for superiority in the world and the goad for spurring the Soviet people into their greatest voluntary effort to achieve that goal.

In this spirit he embarked on his first visit to the United States, his combativeness boosted by the fact that a few days before his departure a Soviet space rocket had planted a pennant on the moon (just as Harry Truman was emboldened to face Stalin at the Potsdam Conference in 1945 by the news that the atomic bomb was a working reality). In the United States Khrushchev, lauding Soviet achievements, asserted bluntly: "we will bury you." He defended his down-to-earth phrase as a novel term for "overtaking and surpassing." He wanted his country to escape from being buried under the reproaches of backwardness. The escape, he argued, was measured foremost in terms of economic productivity.

Already in 1956 Khrushchev had boasted of higher growth rates in the Soviet than the American economy. As a follow-up he launched, in 1959, a new economic plan running seven rather than the customary five years and setting extravagant goals.[6] Citing *sputnik* as "a majestic event in the epoch of building communism," the plan set "the historic task of surpassing the most highly developed of the capitalist countries." Khrushchev was careful not to weaken his case by admitting Soviet dependence on the "capitalist" model. "While competing with America," he argued, "we do not regard America as our yardstick of economic development"; it suffered from unemployment, poverty, and other flaws of "capitalism." Yet in the next breath he was more realistic: "If America's production level is taken as a yardstick for the growth of our economy [as it obviously was], it is only in order to compare this economy with the most developed capitalist economy."[7] Overtaking America economically was a way station on the road to communism, not yet the culmination. There were other achievements along the way of which the Soviet peoples could be proud, their cultural institutions and advanced level of education (in 1962 Khrushchev boasted: "Soviet society is the most highly educated society in the world").[8] Yet these triumphs notwithstanding, the state would not wither away, as Marx had stipulated for communism, until the danger of an "imperialist" attack was ended.

As for overtaking America, the new plan provided thrilling specifics. Within five years after its completion, so Khrushchev assured the extraordinary party congress assembled to enact that plan, the Soviet Union would surpass the United States in industrial output. "Thus, by that time or perhaps sooner, the Soviet Union will emerge first in the world in both physical volume of production and per capita output. This will be a world-historic victory for socialism in the peaceful competition with capitalism in the international arena. (Stormy applause)"[9]

Two years later Khrushchev could boast of further Soviet triumphs in the field of space exploration, unmatched by the United States. In 1961 a Soviet space capsule, appropriately named *vostok*, "the East," had lifted Yuri Gagarin, the first human being, into space. A few months later *vostok* II carried another cosmonaut around the earth seventeen times. Obviously, the space age was dawning in the East, not the West—what better proof that the Soviet Union was surpassing the "capitalist" West in science and technology, key components of "scientific socialism"?

At the next party congress, in 1962, Khrushchev submitted still more encouraging evidence.[10] Not only had the Soviet Union outstripped the United States in its rates of growth, but also in absolute terms; it produced more coal, pig iron, cement, locomotives, lumber, woolens, sugar, butter, fish, and other items. The new party program of the same year clinched the argument, predicting that by 1970 the Soviet Union would have surpassed "the strongest and richest capitalist country, the U.S.A., in per capita production." Even better, by 1980 "the material and technical base of communism will be created that will ensure an abundance of material and cultural benefits for the whole population; Soviet society will come right up to the stage of application of the principle of distribution according to need. . . . Thus a communist society will be built . . . in the U.S.S.R."[11]

What a change of tune since 1956! At the party congress of that year Khrushchev had warned of "some hotheads" who had "decided that the building of socialism had already been fully completed and begun to draw up a detailed timetable for the transition to communism." Such optimism, he feared, would lead to "complacency and self-satisfaction." Why, then, did he subsequently hold out such extreme visions himself? His answer was given in his report at the 22nd Party Congress (1962).[12] "Consciousness of the grandeur of the tasks we pursue is multiplying the efforts of Soviet people tenfold, causing them to be more exacting of themselves and more intolerant of shortcomings, stagnation, and inertia. We must take maximum advantage of the enormous motive forces inherent in the socialist system. (Prolonged applause)" Yet could these grand visions of superiority overcome the pervasive "shortcomings, stagnation, and inertia" in Soviet life? Could they be matched by the results of Khrushchev's policies?

He certainly did not rely for results on the momentum of his exalted vision alone. The appeal to collective pride was part of the familiar litany of stern commands. As he said,[13] "the entire effort of the people must be directed toward fulfillment and overfulfillment" of the new plan. "We must strive to accelerate technical progress. . . . We must strive for a level of industrial and agricultural development. . . . We must advance along the entire front of cultural and social development. . . ." We must, we must—the road to communism was lined with admonitions and warnings. "Bourgeois influences" were still abroad, appealing especially to Soviet youths. Obviously, the Soviet peoples were not spontaneously rallying to the tasks set by Khrushchev; they were not rushing to recast their identity.

Not surprisingly, then, the new party program of 1962 put special emphasis on the necessity for "the socialist upbringing of the masses." It was the party's paramount responsibility, it declared, to "rear all working people in a spirit of ideological integrity and devotion to communism and a communist attitude to labor and the public economy." Echoing Lenin's statement of 1902 that what in a politically free country was done largely automatically had to be done in Russia by organization and deliberate effort, the party now announced that "spontaneous economic development has given way to the conscious organization of production and of all public life." The party assumed responsibilty for "transforming the minds of people in a spirit of collectivism, diligence and humanism, in short, of "scientific communism."[14]

That enlightened *dirigisme* flattered human ingenuity and willpower; hence the intellectual pride—the hubris—of the Communist Party. Yet could its body of knowledge and range of insights match the invisible hand of cultural conditioning that guided "capitalist" development? Apparently not, to judge by the continued use—though not in Stalin's dimensions—of repression and force by the secret police (unmentioned in Khrushchev's speeches).

In any case, by 1964 Khrushchev's accomplishments had not borne out his promises. Productivity in agriculture and industry had not increased according to expectations. His power-oriented testing of Western positions, in Berlin or the Mideast, had not reduced tension in international affairs. His duelling with the United States had gone against him, most signally so when he was forced to withdraw Soviet missiles from Cuba. In addition, communist China, whose revolution in 1949 had been hailed as completing "the process of Socialism's emergence as a world system," had drawn away from Soviet Russia. Most crucial, by his impulsive reorganizations he had dangerously disorganized and antagonized the party cadre, leaving him, exposed and isolated, at the mercy of his vanishing popularity. When in October his closest associates, whom he had raised to prominent positions, confronted him with his failures, he quietly resigned. Worried about the rising gap between Khrushchev's extravagant expectations and Soviet realities, as well as wishing to protect the party's pretension to omniscience, his successors blamed him for his "harebrained schemes." But moved perhaps by compassion for his daring, they allowed him a peaceful retirement.

Compassion he certainly deserves. His years in power may be considered a bold experiment—the most civilized yet in Soviet experience—by a naive patriot working with the simplistic resources of Soviet experience and Marxist-Leninist ideology. He employed his buoyant optimism as the best weapon available against the regime's greatest enemy, discouragement and disillusionment in the inhuman uphill struggle, with little success. His successors, favored by world conditions, wisely sought a humbler and more knowledgeable course.

VI

The new phase in the Soviet experiment opened in unprecedented smoothness. Khrushchev's closest associates, Alexei Kosygin and Leonid Brezhnev, jointly took over the leadership, backed by consensus in the highest party circles; gone were the days of bitter succession quarrels. The new men set a new style of leadership, emphasizing virtues hardly known yet in Russian statecraft: collective leadership, collegiality, and technical expertise. Kosygin was a technocrat bent on efficiency, incorruptible, remarkably selfless, and known for his patience with subordinates. Brezhnev was the stronger man, widely experienced in economic, political, and military work, having been a successful troubleshooter with an impressive record of achievement, most recently supervising Khrushchev's virgin-land project in Kazakhstan and the development of the Soviet space program based there. He too stood for a humane approach to leadership, as may be seen from a passage in his *Memoirs*:

> Remembering people, especially good people, is for a party worker not only a human obligation but also a professional duty. It is always necessary to cultivate human relationships. They give weight to a party worker's quality and strengthen his ties with life; they help to get to know firsthand the intentions, interests, and needs of people. Finally, it is downright pleasant to have a good person—whether worker, farmer, builder, agronomist, artist, or scientist—open himself to you. I never regretted the time spent on that; and, fortunately, it improves the effectiveness of political work.[15]

Brezhnev's *Memoirs* reflect his pleasure in working with all kinds of people and eliciting their best qualities. Whether ghostwritten or genuine—whether showing a real or an idealized portrait—the message was clear. He softened the strident authoritarianism of Russian officialdom, often aggravated through communist zeal, by a touch of Russian heartiness and, in turn, stretched that capacity to greater social and political awareness. Clearly the more powerful personality in the new leadership tandem, Brezhnev loyally shared the work and responsibility, holding the office of party secretary while leaving the management of the state organization, including foreign affairs, to Kosygin as prime minister. Only gradually did he move into the foreground, without public argument with his less forceful partner.

Proving that some measure of human decency had persisted in their training under Stalin and Khrushchev and conforming to contemporary worldwide ideals, the new leaders emphasized collective leadership combined with "comradeliness," "trust in cadre," and "respectful relations to people in the party." The keynote was "scientific management" with a strong accent on technical competence. Developing competence in turn required job security for administrators and their staffs, a further innovation. It also called for expanded technical training and professionalization throughout society, for greater collaboration between specialists and administrators, and for greater flexibility and openness all around. At the same time, ideological dogmatism declined. What was needed, obviously, was pragmatism in the handling of state affairs and even in the assessment of world trends. The requirement of efficiency prepared the way for a new consensus on practical necessities. Meanwhile, the divisive political issues of the past were downgraded; Stalin and the cult of personality disappeared from public discussion. Attention was focused on the present and foreseeable future.

At the same time, material conditions improved noticeably, despite initial uncertainty about the direction of economic policy and subsequent sharp crop failures. Productivity increased, and so did standards of living. Brezhnev hardly boasted when he said in early 1976 that "the history of our country has not known such a broad social program as that fulfilled in the period for which I . . . report [1971–75]."[16] The benefits were felt conspicuously by the new managerial elite, above all in the ranks of the party, where observers noted a rampant new materialism. But the most significant gains were scored by the lowest-paid strata of the population. In any case, income differentials (however measured) were far less pronounced than in Western countries.

Under these conditions the goal of broadening the popular base of the regime made good progress. Party membership increased by 50% between 1964 and 1975. Now the party comprised almost one-tenth of the adult population, with higher percentages among professional people, reaching almost one-half among men with over ten years of education. Public participation in primary organizations like trade unions or administrative councils in towns and cities also expanded. Halfhearted as such participation may have been, it nevertheless indicated some progress in training a larger sociopolitical awareness among the run of the people.

Encouraging constructive criticism, the new level of mass participation also minimized the intensity of dissent. The agitation set off among intellectuals by "the thaw" and Khrushchev's attack on Stalinism came to a head in the late 1960s and then died down. The most outspoken critics (including

Solzhenitsyn) were expatriated, others discouraged by surveillance or imprisonment; there was no evidence of massive support to sustain the protestors—their heroism went unrecognized. More loyal critics among artists and writers skillfully took advantage of the trend toward moderation. Obviously, the regime enjoyed a new legitimacy; its newfound cautious tolerance created its own rewards. Thus began, after more than half a century of brutal crisis, a spell of well-deserved stability and routine. Within the constraints of their cultural conditioning, the peoples of the Soviet Union had never fared so well.

VII

That normalcy was intimately related to the new external security of the Soviet state which emerged in the late 1960s. The United States, the country's chief opponent and measure of achievement, was increasingly immersed in the Vietnam War and absorbed in the domestic crisis caused by it. As part of extricating the United States from Vietnam with honor, President Nixon, with Henry Kissinger's help, was determined to ease relations with the Soviet Union. Thus began the era of détente, of reduced tension between the superpowers; it produced the treaties eliminating anti-ballistic missiles and limiting the production of nuclear weapons (the ABM treaty and SALT I). Even more important for the Soviet sense of security was the fact that the Soviet Union was attaining parity with the United States in nuclear weapons. President Nixon's visit to Moscow in 1973 dramatized the new balance in Soviet-American relations; it was highlighted by a show of cordiality between Nixon and Brezhnev. Brezhnev subsequently regretted Nixon's downfall in the Watergate scandal, even though it tilted the scales of power politics still further in the Soviet's favor.

Meanwhile, Chancellor Willy Brandt of West Germany had helped to ease Soviet apprehensions by taking the initiative for regularizing the relations between his country and the Soviet bloc. His country now accepted the political settlement that had come about after the war, including recognition of the German Democratic Republic. It also expanded its commercial relations with his eastern neighbors, thereby benefiting also the Soviet Union. The process of détente reached its culmination in the Helsinki Agreements of 1975, in which the Western European countries, together with the United States, gave formal recognition to the political boundaries of Eastern Europe. In effect, these agreements were a delayed post–World War II peace treaty, legalizing Soviet domination over eastern Europe.

In one important respect, however, the Helsinki Agreements overstrained the new Soviet confidence. The Soviet government, like the other signatories, agreed not only to promote free cultural relations with foreign countries but also to respect human rights. Did Brezhnev really think that in these critical issues the Soviet Union could live up to Western standards? He permitted an unprecedented number of Soviet Jews to emigrate, but would not allow Andrei Sakharov, the famous physicist who had protested the violations of human rights in his country, to emigrate. And what of human rights in the satellite countries?

Reflecting the new sense of security, Soviet management of its satellites certainly was softening. Admittedly, in 1968 the Soviet government had to shore up its key dependents, above all East Germany, by taking the lead in suppressing "socialism with a human face" as propagated by Dubçek in Czechoslovakia (cleverly using for its justification the excuses given in 1965 by President Johnson for toppling the unwanted government of the Dominican Republic). But thereafter Eastern European regimes enjoyed a measure of autonomy, entering, even more than the Soviet Union itself, into commercial agreements with Western countries. Hungary particularly profitted from the new trend, allowing a revival of private enterprise unthinkable under Soviet planning. Rumania in turn showed considerable independence in its foreign policy.

But the Soviet Union too engaged more freely in contact with the outside world, allowing some of its citizens to travel in the West, attracting foreign experts and industries for implementing "scientific management" and advancing its technology. All told, foreign realities entered more freely than before; Soviet experts on world affairs became more knowledgeable, Soviet diplomats more professional. Though still woefully ignorant of American politics, even top Soviet leaders were willing to adapt more readily to the new globalism. As they felt more secure, their attitude toward Americans became less combative. They needed Western technical know-how and, after their crop failures, foreign and especially American grain. More than before, Soviet economic development was tied into the global economy.

Correspondingly, the Soviet presence in the world increased. The Soviet navy, following the American example, appeared on all oceans; the Soviet commercial fleet significantly entered the world's carrying trade. Soviet emissaries frequented the countries of the Third World, no longer preaching revolution or socialist solidarity but seeking opportunities for trade and nonideological political influence; they invited official delegations and students to their own country or their Eastern European satellites. The economic ties of the Soviet Union and its satellites to Third World countries caused some dismay among Western firms, which had not yet felt any Soviet challenge in these parts, but offered no serious competition. Yet the Soviet Union did not trade from an empty wagon. It offered expertise in economic planning, in security and intelligence management, and in simple technologies. More important, it could give advice on how to build a modern state from the top down, from an elite into the masses, spreading in the name of anti-imperialism the essentials of Westernization. With time Soviet advisors (who found it difficult to adjust to the tropics) toned down their ideological appeal, conforming to prevailing world market trends without, however, disregarding Soviet self-interest.

Admittedly, Soviet efforts to expand Soviet influence in the Third World encountered some notable failures. Its military advisors, for instance, were sent packing from Egypt in 1972. And China remained hostile, suspected of collusion with the United States and widely feared among Russians almost like the source of another Mongol invasion. But all told, the Soviet leaders had cause for satisfaction, particularly during the decline of American prestige around the world following the American withdrawal from Vietnam.

Now the brash ideological self-assertion born of weakness, so familiar from the past, was tempered with a genuine sense of confidence. By the mid-1970s the Soviet government had completed, both in domestic and foreign policy, the most successful decade in Soviet history. For the first time it felt ready, however cautiously and self-seekingly, to open itself to the world and to join, in carefully controlled ways, the new globalism, shifting its arena of competition with the first superpower from ideology to economic and political leverage. At the height of his crisis-studded career and proud of his achievements, Brezhnev showed true statesmanship in combining Soviet ambition with a note of conciliation that lent some credence to the Soviet claim to leadership in promoting peace. Yet he was well aware that the main goal of catching up with the United States, let alone attaining communism, was not yet within reach.

VIII

At the 25th Party Congress, early in 1976—to offer a close look at the Politburo's hortatory assessment of the state of the Soviet Union—Brezhnev took stock of the condition of his country at the stage of "developed socialism on the way to communism" (as the party defined its then current halfway station in Marxist-Leninist terms).[17] He outlined the ideological framework, radiating optimism. "Capitalism," he said, "is a society without a future"; the socialist countries are "the most dynamic economic force in the world." And détente did not interfere with the logic of history; while promoting peaceful coexistence between states, it would not stop the class struggle. Indeed, he emphasized, "we make no secret of the fact that we see détente as a path leading to the creation of more favorable conditions for peaceful socialist and communist construction." In other words, the competition between the United States and the Soviet Union continued unabated, down to the details of the nuclear arms race; Brezhnev reiterated, although in less detailed fashion, his predecessors' promise to match all American advances.

Brezhnev's report was replete with the familiar calls for more rapid growth of labor productivity and efficiency, especially in the light of a foreseeable labor shortage. Needed also were more energy and more raw materials. To these ends the new Five-Year Plan, the tenth, called for hard work rather than raising the volume of goods and services. Among the shortcomings of Soviet labor he listed loss of work time, poor workmanship, a heavy labor turnover. He also spoke of the higher demands yet to be made regarding the quality of consumer goods and public services. As for planning and capital investment, more radical changes were required, more application of science and technology, which in turn called for more changes in the style and methods of economic activity. As for overall directions for the economy, he pressed for Soviet participation in international trade for the sake of a fair distribution of essential raw materials: "the USSR and the socialist countries cannot stand aside from the solution of these problems which affect the interests of all mankind." Such expansion of activity in turn demanded the intensified introduction of scientific methods into planning. In order to meet the international competition, better management was urgently needed and also more effective incentives for individual performance (always including "socialist competition").

But economic reforms were not enough. Brezhnev also called for intensified "organizational work among the masses," for the encouragement of criticism and self-criticism, stressing throughout the need for the "patriotic upbringing of the masses." In the struggle between "capitalism" and "socialism" an attitude of "neutralism" was not permissible. A proper consciousness was especially needed among the younger generation; the army, he said played an important role in its moral education. More generally, he considered it necessary "that the growth of material prosperity be constantly accompanied by a rise in the people's ideological, moral, and cultural levels." Otherwise they would relapse into "petty bourgeois" ways and "philistine psychology." At present he saw too much of "money-grabbing, hooliganism, private ownership tendencies, red tape, and indifference" in Soviet society.

In Brezhnev's style, these admonitions were spiced with an occasional human touch. "It has long been noted," he said,

> that the uninterrupted succession of days that resemble one another, of routine, day-to-day work—and all of us are engaged in such work—often prevents us from fully comprehending the significance and scale of what is taking place around us. . . . Every morning tens of millions of people begin another very ordinary working day: they take their places at the machines, go down into mines, go out into fields, or bend over microscopes, computations or charts. They probably do not think about the majesty of their work. But it is they who . . . are lifting the Land of the Soviets to ever newer heights of progress. . . . We pay tribute to the working people.

Yet, after paying tribute to Russia's working masses in their dull routines, he resumed his exhortations, echoing the Stalinist sternness voiced at the party congress of 1952: "we pay due attention to the problems of strengthening social discipline and to the observance by all citizens of their obligations to society. After all, without discipline and a stable social order democracy is unrealizable."

The party, in short, continued to preach in its peremptory style the message of individual and collective responsibility which the Catholic saints and the prophets of the Protestant ethic—together with kings and parliaments, a plethora of voluntary civic associations, and, not to forget, a string of wars and victories—had instilled into the subconscious realms of "capitalist" society (and which, despite all contrary pressures, still hold it together). Strengthening social discipline in the Soviet Union, however, might mean violating human rights as defined in the West. Even under Brezhnev, the Soviet regime had not outgrown the necessities of compulsion. Reculturation for the sake of matching the "capitalist" model still required the whip.

IX

How successful, then, viewed within its own set of circumstances, was the Soviet experiment in the thirty years after the end of World War II? Recognizing the adversities guiding the Soviet experiment and its built-in limitations, we cannot withhold admiration for its achievements. Building on the work of its founders, it has made remarkable progress under the expanded incentives and opportunities of the new globalism. While the United States was forced during the postwar years into a drastic readjustment of its historic identity, the Soviet Union followed its Leninist course in a steady advance toward the elusive goal of equality with the American model. Yet it too changed its complexion, becoming less compulsive, less militant, in short, a sounder polity.

Foremost, it completed the transition to an industrial service state, mobilizing the diverse and reluctant peoples of the Soviet Union into a working force geared to productivity under large-scale organization and equipped with the essentials. It provided a minimally adequate infrastructure of transportation, education, and basic industries; it even commanded worldwide respect in the prestigious technology of space exploration and nuclear weapons. The new industrialism raised the standard of living far beyond the levels of the past, displaying, as in the Moscow subway, an occasional showpiece of modern efficiency combined with aesthetic appeal.

The regime accomplished this advance despite continuous adversity stemming from a harsh climate, widespread material hardship, and the large size of both the population and the country (let alone the pressure of foreign affairs). It was easy for Americans, benefiting from abundant public and private services and conveniences in a smaller and gentler land, to forge ahead. The Soviet people had to work far harder for slimmer rewards and lesser results. Yet the Soviet system provided them, the newcomers to the insecurities of a far-flung industrial society, with much-appreciated support. As Khrushchev had boasted, unemployment did not exist in the Soviet Union; in addition, the hours of work were reduced below "capitalist" norms. These boons, unfortunately, did not increase productivity; but secure employment and reasonably short hours provided a convenient regime-supporting bridge from the slow-paced preindustrial past to the competitive and demanding present.

The regime also made remarkable progress in transforming a fractious multinational empire into a reasonably unified polity. It dramatically raised the standard of living among non-Russian minorities, especially in critical border areas. As Brezhnev remarked, the more intensive the growth in the separate national republics, the more solid the feeling of unity in the country as a whole. The government also pursued an enlightened policy of both encouraging indigenous cultural creativity and disseminating its accomplishments around the country as incentives for all its artists and writers. The advance of modernity was fastest in central Asia with its Muslim population—where in the neighboring countries did Muslims live so well? (Even Russians had cause to envy the inhabitants of Tashkent, the capital of the Uzbek Republic.) Traces of anti-Russian nationalism persist to this day, especially in the Baltic region or in the Ukraine; but farther east the evidence points to progressive consolidation. "Socialist internationalism" was tried out at home; it shaped a progressively consolidated multinational state always advertised American-style as a model to the world.

More generally, the regime had fashioned a minimal loyalty among its people toward their Soviet motherland; it had transformed them into more docile citizens. Concerned with the masses in their diversity and comparatively low political acumen rather than with a favored minority of intellectuals (who so largely have determined Western understanding of the Soviet Union), the party had succeeded in creating a pervasive common outlook. The bulk of the population showed a patriotic pride in the country (cultivated especially by the grandiose monuments commemorating the suffering of the war). The people also shared a common fear of the unknown and threatening world outside the country; even simple people tried to escape from the humiliations of invidious comparison—if necessary by deliberate ignorance. They accepted the harshness of the regime as a protection against disorder or chaos—there had been too much of that in recent memory. Moreover, they might well ask: what would happen to their slender prosperity if the country fell apart? Soviet people would admit their inability to handle freedom or the pluralism of "capitalist" society constructively; they did not see the hidden discipline underneath the Western practice (or if in emigration they did recognize it, they often resented it as superficial and uncongenial to the best qualities in Russian life).

In foreign relations too we have to give some credit to the Soviet regime. It pursued an embattled but cautious adjustment to the new globalism and to its role as The Other Superpower. In its competition with the United States, the Soviet government exchanged the territorial insecurity of the Russian past for the potentially infinitely worse insecurity of nuclear deterrence; but that insecurity it shared with its adversary. Under these conditions it felt safe to admit, with proper precautions, more of the realities of the non-Soviet world into its hitherto tightly closed universe. Its leadership became gradually more sophisticated in global affairs. It compensated for its disadvantages by its heightened ambition to escape from the penalties of backwardness by any means short of war, trying to harness all anti-American sentiments around the world to its advantage—in the name of peace.

In this endeavor, to inject a note of doubt, did the Soviet Union not pursue a contradictory course that flawed its effectiveness? Its desire for avoiding nuclear war was sincere, as was its desire to stop the nuclear arms race; it was a heavy economic burden. But its self-appointed role as leader in the worldwide peace movement stemmed from a patently competitive motivation. And while the Soviet government preached peaceful coexistence, it asserted that

the rivalry for setting a superior world model continued unabated; in promoting peace, it undermined conciliation by insisting that in the fundamen-
tals of social and political organization—at the deepest source of conflict—
there could be no compromise. According to Soviet perspective, history followed the inexorable laws of the class struggle over the entire world, even in
Latin America, the special preserve of the United States.

In its duplicity, however, the Soviet approach to global politics presented
merely a skillfully contrived mirror image of the Western—and after 1945,
the American—approach. Western governments traditionally favored a policy of peaceful coexistence, which allowed the cultural revolution of Westernization—the ultimate, if invisible, form of power—to operate unhin-
dered; they could afford to let history take its course, generally to their
advantage. Their more realistic Soviet counterparts have endeavored ever
since Lenin to assist deliberately and consciously the inexorable laws of history presumably running in *their* favor. Stalin's successors practiced this
Soviet substitutionism in foreign relations with new sophistication, even
though the tides continued to run patently against them. Decolonization
proceeded under Western rather than Soviet auspices, and the "proletariat" in "capitalist" countries followed social-democratic rather than
communist leaders (if it did not turn anti-communist, as in the United
States).

Meanwhile, the Soviet leaders successfully updated the ultimate—the
metaphysical—justifications of their political ambition, substituting a carefully designed radiant goal for the vaguer and more implicit promises of
"capitalist" development. While the prophets of Western superiority derived
their arguments largely from current practice supported by a vague promise
of "progress" in the future, the Soviet leaders held out the polished (if distant) vision of communism. In the West, to be sure, communism has been
thoroughly discredited by the atrocities especially of early Soviet rule. Even
in Brezhnev's Soviet Union the message of communism has been tarnished,
deprecated as "stale bread." Yet it still filled an obvious need, setting long-
range perspectives and raising a lodestar for a polity trying to liberate itself
from external dependence, with a heavy populist emphasis on the welfare—
and the obligations—of the working masses.

Who can quarrel with the—rather conservative—ideals listed under the
heading of communism? Communism, so the Party Program of 1962 stated,[18]
required:

> Conscientious labor for the good of society—he who does not work, nei-
> ther shall he eat;
> Concern on the part of everyone for the preservation and growth of pub-
> lic wealth;
> A high sense of public duty, intolerance of actions harmful to the public
> interest;
> Collectivism and comradely mutual assistance; one for all and all for one;
> Humane relations and mutual respect between individuals—man is to
> man a friend, comrade and brother;

Honesty and truthfulness, moral purity, modesty and guilelessness in social and private life;

Mutual respect in the family, and concern for the upbringing of children;

An uncompromising attitude to injustice, parasitism, dishonesty and careerism.

Except for an occasionally alien idiom, these ideals applied also to American society, where they have been comparatively more fully realized. Other ideals on the same list promoted patriotism. They called for:

Devotion to the communist cause, love of the Socialist motherland and other Socialist countries;

Friendship and brotherhood among all peoples of the U.S.S.R., intolerance of national and racial hatred;

An uncompromising attitude to the enemies of communism, peace and the freedom of nations;

Fraternal solidarity with the working people of all countries, and with all peoples.

Yet, if viewed as an expression of national solidarity spurred by a world mission, these ideals likewise constitute a mirror image of Western patriotism, certainly of the "America-first" variety.

The communist vision of a social order free of the flaws of capitalism filled a practical need for the "moral upbringing of the people"; it also helped to overcome the ethnic (or national) differences among the Soviet peoples. Yet it also catered to a deeper if counterproductive emotional yearning. As in Marx's case, it seemed to offer the possibility of a shortcut from a romantic, preindustrial vision of the good life to an industrial society so productive as to end the demeaning pressure of material needs—needs which in the complexity of the modern age have become ever more oppressive. When all material requirements have been filled, so the communist vision suggests, the potential of human creativity, as sensed in romantic yearning and Russian spirituality, will have its full opportunity. Then, at last, with the human capacity raised to the utmost individual and collective discipline of communal work, the repressive state will wither away—so that people can enjoy their spontaneous preindustrial undisciplined ways. In its very core, alas, the communist vision is flawed. Meanwhile, on the road to communism, the party relentlessly insisted on more intensive civic discipline.

Whatever its flaws, for a patriotic believer (and there seemed to be true believers) communism was paradise brought down to earth, a magnificent idealistic inspiration shaped by a powerful spiritual past trying to adjust itself to the material complexity of modern life. One should not underestimate, therefore, its hold, especially over the Soviet elite. It served as a substitute religion, as a support against the discouragements of insufficiency or the weakening of the crucial will to persevere; it provided a philosophical justification for hating the United States as the source of all Soviet troubles. Above all, communism supplied the faith suggested by Dostoyevsky's maxim that "every great people believes, and must believe if it intends to live long, that in it alone resides the salvation of the world; that it lives in order to stand at the head of the nations."

All told, then, the Soviet rulers after Stalin, though saddled with a difficult heritage, have succeeded in creating a framework of order and peace within their vast territories; they have provided the essential conditions for greater refinement in human cooperation. They also could proudly point to high achievements, collectively and individually, in space technology, industry, art, literature, as well as gymnastics and sports—achievements attained by comparatively greater effort or even heroism than in the West.

And yet, there hung over the more cheerful Soviet countenance an ominous—and tragic—shadow, the shadow of unsuccessful comparison with the West. In that basic respect, in the largest contexts of global power politics, little had changed. Soviet "communism" remained an idealized mirror image of Western practice for use in a backward country. It offered ample opportunity to discredit "capitalism" for its shortcomings; it superficially disguised Soviet dependence on an alien model. But it remained a paradoxical, conflict-ridden experiment in anti-Western Westernization. Despite its undoubted progress, the Soviet Union did not yet enjoy the true freedom of spontaneous collective development, a key ingredient of Western ascendancy. In the competition with its rival it was still victimized by its geopolitical heritage deeply embedded in its political system. It still had to admonish its peoples to lead more responsible lives; leaving them to their own devices spelled inefficiency and weakness. However impressive, the Soviet advance was limited.

X

To look, then, at the Soviet Union in the broader contexts of its existence, in the competition of the global state system in which the guiding standards of statecraft and collective organization were set by the West. By Western standards, we are compelled to say, the run of Soviet peoples were still geared to small-scale, slow-paced social relationships and taught by experience to avoid taking risks; they were ill at ease and overstrained in their globally oriented industrial service state. They possessed little capacity for coping with strangers, with people of different physiognomy, skin color, or culture. Isolated from the outside world and preoccupied with their own routines, they had little sense of global reality (except as portrayed by party policy).

They were still untrained also for spontaneous civic cooperation in their capacious lands; the Soviet system therefore was constrained to make excessive demands on their energies and mental capacities. In the absence of transcendent cultural bonds, state, society, and the economy were held together by rigid and compartmentalized bureaucracies, which themselves were overburdened. Having politicized all aspects of life in its sweeping push for reculturation, the government had immensely extended the scope of its responsibilities. But it was not only its hugeness that made it rigid; its bureaucrats, however well intentioned, lacked the capacity for outreach and overall grasp, for distinguishing between the important and unimportant. Even the Politburo was overloaded with what in Western practice would be considered trivia.

How under these circumstances could there arise the flexibility needed for effective cooperation and rapid economic growth? Not innovative enterprise (which was unsettling), but security in the vast administrative apparatus was the first consideration even for intelligent managers, all novices to the new political-industrial order. The tradition of secrecy in official transactions—the heritage of troubled times and the pettiness of localism—likewise impeded the effective conduct of business or the development of the larger perspectives needed for realizing the party's goals. Admittedly, the management of large bureaucracies poses problems even in the United States, but in the Soviet Union the available cultural resources needed for the job were far more limited.

And, paradoxically, despite its shortcomings, the government was driven to extend its controls even while trying to encourage greater initiative among the population and the bureaucrats. Initiative from below was permitted only within prescribed guidelines for prescribed purposes. Spontaneity was suspect or even prohibited—for good reason. It diverted popular energies from overall objectives which, however justified in the common interest, had little grounding in popular understanding. Or worse, given the centrifugal tendencies in Soviet life, it might run counter to public interest as interpreted by the party. The common people—"the masses"—could not be trusted in their ignorant diffidence, however creative their role according to the official ideology. The government—or the secret police—had to keep a close watch.

Even so, despite its pretensions to power, the government had to tolerate a wide range of activities incompatible with its aims. An informal "second economy" provided essential goods and services in the interstices of the Five-Year plans. It promoted a more primitive socialization than that envisaged by the party, but smoothed the routines of ordinary life. In addition, even more informal arrangements, under various nicknames but all adding up to "corruption," achieved a flexibility in Soviet life not otherwise attainable; incompatible with the ideals of communism, they provoked an occasional burst of punitive "communist discipline," but continued all the same.

Wherever one looked in the Soviet Union—and especially in the big cities—one could observe a conflict between the high-strung, globally oriented, and contrived activities of the government, often conducted in a surly or rigidly authoritarian manner extending down to the lowest ranks, and the narrow confines of private life. Here one found human warmth and spontaneous intimacy, or intense and often defensive self-indulgences ranging from immersion in artistic creativity or scientific specialization to alcoholism. An ancient pastime much advanced by the new prosperity, the latter rose to worrisome proportions, illustrating how little "communist discipline" had taken hold among the public.

Prosperity and increased contact with the West introduced further variants of self-centered escapism. Rather than set an example of social discipline, the West—with the United States in the lead—increasingly radiated abroad its most superficial luxuries, its fashions and popular entertainment, which proved only too attractive to spoiled youngsters disaffected by official insistence on austerity. The Soviet authorities had good cause to fear contact with the "capitalist" West.

Defeat in invidious comparison still haunted the government day and night, and with it the fear of disloyalty. It could not afford to return to the isolationism of Stalin's days; it needed interchange with the foreign world. But at the same time, it had to prevent its peoples from knowing "how good it is in foreign lands." If it painted too dark a picture of "capitalism," it ran the danger of discrediting itself. Reports from Soviet travellers to the West circulated by word of mouth, persuading some to defect when possible; others with permission to go abroad abused that privilege by defecting. As victims of popular prejudice, Jews were especially prone to seek escape.

But even privileged officials could not be trusted. While Brezhnev was speaking to the party congress in 1976, the highest-ranking Soviet diplomat, stationed in New York, UN under-secretary-general Arkady Shevchenko, was secretly working for the CIA; and the top KGB official at the Copenhagen embassy, Oleg Gordeevsky, was reporting to the CIA's Danish counterpart. Both men had seen "how good it is" to live in the West. There were other cases of defection, all of them highly embarrassing to Soviet pride. Prominent Americans did not defect to the Soviet Union; and American tourists unconsciously paraded their sense of superiority while travelling there. Such humiliation strengthened the resolve of Soviet patriots to get even with the humiliators (how would Americans have reacted under similar circumstances?).

The most troublesome category of citizens under the care of the Politburo and the secret police were the professional elites (including many party members) who by the nature of their function craved access to the West. From their work and travels, from government sources, and from Western broadcasts eagerly listened to, they knew about the differences between their country and the outside world; they were also aware of the gap between the myth of Soviet superiority and Soviet reality. Trying to be good citizens, they were torn between the elemental urge, common to all elites in developing countries, to rise to Western standards of personal comfort and their devotion to indigenous tradition and the regime's vision. Between these conflicting pressures they had to find their identity by choosing from a wide range of compromises and degrees of ideological commitment—a tragic condition and a source of immense anguish of a depth unmatched in the West—as well as of a nagging disloyalty.

What tragic challenge also for the regime's leaders! They found their most valuable people existentially caught in the crunch of cultural and political differences between their Soviet motherland and the West, and most exposed to the temptations of disloyalty. But what alternative was there but vigilance and indoctrination in the saving fervor of party militancy? The only alternative to doubt was the uncompromising, unfeeling hardness of belief and conduct, of *partiinost'* (party-ness), enabling the inner core of the party to view Soviet reality through the Marxist-Leninist myth. Unfortunately, such measures tended to increase the troubles of that professional elite and to limit its contribution. In this respect, as in all others, the government's efforts to cope with its problems merely perpetuated them. There was no escape from the curse of living in the shadow of a superior model which discredited all efforts to match it. But there could at least be more skillful management.

The key job of controlling the morale of the Soviet peoples, entrusted to the Committee on Government Security (KGB), was not repression, as in Stalin's days, but engineering human souls, as in Lenin's vision. The secret service had to sense public opinion, shape and reshape it according to government policy, watching over morale and morals, assessing loyalty and social cohesion, and adjusting the official tolerance of diversity and dissent accordingly. Repression was a disagreeable and often embarrassing incidental necessity, to be avoided if possible; the emphasis now rested on positive measures and more sophisticated methods—another improvement of the post-Stalin era. Because of its crucial role, the KGB became the political brain center of the Soviet system, the training ground of future leaders more sensitive to the issue of human rights and public opinion, yet still bound to the adverse conditions of their country. The conscious shaping of a superior society by human intelligence in the service of the highest ideal could not evoke the subliminal unguided social discipline produced by the free play of sociopolitical interaction. Substitutionism could not provide the voluntary discipline needed for freedom.

But why, outsiders ignorant of the power of invidious comparison often ask, could the Soviet regime not take a more honest and open view of its conditions, explaining to its subjects the problems they faced in nonideological terms and setting a slower and more humane course of development? Yet, as the insiders know, an admission of weakness adds to weakness itself. It encourages among the Soviet peoples their ever present resistance to cultural change. Above all, it establishes the superior Westerners more firmly in their superiority. Compassion comes easy to people secure in their preeminence. It is not compassion, then, which the Soviet leaders seek, but equality—equal superiority with the Americans. Only victory—or at least a reasonable balance—in the contests of invidious comparison can cure the ills of their polity. Escape from the Soviet predicament is a matter of power measured by the example of the First Superpower, the United States. In this sense, Kremlin policy is shaped by Washington and, more broadly, by the people of the United States.

XI

To complete, then, this sketch of the Soviet condition in the comparative perspective with a few comments on U.S.-Soviet relations. In the light of the foregoing analysis it is clear that the source of hostility between the superpowers lies in the clash of incompatible political systems, products of sharply divergent historical evolution competing for domination in the global state system. When toward the end of World War II the two polities for the first time confronted each other not only ideologically but also politically and territorially, across the full range of power rivalry, the issue was joined. It was global confrontation short of nuclear war. Each country strove, for its own sense of security, to create a world order conforming to its political institutions and ideals; there could be no compromise on this point.

According to this deterministic and pessimistic view of American-Soviet relations, the insights and intentions of individual statesmen or their advisors in the evolution of the Cold War, much studied on the American side, were of relatively little consequence. Admittedly, we observe considerable disagreement between American liberals and conservatives, the former perhaps more knowledgeable and perspicacious, the latter more rigidly America-centered; yet even liberals hardly deviated from their conviction of American superiority, though aiming at a lower intensity of conflict. And in the competition for votes, the liberals generally lost out; American public opinion, representing entrenched isolationist tradition, favored a hard line. On the Soviet side too we observe, more obscurely, a variety of positions, none of them implying a surrender of traditional anti-foreign conviction. How under these conditions could there emerge sufficient common ground for agreement? The results of the compromises reached in the thirty years after 1945 certainly were insignificant, considering the ever growing ferocity of the weapons and weapons systems under discussion. The escalation of military preparedness increased by leaps and bounds.

In the power contest the United States was by far the stronger partner. It could count on a loyal and knowledgeable population and a highly productive economy drawing on advanced science and technology; it could afford to give its people a large measure of freedom in all aspects of life, knowing that independent initiative would never undermine the basic consensus. It was assisted, despite occasional disagreements, by the countries of Western Europe, which buttressed its claim to leadership of the West; in the Far East Japan was its loyal ally. And in its relations with the rest of the world, it could draw on its unrivalled material wealth and human experience; for that reason it enjoyed ready access everywhere. It was the most expansionist country in history, as much by merit as by design. In short, in all invidious comparison it could hold its own, whatever the much-discussed flaws in its midst.

Yet even the United States was not invulnerable. Its major weakness in the power contest was psychological. Startled out of its secure hemispheric isolation and saddled with a new responsibility for upholding and advancing the Western-oriented global order built up by Western Europe, it succumbed to a profound sense of insecurity, an insecurity enhanced by national pride. For carrying out its obligations or even preserving its identity it craved a world shaped after its own image. Upholding freedom and democracy in these tumultuous, rapidly changing times required ever more effective instruments of power. Not surprisingly, its psychological insecurity quickly gave rise to a sense of military insecurity. Considering that the country lacked effective knowledge for managing the new globalism peacefully, it felt compelled to deal with the world and its adversaries from a position of strength, above all military strength. For that reason, the United States took the lead in the arms race, the only phase in the competition for power in which it might conceivably be deficient.

By contrast, the Soviet Union was weak in all aspects of power apart from military strength. Its armed forces, designed for internal as well as external use, seemed impressive on the surface. In the nuclear arms race the Soviet arsenal had risen to rough parity by the mid-1970s, barely enough to satisfy Soviet leaders keenly aware of their country's overall political weaknesses. Their strength lay in their proud resolve to achieve equality with their more powerful rival and in their skill to display that resolve as a source of power. They conned apprehensive American leaders, so unaware of their superior assets, into accepting the Soviet assessment of strength at face value, thereby acquiring a measure of power over their opponents. In matters of defense, Washington policy was made in Moscow.

Having acquired a basic hold over each other, the enemies became more like each other; their hostility forced them into a collision course of convergence. Americans, following the logic of Nitze's advice that they must employ Soviet Cold War methods themselves, gradually built up the "national security state," a state within the state operating under cover of secrecy and infiltrating even into domestic politics, restricting traditional civil liberties and subverting constitutional rights (as proved in the Watergate scandal). At the same time, the subtle militarization of American life proceeded apace, accepted as a necessity by public opinion. And in its management of international relations American practice readily adapted to the revolutionary heritage in Soviet foreign policy, sometimes even resorting to clandestine terrorism (as in Central America).

On the Soviet side, where the instruments of raw power had long been in place, convergence introduced more American accomplishments, above all technology and administrative skills, but also consumer goods and styles of living. Legality, civil liberties, and human rights were likewise stressed, especially in the new constitution of 1978; more than under the Stalin constitution of 1936, the Soviet Union was given a democratic facade. In all their activities Soviet leaders tried to impress the world by their country's modernity, offering further evidence of convergence toward worldwide norms set essentially by the United States.

Convergence, however, did not provide common ground for cooperation. On the contrary, by making the enemies more alike, it magnified their still considerable differences and intensified their competition, which reached dimensions of hostility unprecedented in peacetime international relations. War and preparation for war became daily topics in the news. More ominously, there raged a vast and vicious outright war, well hidden from the public, of electronic surveillance and ceaseless probing of the enemy's defenses, of espionage and subversion, of intelligence gathering and outwitting the enemy's counterespionage precautions. From this hidden war powerful networks of vested interests arose, the military-industrial-intelligence complex on the American side, dominating the national economy. Its Soviet counterpart, more central to the survival of the regime, was even more deeply entrenched in the body politic; out of weakness and cultural conditioning, it projected the cynical refinements of revolutionary tradition into international relations.

Under these circumstances, national policy in both countries was predetermined in favor of conflict, and even of escalation of conflict. What counted were the psychodynamics of collective ambition deeply embedded in cultural tradition, of patriotic pride wedded to ignorance of the world beyond the country's boundaries. In the most basic respect, both superpowers pursued the same irreconcilable aims: their ultimate security, metaphysically and politically, lay in a world order conforming to their own interests and patterned after their own historic experience. There could be no compromise on that score.

How could the deadly trend be reversed? Dealing with two incompatible political cultures, should the incompatibilities be accentuated? This approach, recommended by Nitze as the guiding rationale of American policy toward the Soviet Union, was bound to force the latter into an equally uncompromising self-assertion. Or should the incompatibilities be minimized by a search for common ground? The latter course, however, was not a matter of changing government policies (an unlikely prospect in any case), but of first understanding the nature of the conflict and then recasting collective thought, of enlightening public opinion and changing national consensus. What was needed at the outset, one might argue, was a humbler, more realistic view of this generation's limited capacity for the management of international relations, a recognition of the dangerously high price of pride and ignorance, and, above all, a more inclusive view of the world, covering the conditions of the majority of humanity neither American nor Soviet.

Propelled by the culmination of the world revolution of Westernization, that majority, the Third World so-called, added further incentives for the competition between the superpowers. In Soviet eyes the developing countries repeated, in some respects, Soviet experience in state-building; they held out a chance to expand the global counterrevolution. For the United States and the other countries of the West, they provided added opportunities for extending their customary sway. The newcomers, however, were determined to escape the polarization of global politics because it threatened to diminish their own opportunities in the emerging world system. It is to them, to their emergence and search for wealth, power, and identity in the high tide of Westernization, that we now turn.

VI

The Third World:
Decolonization, Independence,
and Development

19

The Bandung Generation

I

Ten years after the conclusion of World War II, the newcomers—twenty-nine states from Asia and Africa—gathered at Bandung, Indonesia, to announce their presence in the world. The Bandung Conference of April 1955 marked a milestone in the evolution of postwar globalism. It established the developing countries, aligned and nonaligned, as a force to be counted in world affairs.

Global politics were then in full swing. Superpower rivalry had heated up during the Korean War in 1950. After Stalin's death in 1953 that war was ended, but not the talk of war. Hostility between the United States and the People's Republic of China (established in 1949) simmered as the United States blocked the extension of communist rule over the offshore islands of Quemoy and Matsu. At the same time, the United States expanded its military alliances into Asia, from the east through the Southeast Asia Treaty Organization (signed in September 1954), and from the West through the Baghdad Pact (concluded on the eve of the Bandung Conference). The American moves stimulated Soviet countermoves; North Vietnam was becoming an outpost of Soviet influence. Communist China, in turn, was pressing alarmingly against its southern neighbors. That was why Zhou Enlai, Mao's foreign minister, had been invited to Bandung.

Meanwhile, the agitation for decolonization, stirring since World War I and intensified during World War II, was rising to its climax. Independence had come to India in 1947, to Burma in 1948, and to Indonesia in 1950. It was on its way in Africa, after Libya had become a sovereign state in 1951. It came first to the Sudan (1956) and then, more spectacularly, to the Gold Coast, renamed Ghana on Independence Day in 1957. Ghanaian independence cleared the path for other African colonies, whether French or English. Everywhere independence generated a new sense of power, an ambition to be heard and to be admitted to participation in world affairs. As President Sukarno of Indonesia said in his opening speech at the conference: "We have been the un-regarded, the peoples for whom decisions were made by others whose interests were paramount, the peoples who lived in poverty and humiliation. Then our nations demanded, nay fought for independence, and achieved independence, and with that independence came responsibility."[1]

And in what turbulent times! As Sukarno observed in his speech:

> There has been a "Sturmüber Asien"—and over Africa too. The last few years have seen enormous changes. Nations, States, have awoken from a sleep of centuries. The passive peoples have gone, the outward tranquillity has made place for struggle and activity. Irresistible forces have swept the two continents. The mental, spiritual and political face of the whole world has been changed, and the process is still not complete. There are new conditions, new concepts, new problems, new ideals abroad in the world. Hurricanes of national awakening and reawakening have swept over the land, shaking it, changing it, changing it for the better.
>
> This twentieth century has been a period of terrific dynamism. Perhaps the last fifty years have seen more development and more material progress than the previous five hundred years.... But has man's political skill marched hand-in-hand with his technical and scientific skill?[2]

It was a pertinent question.

The storm described by Sukarno, however, was even bigger than he imagined. The world revolution of Westernization was reaching its culmination. Decolonization meant the end of the first phase of that revolution, accomplished through colonialism and imperialist outreach. Now the revolution proceeded on its own, in new states patterned after the European nation-states, through the newcomers themselves, who subtly and for their own self-respect transformed "Westernization" into "modernization" or plain "development." Liberation meant a far more intensive submission to the Western model than had been possible under colonial rule. In some respects, colonial rule had acted as a buffer to Westernization; it has preserved a semblance of indigenous authority even while transforming indigenous society; it had provided internal peace and order plus protection from the pressures of global politics. Now the emerging states had to fend for themselves, taking over more than the functions of the departing colonial masters. They had to build their polities after alien patterns while holding their own amidst the ominous tensions of world politics.

Building an equivalent to the Western nation-state out of unsuited and resisting human raw material was an excruciatingly difficult experiment following, in a new era and in new settings, the pattern of the Westernizing counterrevolutions set off after World War I. The new experiments had to be carried out among people even less well prepared and subject to the rising pressures of the global state system. Those pressures operated both from within and from without.

From within they proceeded through intensified invidious comparison. "The colonised man," according to Fanon, "is an envious man";[3] the ex-colonised man was even more so. He wanted—and as soon as possible—all the boons which the West, and the United States foremost, radiated into the world: a civilized standard of living, a proud ego individually and collectively, and above all power. And he wanted it on his own terms, affirming his own identity like the old colonial masters, like the Americans now. Obviously, he was anti-Western, carrying with him the memories of endless past humiliation. But the realization of these hopes depended on learning the skills that produced power. That learning meant Westernization, reculturation, the most difficult and exasperating revolutionary transformation, conducted without guidance, without insight, in the face of Western incomprehension and moral disdain. Reculturation, in turn, radiated the internal tensions back into the tension-ridden world at large.

From the start the newcomers were immersed in quarrels with their neighbors, drawn into the rivalry of the superpowers, and involved in the politics of the United Nations, their most congenial forum for airing their needs. Lacking adequate means to satisfy domestic demand, they had to support a foreign policy establishment as well, complete with diplomatic service and foreign embassies. However powerless, they had to compete in the arms race, building up armed forces for both domestic politics and foreign relations.

Viewed in full context, their own experiments in state-building were part of the larger experiment affecting all humanity. How in the age of nuclear weapons could the precipitous global confluence of the world's disparate cultural and political traditions be channeled into orderly cooperation reconciling human wants with the earth's resources? Could humanity's political skills be advanced to match its scientific and technological skills?

II

The participants of the Bandung Conference, though hardly aware of the magnitude of the tasks before them, had had a taste of the problems from their own experience. Consider Jawaharlal Nehru, then in his mid-sixties, the prime minister of India and the most influential among the statesmen assembled at Bandung, self-confident and perhaps even overbearing in his pride. Under his guidance India had achieved independence in 1947, the first major country to emerge from colonial rule, the self-appointed leader of the newcomers on the world's stage.

But at what price! On the Indian subcontinent the ideals of freedom and self-determination had circulated during the war with renewed intensity, accentuating the diversity of creeds and cultures among its peoples. The division between Hindus and Muslims, long simmering as a source of conflict, had broken into the open as independence approached. As a result, not one but two states emerged in 1947: India and Pakistan. The separation was accompanied by murderous clashes between Hindus and Muslims, the carnage taking the lives of many hundreds of thousands (and Gandhi's life too, because he had counselled conciliation). In addition, over eight million people scurried for safety to and from the lands assigned to Pakistan, in indescribable confusion and hardship. Over fifty million Muslims remained in India, a troublesome minority. The neighbors started state-building in mutual hostility, each in turn coping with internal divisions prompted by the postwar political activism in the name of freedom and self-determination.

Nehru's India proved to be the more successful experiment in state-building. But consider the obstacles. To start with, the new India required unity under a federal constitution. The princely states, holding roughly a quarter of the population, had to be absorbed; out of the disjointed mosaic of languages, religions, and cultures reasonably coherent member states had to be formed. The peoples of India were also deeply divided by caste, the "scheduled" castes and tribes at the bottom of the social structure comprising about one-fifth of the total. Four-fifths of the population lived in villages raising food by traditional methods; 60% of all children went without school; the economy stagnated. And yet the population, numbering 357 million in 1951 (and twice that figure by 1985) and making the country the second most populated country in the world, was growing rapidly. How under these conditions could the promise of a better life in independent but fragmented India be realized?

State-building—India's "tryst with destiny," as Nehru called it—got off to a good start, partly because of Nehru himself. The quintessential representative of the anti-Western Westernized nationalist intelligentsia, he combined the proudest indigenous tradition with the most privileged British education. He has been seasoned in the long struggle for Home Rule, matured by lengthy terms in jail, and apprenticed to political action in the Indian National Congress, the core organization of Indian nationalism. Released from jail at the end of war, he emerged as India's leader when independence approached, a high-caste Indian steeped in British tradition both in England and in India, as well as in Western perspectives generally.

For building the Indian state he possessed a ready instrument in the Congress Party, a sufficiently cohesive mass-based political organization guaranteeing a measure of unity for the new state. He also had at his disposal the administrative machinery left behind by the British. More important, he possessed the charisma of a popular leader. The idol of the Indian masses second only to Gandhi, he knew how to reassure them and fuse them into loyal followers. "I go out and see masses of people . . . and derive inspiration from them. There is something dynamic and something growing with them and I grow with them. I also enthuse with them." The enthusiasm provided a much-needed sense—or at least an illusion—of a common "Indianness."[4] The vitalizing interaction between leader and followers, familiar from the politics of central Europe in the 1920s, worked in India as well; in both cases the state was consolidated around a charismatic leader.

In India too the state was given a central role in providing a better life for all. A Planning Commission was established in 1950, examining precedents elsewhere. The next year it submitted its first Five-Year Plan, outlining "a process of development which will raise living standards and open out to the people new opportunities for a richer and more varied life."[5] In the year of the Bandung Conference the aims of the second Five-Year Plan were defined more inclusively: "The task before an underdeveloped country is not merely to get better results within the existing framework of economic and social institutions, but to mould and refashion these so that they contribute effectively to the realization of wider and deeper social values."[6] The aim, in other words, was reculturation; the "wider and deeper values" obviously were less of Indian than of Western derivation.

As in the 1920s, planning implied some form of socialism. In 1956 the Indian government made "the adoption of the socialist pattern of society"[7] the national objective. The state had to assume direct responsibility for the future development of industries over a wider area. As Nehru was aware, the British Labour government after 1945 had set a suitable example. Planning in India, admittedly, took a relatively mild form; its socialism was essentially democratic socialism. The private sector of the economy remained powerful, but its opposition to state control was restrained. A laissez-faire economy, admittedly, would not as quickly provide the "richer and more varied life" expected from independence.

Expectations ran high, always ahead of results; planned development was a difficult experiment. As for the first Five-Year Plan, Nehru himself admitted: "We just took what was there and called it a plan."[8] And as for his country's accomplishments in the first three years of independence, perhaps it was true that the new government functioned rather like the British *raj,* although possibly with less efficiency. Nevertheless, the pace of accomplishment was impressive; the basic structures of government were expeditiously put in place.

In any case, the course was set for deliberate Westernization. Nehru welcomed "the powerful impact of the industrialized West." "There can be no doubt," he was to write in 1962,[9] "that this new urge for industrialization or the adoption of modern techniques will succeed and change the face of India"—without loss, he pleaded, of India's individuality.

Yet the fusion of Western and Indian cultures remained problematical. State-building and economic development in the Indian setting could not be accomplished as rapidly as was assumed in the optimistic years after 1947, even though it was favored by a variety of propitious circumstances. These included a ready receptivity of Western cultural skills based on indigenous capacities, the presence of a Western-trained elite and a core of capitalist entrepreneurs, the continuity of British legal and political institutions as well as of English as the only all-India language (spoken by 1% of the population), and—at the outset, at any rate—a safe distance from the storm centers of world politics.

State-building, however, meant more than internal reconstruction. With a hostile Pakistan next door and communist China looming in the north, India had to prepare its defenses, eventually to the point of testing its own atomic weapon in 1974. More rewarding—and highly useful for rallying domestic support—was putting the new state on the world stage and catching the world's respect in the manner of the former colonial masters (though with a Gandhian pacifist slant). Here Nehru was in his element. Even before independence, in the midst of the communal violence, Nehru had staged his debut as a world figure hosting the first Asian Relations Conference. That occasion brought together Arabs, Jews, Uzbeks, Kazakhs, Burmese, Indonesians, Chinese, Koreans, and still others.

That conference in 1947 resumed, in a new age, the tradition of non-Western anti-Western gatherings inspired by the emerging globalism at the start of the 20th century. It also enhanced Nehru's status at home and abroad; it gave prestige to his country. Its rhetoric echoed, in the humbler voice of powerless newcomers in distant parts of the world, the aspirations of the great Westernizing counterrevolutionaries of the previous generation. "Asia, after a long period of quiescence," Nehru proclaimed, "has suddenly become important again in world affairs. . . . We live in a tremendous age of transition and already the next stage takes shape when Asia takes her rightful place with other continents."[10] With the help of other Asian states India was to become a force in global affairs. In this spirit Nehru had laid the groundwork for the Bandung Conference.

III

At Bandung in 1955 he encountered another aspiring Asian nation-builder, President Sukarno of Indonesia, the host of the conference. Sukarno too was a charismatic anti-Western Westernizing nationalist. His pronounced hostility to the West—he lumped all western European countries and the United States together under that term—testified to his relentless preoccupation with Western power and culture; both inspired his political career. Born in 1901 to a privileged Javanese family and graduating with an engineering degree from a Dutch technical college in Bandung, he was soon absorbed in the agitation for Indonesian independence; by the late 1930s he was its acknowledged leader, eager to find a proper ideology for unifying the diverse and scattered peoples of the Dutch East Indies, a formidable venture.

The Indonesian archipelago, as he used to call it— "even a small child, when he looks at a map," he said, "can see that the Indonesian archipelago forms a unit"[11]—was a motley collection of a few big and many small islands scattered over an area almost as large as the Indian subcontinent yet lacking effective internal communications. It had been brought together by Dutch rule, but otherwise possessed few common bonds. Among the population, the third largest in Asia after China and India, the indigenous culture was mixed with Hindu and Buddhist as well as with Muslim elements. For patriotic intellectuals like Sukarno, that cultural brew was further diversified by Western influences with a strong Marxist twist and laced with all the other intellectual and political currents that had drifted into his world. Further disunity stemmed from the diversity of languages—over three hundred of them—from local subcultures, climatic and geographical differences, and intense parochialism everywhere. In addition, a large Chinese minority had assumed a position of economic power among a resentful population generally lacking in economic enterprise. How could there arise a modern nation out of that centrifugal conglomerate?

That question had haunted Sukarno from the start of his political career in the late 1920s. By 1945, at the very end of the Japanese occupation, when independence seemed assured, he stepped forward with his answer, an odd conglomerate called the Five Principles, the *Pantja Sila*.[12] The first was Indonesian nationalism, "all for all, and all for one" within the Indonesian archipelago, whose unity was an article of faith for him. Second came humanitarianism equated with internationalism, also defined as pan-Asianism or inter-Asianism, even as world unity and God's will. The third principle was democracy, which meant not parlimentary democracy Western-style, but unanimous consent representing all sides in a manner mutually satisfying. Social justice, the fourth principle, grew out of the third. Presumably, it meant a fair share for everybody, but it had no room for Western individualism. The crowning principle was belief in an Almighty God, the ultimate guarantee, he said, of unity through mutual tolerance.

Viewed as a whole, these Five Principles resembled a flimsy ideological net designed to catch bits of each of the major religious and political currents abroad among influential sections of the population. Confronted with a doubting Muslim believer, their author suggested that Islamic laws were "flexible as rubber"[13] (rubber being one of the major products of Indonesia). Indeed, providing unanimous consent representing all sides in a manner mutually satisfying meant stretching every creed, every concept, every conviction—or everybody's credibility—to the utmost. How else could unity for Indonesia be achieved?

Yet Sukarno's thoughts stretched even further. "We, the Indonesian people," he once said in a radio address to "the peoples of Asia" (no less), "have learned to think not in centimeters or meters, not in hours or days. We have learned to think in continents and decades."[14] Yet, for all his grandiloquence, he fashioned his ideas into a surprisingly effective political force. Expressed with utter self-confidence, they gave him the aura of "the Chosen One," of charisma. A German observer credited him with a "unique ability to address the masses and to carry them with him, whether he promised them paradise or proclaimed a life-or-death struggle."[15] In this respect he resembled Nehru—or the spell-binding dictators of the 1920s. Like them he could say: "Why is it that people ask me to make speeches to them? . . . The answer is that what Bung Karno [the Chosen One] says is in fact already written in the hearts of the Indonesian people. The people wish to hear their own thoughts spoken, which they cannot voice themselves."[16] As a charismatic spellbinder he stood a notch above other spokesmen for Indonesian independence, and with greater personal ambition.

Immediately after the Japanese surrender, the Indonesian patriots unilaterally declared independence, under a constitution which made Sukarno president with nearly unlimited power. He was to be "the true leader of the nation and of one mind with the whole people."[17] His ideal was "centralized democracy" (or democratic centralism) with but a single party, a *Führerstaat* according to a plan he had formulated during the Japanese occupation. And, like Hitler, he reached far and wide for territorial dominion. His Greater Indonesia, as envisaged in the summer of 1945, was to cover all Dutch, English, and Portuguese possessions in the area, plus the Malay peninsula with the Strait of Malacca, plus the Philippine Islands. As the war ended, anything seemed possible; globalism had come to the Dutch East Indies.

Events, however, took a contrary course. Within a month Sukarno's power was limited by his political rivals. Then came the restoration of Dutch rule, lasting until December 1949. When through the good offices of the United Nations and American pressure on the Dutch government independence was at last secured, the new constitution installed Sukarno as figurehead president without real power (yet with his ambition unclipped). Indonesia started independence as a Western-style parliamentary democracy within more limited boundaries than Sukarno had wished. West New Guinea, for instance, was still under Dutch rule, a handy nationalist grievance proving that the revolution for liberation was still incomplete.

Meanwhile, the promise of liberation grew stale. After the unifying impetus of the struggle for independence had spent itself, the old diversities reasserted themselves; the country's deficiencies grew more prominent. The administration, reasonably efficient and honest under Dutch rule, was now staffed with Indonesians. Lacking the cultural discipline of European civil servants integrated into the complex interrelationships of large-scale organization, the indigenous officials were guided by traditional incentives built around small-scale perspectives; government became corrupt, far more than in India. The Korean War provided a transient economic boom, but the marked backwardness of the economy persisted. In 1951 the government launched an Emergency Industrialization Program, followed by a Five-Year Plan for the years 1956–60 (prepared in the year of the Bandung Conference), with little effect. The first elections for parliament were held soon after the conference, accentuating the growing disunity; none of the forty parties in the running gained a majority.

Not surprisingly under the circumstances, effective power in the country drifted toward organizations with a sense of discipline, the armed forces and the Communist Party (the latter, however, checked by anti-communist Muslim opinion). Asserting himself, Sukarno moved again into the foreground of politics, allied with both communists and the soldiers, soon courted by the Soviets. The Bandung Conference provided him with worldwide visibility and new prestige at home. It revived his earlier ambition for a national revolution, with himself leading a Greater Indonesia while thinking in terms of continents and decades. Buoyed by hope and ambition, he welcomed his foreign guests to Bandung, among them Colonel Gamal Abdul Nasser from Egypt, another rising star among the anti-Western Westernizing state-builders.

IV

Colonel Nasser (born in 1918) was the youngest among the conference's luminaries and a newcomer to global notoriety. He carried weight because of his country's geopolitical significance, "the geographical crossroads of the world,"[18] as he boasted. Egypt bridged Asia, the Mideast, and Africa, with strong ties to the Mediterranean basin. In addition, Nasser was seen as the coming strong man in a strategically important country, a hardworking, intelligent soldier with a good understanding of the needs of the times. He had brought with him the "Six Principles" proclaimed in January 1953, after the "revolution" of July 1952 had ousted King Farouk and initiated a republican regime.

These Six Principles,[19] vague and as yet unapplied, fitted well into the Bandung rhetoric. Foremost they proclaimed the liquidation of imperialism, and also of feudalism and capitalism (whatever the meaning of these terms). They also proclaimed social justice and democracy (again undefined). Most members of the conference would also agree with another important principle: the need for a strong army for fighting imperialism. It did not necessarily contradict the other principles: they could be realized only after imperialism had been defeated.

Clearly, Nasser was one of the new nation-builders, faced, like the others, with a difficult task. Located along the Nile River, Egypt admittedly constituted a well-defined territory with a long and famous history built around a river-based culture favoring a centralized authoritarian government. But as a crossroads it had suffered from the confluence of many cultures—Arabic, Turkish, French, English, or generally Western, all within a fluid Islamic state of mind—without ever achieving true integration. Observers of its society and politics would hardly call Egypt a nation, even though the country, organized as a constitutional monarchy, had attained nominal independence in 1922.

In Egypt's population of over twenty million the peasants, the true indigenous element, constituted the great majority, poor, illiterate, and dominated by large and often foreign landowners; the urban poor were crowded into miserable slums in Cairo and Alexandria. The cosmopolitan middle and upper classes, though reared under Islamic influence, were open to foreign cultures, often tied to foreign business, the source of Egypt's limited prosperity. The government was under the care of the king and an overgrown, cumbersome bureaucracy administering the country, guided in theory by an elected parliament, but in practice by royal manipulation. Among the elite the Wafd Party, the dominant party in the 1920s and 1930s, had taken an anti-British stance, which occasionally gave it mass support. But it was not open to lesser folk; it had no social message. The void, elsewhere an opportunity for Marxists, was filled with religious fanaticism under the aegis of the Muslim Brotherhood, Islamic fundamentalists. There were smaller parties, including communists, all faction-ridden, lacking discipline and firm conviction. Egyptian politics possessed a peculiarly fluid, indefinite, and surprisingly mellow character; life was not violently politicized. The political crises of Egypt had arisen out of confrontation in foreign, not domestic, affairs.

The chief cause of confrontation, of course, had been the British government. Despite having granted independence to Egypt in 1922, it remained the guardian of the Suez Canal, and thereby of Egypt as well, much against the wishes of Egyptians fired by the ideal of self-determination. In 1936 an agreement had been reached under which the British would withdraw from Suez after twenty years under certain safeguards not pleasing to Egyptian pride. In 1942, with the Germans at the gate, the British had forced a reluctant king to submit to stricter controls by the British army; its soldiers were stationed in the major cities. By 1946 these restrictions were lifted, but the British kept their base at the canal, again to the chagrin of Egyptian patriots. Then, in 1948, came war with the new state of Israel; Israel, so it seemed, was another product of British imperialism. The war was disastrously lost by the Egyptians; it became the major cause of the revolution of 1952.

Who was responsible for the humiliation but the king? Who was responsible for the subsequent raids by British forces bombing and shooting defenseless (though rebellious) Egyptians? After a particularly brutal clash in January 1952, mobs incited by the Muslim Brotherhood attacked British offices and clubs in Cairo, and Western imports like bars, cinemas, and nightclubs as well, burning a large area in the modern center of the city. When King Farouk replaced his prime minister four times in the following half-year without producing a show of strength, a conspiracy of junior officers organized in the Association of Free Officers struck on July 22 and sent him packing. It was a bloodless putsch, subsequently endowed with the aura of revolution in order to lend it dignity. The so-called revolution was subsequently given an ideological content through the Six Principles.

The center of power now lay in the hands of the Free Officers, a small conspiratorial group of military men drawn from the lower classes and trained in British discipline as bequeathed to the Egyptian army. They were able men, open to the portents of the times and keenly resentful of the humiliations inflicted by the British; they wanted to make their country into a nation respected in the world. After their "revolution," however, they proceeded cautiously, aware of their lack of experience. Colonel Nasser, their leader, kept himself offstage while his associates slowly consolidated their power, suppressing opposition within the army and the government, dismantling the opposition parties, and building a military dictatorship by rather conventional means. In the summer of 1954 at last Nasser himself stepped forward in the civilian role of prime minister. His real advance to prominence came only in October, after an attempt on his life.

It was not potential martyrdom, however, that made the crowds (encouraged by the show trials that followed) cheer him as their leader, but rather the assertion of Egyptian pride and independence in the escalating Cold War. Already in the summer of 1954 he had scored a success by persuading the British government to evacuate its base on the Suez Canal within less than two years (with the usual reservations in case of a major international conflict). In October he stood up more forcefully for Arab unity, a spokesman for the League of Arab States founded in 1948. The issue was the Baghdad Pact, linking Turkey, Iraq, Iran, and Pakistan to Britain and the United States. Should Egypt join, as urged by the Western powers? Should the Arabs join? Nasser answered with a decisive No, thereby making himself the hero of the patriotic crowd. Having taken a courageous stand, he found himself courted by Marshal Tito of Yugoslavia and by Nehru, both prominent advocates of nonalignment. How did he interpret his mission as he set out for the Bandung Conference?

As it happened, he had just published a pamphlet, allowing a revealing glimpse into the mind and experience of a state-builder in the Egyptian setting, under the bold title *Egypt's Liberation; The Philosophy of the Revolution.*[20] The boldness, however, was quickly toned down, as the author called his analysis merely "a reconnaisance patrol, . . . an effort . . . to discover who we [the Free Officers] are and what our role is to be in the succeeding stages of Egypt's history."[21] Nasser did not hide the fact that he himself was a novice in revolutionary theory. That air of hesitance and uncertainty prevailed throughout his exposition. Admittedly, he left no doubt about the intentions of the Free Officers. They were a vanguard which in July 1952 had "performed its task and charged the battlements of tyranny." But, unfortunately, the vanguard was not followed by "the sacred advance" of the masses "towards the great objective." When eventually the masses arrived, they came as "disunited, divided groups of stragglers." "We needed unity," Nasser complained, "but we found nothing . . . but dissension. We needed work, but we found . . . only indolence and sloth."[22]

His assessment of his countrymen was discouraging indeed. "If anyone had asked me in those days what I wanted most," he wrote, "I would have answered promptly: to hear an Egyptian speak fairly about another Egyptian. To sense that an Egyptian has opened his heart to pardon, forgiveness and love for his Egyptian brethren. . . . There was a confirmed individual egotism. The word 'I' was on every tongue. . . . I had many times met eminent men— or so they were called by the press—of every political tendency and color, but when I would ask any of them about a problem in the hope he could supply a solution, I would never hear anything but 'I.'"[23]

Subsequently Nasser complained of "the lack of a strong and united public opinion in our country. The differences between individuals are great and between generations they are still greater. . . . We live in a society that has not yet taken form. It is still fluid and agitated and has not yet settled down. . . to catch up with those other nations that have preceded us on the road." He fervently hoped that his people would become "a strong, homogeneous, unified whole." But at present "our position is blown upon by the wind from all directions. We are on a field roaring with hurricanes, dazzled by lightning and shaken by thunder." Under these circumstances, he added with typical Egyptian mellowness, "it would be monstrous to impose a rule of blood."[24] He was not to be a Stalin. Yet what was to be done?

Nasser explained that his country was going through two revolutions at the same time, a political revolution "which wrests the right to govern itself from the hand of tyranny, or from the army stationed upon its soil against its will," and a social revolution. But the social revolution rather obstructed the political revolution. "It shakes values and loosens principles, and sets the citizenry, as individuals and classes, to fight each other. It gives free reign to corruption, doubt, hatred, and egoism."[25] Nasser felt caught between these two revolutions, as between two millstones. In more stylish image he observed, "Our people are now like a caravan which seeks to follow a certain route, but the route is long, and the diversions to be encountered are many. Thieves and highwaymen may hold it up, and the mirage mislead it from the true way. . . . Groups may go astray . . . and individuals scatter in different directions." He even admitted that there were no leaders: "It would be illusion if I thought that we could solve all the problems of our country. . . . We are simply not competent to do that job. . . . " Nor would he let himself be carried away by "glittering phrases."[26]

Nasser was on firmer ground when he discussed the place of Egypt in world affairs. "The age of isolation is gone," he wrote. "No country can escape looking beyond its boundaries to find the source of the currents which influence it, how it can live with others." Egypt, he argued, existed in the center of three concentric circles. The first was the Arab circle, "its peoples . . . intertwined with us by history," a keenly felt personal experience for Nasser. The Arabs were endowed with considerable strength; they possessed a common civilization, were strategically located in the world, and sat on an inexhaustible supply of oil. Nasser stated that he did not seek the leadership of the Arab world; he advocated interaction in a common "experiment, with the aim of creating a great strength which will then undertake a positive part in the building of the future of mankind."[27]

The second circle was Africa, the "Dark Continent." Here it was Egypt's "responsibility to support, with all [its] might, the spread of enlightenment and civilization to the remotest depths of the jungle. . . ." The third and final circle, "which circumscribes continents and oceans," was inhabited by the adherents of the Islamic faith, "our brothers, . . . pious and humble but strong, . . . convinced of their place in the sun, . . . wield power wisely and without limit."[28]

Nasser's attempt to define the theory of the Egyptian revolution ended on an upbeat yet still vague note. Alluding to Egypt's history as a play filled with heroic roles but no heroes to perform them, he concluded: "Here is the role. Here are the lines, and here is the stage. We alone, by virtue of our place, can perform that role."[29] That role, we gather from this document, was to create Egypt as a unified, homogeneous whole like those "other nations that have preceded us," and do so by playing a positive role in the world at large.

Nasser was an exceedingly hardworking, incorruptible, and pragmatic military man with good common sense and an endearing strain of humility, but no political theorist, no presumptuous globalist. Yet he pursued the same essential goal as the other anti-Western Westernizing state-builders: he set his people purposefully into the all-too-suddenly enlarged and open world; he gave them a positive perspective, a source of hope and self-respect, an escape from the ignominy that had oppressed them in the past. At the time of the Bandung Conference he had barely started; the conference gave him both courage and inspiration. It also provided him with a chance to meet an emissary from the "Dark Continent" representing Kwame Nkrumah, prime minister of the Gold Coast, the British colony scheduled to become independent in early 1957.

V

Absent at Bandung but very much present in spirit, Nkrumah signified a fourth and especially fascinating mix of anti-Western Western attitudes with an articulate West African—or, more specifically, Gold Coast—flavor. Under his leadership the Gold Coast colony was scheduled to become the state of Ghana, named after a fabled ancient African empire. In his "Motion of Destiny" speech of 1953, which prepared the way for independence, Nkrumah had given the reasons for this choice: "In the very early days of the Christian era, long before England has assumed any importance, long even before her people had united into a nation, our ancestors had attained a great empire. . . . Thus may we take pride in the name of Ghana, not out of romanticism, but as an inspiration for the future."[30] Ghana's past greatness, dwarfing Europe's achievements of that time, was to be the guide for the new Ghana, the core, perhaps, of a united Africa. That aspiration was the reason why Nkrumah had been invited to Bandung; he was the coming man in black Africa.

In his "Motion of Destiny" speech he had boldly advertised the Gold Coast colony's assets for independence and greatness: a stable society and a healthy economy. "Our people are fundamentally homogeneous, nor are we plagued with religious and tribal problems . . . we have hardly any colour bar."[31] Yet, as he was reminded soon afterwards, the realities were different. The Gold Coast, a smallish territory with less than eight million inhabitants, was much divided by tribal loyalties. The majority of the people lived by traditional ways, guided by family or lineage ties, by spirits and local deities, untrained in the mechanical arts, and confined in small communities even after they had come under European influence. An enlightened British administration, with the help of the missionaries, had introduced stability to the area by minimizing tribal conflict and, thanks largely to cocoa, bringing a measure of prosperity unequalled in West Africa to the coastal districts and the lands of the Asante tribe. The Gold Coast was a model colony where black and white had long worked together; but what sense of unity and common progress there was came from colonial rule. When it faded, traditional rivalries revived, aggravated by the agitation for independence.

Into that colonial world Nkrumah was born (he said) in 1909, destined to rise far above it. He began his schooling under a German Jesuit who, recognizing his unusual ability, launched him into a Western education. Attending Achimota, the Gold Coast replica of a British public school, for a teacher-training course, he heard about Dr. Aggrey, the famous black American educator who had advocated black-white harmony in the manner of music played on black and white piano keys, and about America too. While teaching at Catholic schools he absorbed the early messages of African liberation emanating from England or colonial centers along the West African coast. After toying with plans for a life as a Jesuit, he determined to prepare himself for a political career via an American education. When he left for the United States in 1935, his mother, a Catholic convert, gave him a half-Christian half-African blessing: "May God and your ancestors guide you."[32] His own mood at the time—and his ambition—he described in Tennyson's words: "So many worlds, so much to do, / So little done, such things to be."[33]

On his way he stopped in London, his first encounter with the white world; it chanced to be the day Mussolini invaded Abyssinia. "That was all I needed," Nkrumah recollected later. "At that moment it was almost as if the whole of London had suddenly declared war on me personally. For the next few minutes I could do nothing but glare at each impassive face wondering if those people could possibly realize the wickedess of colonialism, and praying that the day might come when I could play my part in bringing about the downfall of such a system. . . . I was ready and willing to go through hell itself, if need be, in order to achieve my object."[34] Such things to be!

Then followed ten years (1935–45) in the United States. He was a successful student at Lincoln University (founded by a Quaker for the advancement of Negroes), located not far from Philadelphia, and subsequently at the University of Pennsylvania, studying philosophy, theology, and education, teaching and preaching while advancing toward a Ph.D. degree. In other respects those years proved hellishly hard times for a pennilness, solitary young African relying on his wits and menial summer jobs for survival in depression-ridden and wartime America.

Africa, of course, was always on his mind. He helped to set up an African Studies section at the University of Pennsylvania; he founded an African Student's Association—the beginning of his political career. He also became interested in the ideas of Marcus Garvey (by then expelled from the country). During his last four years in America, the country—the whole world—was at war, though not Nkrumah, the African outsider wrapped up in his own pursuits; there is no mention of the war in his autobiography, composed eleven years later. When the war ended, in May 1945, he left New York for London. As he passed the Statue of Liberty, "with her arm raised as if in a personal farewell to me," he vowed: "I shall never rest until I have carried your message to Africa."[35]

His political career began during the next two and a half years in England. Giving up his academic interests, he joined like-minded blacks, including George Padmore, who had once toyed with the Comintern, other African students (including Jomo Kenyatta), and African manual workers, all clamoring: "Africa for the Africans." He helped organize the Fifth Pan-African Congress, the first to stress indigenous African nationalism; he acted as secretary of the newly founded West African National Secretariat for liaison with West African colonies, travelling to Paris to meet Senghor and other representatives of francophone West Africa; he founded a monthly newspaper, *The New African,* and wrote editorials preaching African unity.

He also ostentatiously read the communist *Daily Worker* and organized a society of like-minded African nationalists called "the Circle." That organization, following well-known precedents, was dedicated to Service, Sacrifice, and Suffering, considering itself the "Vanguard of the struggle for West African Unity and National Independence" with the aim "to create and maintain a Union of African Socialist Republics." Committed on the twenty-first day of each month "to fast from sunrise to sunset" and to "meditate daily" on the aims of the Circle, the members were also pledged "to accept the Leadership of Kwame Nkrumah."[36]

That African nationalism, however, was but a high-flown and vague vision. The African continent, with its more than five thousand different languages and ethnic traditions, its sharp geographic and climatic differences, was not destined for a common government. Even its subunits, virtually all of them organized as European colonies, were conglomerates of tribes which defied integration into nationhood. African nationalism was an abstract Western ideal, inspired by pan-Africanism. In the aftermath of World War II it was raised to political prominence by Africans residing in London.

Life in London certainly did not incite revolution. While attacking imperialism, these Africans lived much like ordinary English people in their postwar drabness. Their Africanness was a matter of their thoughts, not of their lifestyle. Nkrumah subsequently expressed his appreciation of the good-natured tolerance of the English people and of the voluntary clerical help given his organization by young women of good family.[37] One of them later became his devoted secretary, in a strictly professional relationship dedicated to an un-African efficiency.

Having made his mark in England, Nkrumah came to the attention of the leaders of the United Gold Coast Convention, the elite of lawyers, businessmen, and progressive chiefs bent on achieving greater participation in British colonial rule. On their invitation he returned in late 1947, a thoroughly Westernized, proud African, soon sensing a widespread popular desire for ending colonial rule altogether.

Gold Coast people, along the coast and in the towns farther inland, were in ferment. Many men had fought with the British army in East Africa and on the Burma front, pulled away from their roots into the global mainstream with its agitation for freedom, self-determination, and racial equality (of which they had had a taste in the army). The ex-soldiers were joined by young Western-educated intellectuals like Nkrumah and, in greater numbers, by workers now organizing in unions after the British model. The market women too were on the move, claiming a share in public affairs. The discontent cropped up in boycotts of English and Lebanese firms, in demonstrations, and in strikes, incited by news from the Philippines, India, Burma, Indonesia, and China. On the Gold Coast too people clamored for independence. Here beckoned Nkrumah's opportunity.

Breaking away from his sponsors in the United Gold Coast Convention, he founded his own party, the Convention People's Party (CPP), a vanguard and at the same time a mass organization. Its aims were independence, democratic government, and "the Political, Social and Economic emancipation of the people, more particularly of those who depend directly upon their own exertions by hand or by brain for the means of life." As regards international affairs, the party was pledged "to work with other nationalist democratic or socialist movements in Africa and other continents, with a view to abolishing imperialism, colonialism, racialism, tribalism and all forms of national and racial oppression and economic inequality among nations, races and peoples and to support all action for World Peace"; it envisaged the Gold Coast to be a nonaligned emerging country. The party also was "to support the demand for a West African Federation and of Pan-Africanism by promoting unity of action among the peoples of Africa and of African descent."[38] The program of the CPP tied the Gold Coast into the full global context.

Following well-known models, it also sported a slogan (Forward Ever— Backward Never), a flag (red, white, green), and a hymn, the Methodist "Lead Kindly Light" sung, after it had once been spontaneously chanted by a fervent woman supporter, at party rallies. The party contained a Women's League and a Youth Movement; it celebrated Party National Holidays, including Women's Day and the birthday of the life chairman, Nkrumah. In one respect, however, the party's program suffered from an unresolved ambiguity. Independence, so the preamble declared, was to be achieved for "the people of Ghana (Gold Coast) *and their chiefs*"[39] (italics added). Nkrumah obviously recognized the continued popularity of the chiefs, hoping to attract them to his side. Yet, at the same time, he proclaimed the all-around emancipation of the people, denouncing tribalism ("tribal feudalism," he called it on other occasions) as a tool of the British administration. The issue was to haunt Ghanaian politics until the present. Western democracy (especially social democracy) and indigenous tradition centered on chieftaincy would not mesh.

Whatever the long-run problems, Nkrumah proved an indefatigable organizer, driven by the ascetic work discipline required for success in the demanding age of globalism. As a speaker at rallies he led his followers to a feverish pitch of emotion, which in turn convinced him of the lasting devotion of the people; he too stood out as a charismatic leader, providing unity by promising wealth and power for all. Soon his agitation upset the colonial authorities; in 1950 he was clapped in jail for sedition, suspected of communist sympathies.

Recognizing the inevitable, however, the governor-general, with encouragement from London, was determined to work toward a gradual transition to independence through the establishment of representative government. Elections were held in early 1951. The CPP campaigned under the enticing slogan "Seek Ye First the Political Kingdom and All Things will be added unto it,"[40] winning overwhelmingly. The day after the election Nkrumah, "a marxian socialist and an undenominational Christian,"[41] was led from jail to the Governor's Castle, purified along the way by thrusting his foot seven times into the blood of a sacrificial lamb and appointed Leader of Government Business; he became prime minister in the next year and, in March 1957, the hero of Ghanaian independence.

The scope of his mind and ambition in the preparatory years was revealed in his autobiography, published on the eve of independence. Here he listed his models for his campaign against imperialism: "Hannibal, Cromwell, Napoleon, Lenin, Mazzini, Gandhi, Mussolini and Hitler,"[42] an odd collection of nation-builders. He placed the two fascist leaders, America's and England's enemies in World War II, alongside Gandhi, whose philosophy had appealed to him. Endowed, like many coastal people in his country, with an engaging gentleness himself, he had always advocated nonviolence in his agitation against the colonial authorities. And yet he paired Gandhi with notorious conquerors.

His ends as outlined in his autobiography justified his selection of models, echoing Stalin (who had just been made unmentionable by Khrushchev's speech in 1956 and therefore was not included in the list above). "All dependent territories are backward in education, in agriculture and in industry. The economic independence that should follow and maintain political independence demands . . . a total mobilization of brain and manpower resources. What other countries have taken three hundred years or more to achieve, a once dependent territory must try to accomplish in a generation, if it is to survive. Unless it is, as it were, 'jet-propelled', it will lag behind and thus risk everything for which it has fought."[43]

He went on even more boldly: "Capitalism is too complicated a system for a newly independent nation. Hence the need for a socialistic society. But even a system based on social justice and a democratic constitution may need backing up, during the period following independence, by emergency measures of a totalitarian kind. Without discipline true freedom cannot survive. . . ."[44] Escape from backwardness meant reculturation, and reculturation could not be achieved by leaving people to their own devices; reculturation implied socialism or even totalitarianism. Obviously, Nkrumah was well informed about world events, though he never alluded to the Soviet experiment and sincerely cooperated with the governor-general in providing Ghana with a Western democratic constitution.

And yet, while advocating catching up at top speed, Nkrumah the Westernized African familiar with romantic philosophy also rebelled against wholesale Westernization. As he observed in his "Motion of Destiny" speech, "We feel that there is much the world can learn from those of us who belong to what we might term the pre-technological societies. [There are] values which we must not sacrifice unheedingly in pursuit of material progress. . . . In harnessing the forces of nature, man has become the slave of the machine, and of his own greed."[45] Which way, then, was the experiment of Ghanaian statehood to move? For the present, certainly, Nkrumah favored rapid industrial development.

Inspired by Nkrumah's African-made Third World ideology composed of conflicting inspirations, his representative travelled to Bandung and listened as Ali Sastroamidjojo, Indonesia's prime minister, asked: "Where do we, the peoples of Asia and Africa, stand; and for what do we stand in this world dominated by fear?"[46]

VI

Hailing from Asian and African lands that encompassed the bulk of the non-Western world, the delegates to the Bandung Conference brought with them the divergent concerns of their countries. India and Pakistan quarrelled about Kashmir; Indonesia voiced its complaint over West New Guinea. Nasser pleaded not only for the Arabs in Palestine but also for the liberation of Morocco, Algeria, and Tunisia from French rule. Laos and Thailand complained about Chinese penetration into their lands. Zhou Enlai joined the protest against apartheid in South Africa. All delegates agreed on denouncing colonialism, but quickly argued over the targets of their denunciation. Should the Soviet Union and China be included?

The Ceylonese prime minister unexpectedly raised the issue of Soviet domination of eastern and central Europe, backed by the delegates from countries allied with the United States, Turkey, Pakistan, and the Philippines. The representatives from Thailand and Laos pointed accusingly at China. Nehru was hard-pressed to minimize this divisive issue. But the anti-Soviet sentiments represented at the conference hardly escaped Zhou Enlai's attention. Communist China's close association with the Soviet Union won his country few friends in Asia. And Nehru above all wanted to draw China—a peaceful China—into the Third World orbit.

The final communiqué[47] summed up the common interests. First came the urgent plea for economic development, admittedly with the help of foreign capital from countries outside the region. The wording of that plea was kept vague—obviously, only Western countries, and the United States foremost, could furnish that capital—with repeated reference to the United Nations and to the World Bank. In addition, better terms were needed for essential services: shipping fees, banking, and insurance; for more intra-regional consultation and trade. Nuclear energy in particular was an item of high priority. Asian and African countries wanted representation on the International Atomic Energy Agency; they wanted access to nuclear energy for peaceful purposes.

The delegates also endorsed a strong condemnation of colonialism "in all its manifestations" (a formula evolved after long dispute whether or not to condemn the Soviet Union too). On the basis of their common anti-colonialism and the cultivation of their national cultures, the participants agreed on cultural cooperation among their countries. They furthermore strongly backed the Charter of the United Nations and the Universal Declaration of Human Rights in the name of self-determination and the assertion of the sovereignty of their countries. They denounced racial discrimination and pleaded for universal equality.

For the sake of "social progress and better standards of life in larger freedom" the countries of Asia and Africa also opposed the costly arms race. They stood for world peace and cooperation, for all the generous ideals of the Western liberal tradition, yet insisting, for their own protection, on non-intervention or noninterference in their internal affairs (where the endorsement of human rights hardly applied). In principle, they recommended "abstention from the use of arrangements of collective defence to serve the particular interests of any of the big powers." An effective policy of nonalignment, however, was nowhere in sight. Subsequent conferences of nonaligned states faded into obscurity.

Whatever the declaration of common interests, these countries could not escape from their helplessness. The high-flown aspirations and grand words that gave distinction to their conference and sustained them in their subsequent trials were but a bold show. Indeed, aspiring to play an active role as independent sovereign states, they increased their vulnerability. They could not aspire to form a bloc of their own; several of the states here assembled eventually found themselves at war with each other—India with Pakistan, India with China, North Vietnam with South Vietnam, and Iraq with Iran. While calling for cooperation, they faced the future singly and in continued fear, exposed to all the tensions of the global state system and, in their fragility and dependence, further contributing to them. Yet in 1955, inspired with a new sense of importance, the delegates went home to their separate experiments in state-building and to their separate personal destinies.

What happened subsequently to Nehru, Sukarno, Nasser, and Nkrumah? At the time of the Bandung Conference these four leaders of developing states had looked boldly toward the future. As their subsequent fate proved, however, they did not possess the superhuman political skills needed to revamp the customs and habits of their peoples; they did not live up to the high hopes on which they had ridden to prominence.

VII

Of these four leaders, Nehru was the most favored. In the two years after Bandung his influence rose to its culmination. He was respected around the world, the senior statesman by age among impressive peers, attempting the role of peacemaker in the Cold War among the superpowers. His prestige abroad added to his charismatic role at home. He was irreplaceable; his occasional threats of resignation consolidated his power even more. He ruled paternalistically, unable or unwilling to delegate responsibility, devoting himself diligently to his duties, a model of selfless leadership. At the time of the second general election in 1957 optimism ran high in his country, giving the Congress Party a large majority in the Indian parliament. Then, suddenly, after ten years, the exultation of independence began to wane.

The promise of quick ascent to modernity could not be fulfilled. The second Five-Year Plan, for instance, did not live up to expectations; its goals had to be reduced. Jobs did not grow as rapidly as the population; the terms of trade deteriorated; the inflow of foreign capital was insufficient; harvests failed. Under the third Five-Year Plan, started in 1961, progress again was disappointing; unemployment increased; more food had to be imported; stark poverty persisted. The planning process itself proved inadequate. The economic stagnation cast a shadow over Nehru's last years.

In politics, too, his appeal declined. Resistance to planned reculturation mounted. A new conservative party gained votes, reducing Nehru's authority within the Congress Party. Decentralization and the extension of democracy to local bodies, rather than promoting the changes desired by Nehru, strengthened the forces of parochialism and separatism, benefiting the entrenched traditional elites and giving Indian politics a devisive and occasionally violent turn. In 1959, rather against his will, he was compelled to oust an elected caste-based Indianized communist regime in the state of Kerala. Two years later (in 1961), again over his objections, the Portuguese were driven by force from their Indian colony at Goa, to the plaudits of the Indian public and African opinion. In the fall of the next year, at the time of the Cuban missile crisis, a short war erupted with China over a long-simmering border dispute in the Himalaya Mountains—a minor affair unexpectedly begun and as unexpectedly ended by China, revealing the weakness of the Indian army. The humiliation, however, led to a welcome if brief flare-up of Indian nationalism, cleverly used by Nehru. He offered his resignation, which was again refused—he was still indispensable.

Yet his charisma was impaired; his health deterioriated. In May 1964 he died of old age, peacefully. But his work of molding and refashioning "the existing framework of economic and social institutions" according to modern (or Western) principles was far from complete. The younger generation of Indian politicians, which had come to the fore in his last years, was more Indian, more pragmatic, more tradition-oriented; the idealism of the founders of the Congress Party and of Indian nationalism had quickly evaporated.

Consider, for instance, the lack of progress in the Hindu Marriage Law passed in 1956. Though equality had been promised to women according to Western principles, the position of women in India remained shaped by tradition. Similar prevarication was evident in matters of caste. Though officially abolished, the caste structure remained virtually intact, the untouchables kept in place, by raw violence if necessary. The caste structure also permeated local politics, encouraged by the process of decentralization. Quite logically, rather than advancing the alien Western model, Indian democracy strengthened local practices; it entrenched tradition, counteracting reculturation. In the same manner the Indian bureaucracy, indigenized after independence, became an obstacle to innovation. Cumbersome, corrupt, and caste-ridden, it imitated the splendors of the colonial masters and thereby accentuated the inequality between the ruling elite and the masses.

Wherever one looked, the officially proclaimed ideals of the government and the realities of Indian life diverged sharply. Dirt and filth disgraced government offices, death and poverty the streets even in the capital. In the villages the age-old misery and inhumanity persisted; traditional practices long outlawed by the British *raj* (like widow-burning) revived. The widespread indifference to human suffering, encouraged by the caste system, continued to shock Western observers, while Indian intellectuals turned cynics, another obstacle to purposeful reculturation.

What was lacking above all was the social discipline needed for the effective assimilation of Western—or modern—skills. As Gunnar Myrdal observed, the Indian government ran a "soft state,"[48] proclaiming advanced ideals and yet doing little or nothing to carry them into practice. Nehru himself was responsible for that characteristic softness, advocating "change *with* continuity." He persistently refused to employ the continuity-cracking compulsions demonstrated so temptingly (at least for some Indian radicals) by communist China or the Soviet Union. Reculturation in India was to be accomplished by the molecular processes of democratic—or even Gandhian—methods, which meant at a snail's pace at best. The individuals composing the Indian body politic, especially the most privileged, abided by Hindu tradition, preoccupied with their bodily and spiritual selves, incapable of the community-building outreach implicit in the Western model.

At the time of Nehru's death, in short, little progress had been made toward transforming India into a modern nation. State-building was a much slower and more complex process than anticipated; the hidden socializing discipline of Western political culture could not be easily transferred into the Indian setting. Yet what progress had been made—it was not negligible—had been achieved comparatively peaceably, with reasonable goodwill and tolerance among the elites, and with much long-suffering among the tradition-bound masses. For these reasons the Indian experiment (next to Japan's) was unique in the world.

VIII

By comparison, the post-Bandung career of Sukarno was far more tragic. Reverting to his stated aims of 1945, he soon dismantled the parliamentary government instituted at independence and proceeded to establish what he called "guided democracy." By 1960 he emerged as "the Great Leader of the Revolution" and President for Life, exalted in public monuments like Hitler or Stalin. Yet underneath the pomp of dictatorship he reverted to the style of a traditional ruler. A pleasure-loving womanizer, he lacked drive, discipline, and even a detailed program to give substance to his grandiloquent sloganeering about revolution. Indonesia's standard of living declined despite Sukarno's program of "Indonesianization," which nationalized formerly Dutch enterprises, and despite Soviet industrial aid. He also received Soviet military aid for his policy of building a Greater Indonesia by a persistent guerilla war with Malaysia. On that score he quarrelled with the United Nations, even cancelling his country's membership in that organization in 1965 and launching a rival body called the Conference of New Emerging Forces.

By October of that year, however, he had lost his charismatic appeal. A foiled communist coup against the military leadership led to a popular outburst of anti-communist indignation, culminating in a bloodbath lasting months. It eliminated the communists and their sympathizers, together with untold Chinese; the human toll was almost as large as in the Indian communal riots of 1947–48. Sick and indecisive at the time, Sukarno survived that event, but his ascendancy had ended. Eased out of the presidency by General

Suharto, the leader of the army, he died in 1970, buried in an unmarked grave in his hometown. General Suharto meanwhile had set his country on a pro-Western course, rewarded by Western aid that at last led to a gradual improvement of living standards under continued military rule. State-building continued indecisively with less rhetorical fanfare and more pragmatism under the highly adverse conditions of the Indonesian archipelago.

IX

Pragmatism and more auspicious circumstances meanwhile favored the work of Colonel Nasser. Cheered at Bandung and even more resoundingly at home, he accelerated his ascendancy as Egypt's leader and protagonist of Arab unity, rising to astonishing success before being trapped in the intricacies of state-building and the complexity of Mideastern politics. Regularizing their "revolution" in the year after Bandung, Nasser and his Free Officers provided their country with a Western-style constitution, resigning their commissions and assuming civilian positions. The constitution, ratified by popular referendum, declared Egypt a democratic republic, thus giving some substance to one of the Six Principles. A freely elected National Assembly possessed the right to pass government bills and the budget, although checked by the president's power to dissolve it at will. The president was selected by the Assembly, subject to ratification by a general election. Elevated to the presidency, Nasser received 99% of the popular vote.

His new constitution, however, was but a facade, like its predecessor under King Farouk. It gave a fashionable democratic touch to a presidential regime centered on Nasser himself. Although criticism of the government was permitted under the constitution, all previously existing parties were dissolved, their place taken by the National Union, the official party, which carefully screened its membership. In 1958 that constitution was replaced by a Provisional Constitution for the new United Arab Republic, and changed again in 1962, when that venture had failed. Constitution-making was part of the unceasing experiment of nation-building, part of the relentless process of Westernization, yet without its essence. The opposition enjoyed no protection, though political repression remained comparatively mild. The gentle authoritarianism of Egyptian tradition required no formal dictatorship.

As for social justice, another of the Six Principles, the Free Officers from the start had undertaken a land reform which broadened the base of landownership in favor of the peasants; it assured the new regime of peasant support ever thereafter, while granting the dispossessed landowners generous compensation. Otherwise the principle of social justice remained dormant; socialism as a political system was but a fleeting reference. For the sake of economic development, the Free Officers in their civilian capacities cooperated with the existing elites, which, reduced to political passivity, gladly gave their consent, reassured by Nasser's successes in foreign policy.

Soon after Bandung Nasser concluded an arms deal with Czechoslovakia which, in line with another of his Six Principles, strengthened not only the army but also Nasser's anti-imperialist bargaining power. Nonalignment held out the promise of greater significance for a strategically located newcomer like Egypt; while playing off the superpowers against each other, it might obtain assistance from each, like weapons from the Soviet bloc and food plus other economic assistance from the United States. Nasser's revolution aimed at economic development, foremost through the building of the Aswan Dam. The dam, an impressive experiment in applying Western technology to an unknown setting, was to regulate the flow of the Nile for the benefit of agricultural production. But above all, Nasser aimed at the assertion of Egypt's power.

When in 1956 the United States cancelled its promised support for the Aswan Dam (letting the Soviet Union take its place), Nasser's moment of glory arrived. Upon receipt of the news from Washington, he announced to a jubilant population the nationalization of the Suez Canal (with due compensation for the original shareholders), thereby ending the British presence in Egypt. His action provoked an international crisis, as an Anglo-French expeditionary force arrived to reclaim the canal and Israel advanced toward Egypt, easily defeating its armed forces. Yet Nasser's gamble succeeded. President Eisenhower and the United Nations, backed by public opinion in Asia and Africa, came to Nasser's rescue, compelling the withdrawal of the invaders; the Soviet Union, engaged in suppressing rebellion in Hungary, expressed its moral support. In the upshot Egypt scored a political victory, master at last of that strategic waterway. Nasser was praised as the architect of English and French humiliation, the hero of the Arab world (the Israeli victory was conveniently disregarded). His Egyptians now worshipped him like a god; never did his star shine more brightly. He was the answer to all prayers for deliverance from age-long oppression and humiliation, a true savior. The exultation demonstrated, even more than the Bandung Conference, the intensity of the accumulated anti-Western resentments, as well as the capacity for exaggeration and self-delusion among the newcomers.

After his triumph in the Suez crisis, Nasser briefly emerged as the acknowledged leader of pan-Arab unity, increasingly immersed in Mideastern politics. In 1958 he arranged the union of Egypt with Syria under the name of the United Arab Republic, another widely hailed success. Yet the union did not thrive. Egyptian high-handedness and ignorance of Syrian conditions alienated the Syrian leaders; in 1961 the union collapsed. At the same time, Nasser's leadership among Arabs had come under suspicion. His ambition was resented—he had gone beyond the modesty expressed in his pamphlet of 1955; he would not take the advice of more knowledgeable counsellors. How in any case could anybody achieve unity among the Arab states extending from North Africa through the eastern Mediterranean to Saudi Arabia and Yemen, among peoples so individualistic and fragmented in their political culture? In the 1960s Nasser additionally involved himself in the politics of emergent Africa, another political quagmire.

The year 1961 marked a change in Nasser's fortunes. His health declined (he suffered from diabetes), though he kept his relentless pace of activity, unwilling to delegate power. As a longtime conspirator temperamentally suspicious of his associates, as of political leaders generally, he withdrew into himself while trusting in the adulation of the crowd. He suffered from the corruption of power while yet retaining a trace of that humility and honesty that had made him confess his incompetence early in his political career. He also preserved his sense of humor, that graceful lubricant of human relations. And he never forgot that his revolution was yet unfinished.

Devoting himself again to domestic politics, he turned in 1961 toward socialism in ideology and administration. After the nationalization of the Suez Canal, the public sector of the economy had been gradually strengthened. In 1960 a ten-year plan for economic development had been enacted, envisaging a doubling of the country's GNP by the beginning of the next decade; in the same year, the press had been nationalized. Now, in 1961, came the turn of the big banks, then of the insurance companies and of all heavy industry. In 1963, after another spurt of nationalization, 80% of Egypt's industry found itself in the public sector.

The economic changes were matched by new constitutional arrangements. In 1961 a new National Charter, declaring "war on imperialism and domination" while "laboring to consolidate peace,"[49] stipulated that one-half of all seats in the Assembly and its committees be held by workers and peasants, a provision maintained in the constitutional revisions that followed. Egypt now was declared a "democratic socialist state" on the way to "scientific socialism." Stressing "democracy" and dissatisfied with introducing innovations from above, Nasser even called for a new mass party, the Arab Socialist Union, which was duly organized—without eliciting the desired spontaneous change from below.

Always concerned with the "sacred advance" of the masses as the ultimate success of the revolution, Nasser found himself in a quandary. The Egyptians were not responding to nation-building as he had hoped; they lacked vision, enterprise, and above all the discipline that Nasser and his Free Officers had learned in the army. Some of Nasser's close associates strayed from the austerity of military discipline; the privileged classes splurged on luxuries rather than investing in industry; the bureaucrats and technocrats followed their narrow routines; and the lesser folk continued in their traditional ways. Successes in foreign policy, though contributing to economic growth, did not provide sufficient incentives—too much revenue was diverted to the army in any case.

Nasser's renewed efforts to elicit more spontaneous mass participation, therefore, was a dubious experiment. Fully aware of the unmodern ways of his people—yet appreciating their adulation—he rightly distrusted spontaneity. He even feared, with some justice, that the Arab Socialist Union might turn into a political threat to his rule. As he admitted in 1964 to the National Assembly: "As far as the moral texture [of our society] is concerned, we cannot change it overnight. So far, we have not been able to create a model for the new relationships that should prevail. . . . Our social relationships have not changed."[50] He never solved that central problem of reculturation. Did he really understand it?

In the 1960s Nasser's foreign policy also turned sour as he was caught in the gridlock of Mideastern politics. He vainly intervened in the civil war in Yemen, trying to oust the British from Aden, while posturing as the Arab guardian against Israel, alienating opinion in the West and gravitating toward Moscow. In 1967 he let himself be prodded into the disastrous Six-Day War with Israel that ended in utter defeat for the Egyptian armed forces and the collapse of Nasser's leadership in pan-Arab unity; it was the nadir of his career. Assuming responsibility, he announced his resignation. Then the unexpected happened: upon hearing the news the population of Cairo poured into the streets, apparently quite spontaneously, urging him to stay in office—he was indispensable, his charisma still intact.

His last years, still busy in pursuit of his goals despite his failing health, proved unproductive. Having earlier lost Nehru's friendship, he now also alienated his long-standing ally Marshal Tito. He died (age fifty-two) of a heart attack in 1970, buried amidst a public mourning so intense as to disrupt the funeral procession, the hero of Egyptian deliverance from ignominy. Under his dynamic leadership Egypt had indeed come a long way toward modern statehood, whatever his failures.

Yet the real objective of his revolution, reculturation after the Western model, had not been attained. How could it have been, considering the attachment to the country's traditional identity cultivated in the name of an anti-Western nationalism? As Nasser's successor, Anwar Sadat, one of his closest associates among the Free Officers, observed in 1974: "The real challenge confronting peoples with deep-rooted origins who are facing the problem of civilizational progress is precisely how to renovate their civilization. They should not reject the past in the name of the present and should not renounce the modern in the name of the past. . . . " But as for that creative combination, the crux of the problem of reculturation, Sadat too wavered: "Modernism is knowing the right order of priorities. . . . We should compose the suitable environment and necessary stage of development which will make us capable of invention and creativeness and consequently of a true contribution to human civilization."[51] How was that to be done, and how much of the cherished origins would be left at that ultimate stage?

The cherished origins, incidentally, did not fit into the contexts of Marxist "scientific socialism." Marxism looks forward, repudiating the feudal past as something to escape from as rapidly as possible. Yet Third World countries emerging into statehood, although attached to Marxist mobilization techniques, tended to adopt a positive—almost fascist—stance toward the past. The colonial era, to be sure, was mercilessly denounced. But the precolonial past was viewed as the storehouse of collective identity and the guarantee of cultural continuity. It was often romanticized, its memory cultivated as the source of a distinct and worthy contribution to modern humanity. The problem remained: how to reconcile the nostalgia for the past with the demands of the times in which the West still set a fast pace.

X

Or rather—to shift now from Nasser to Nkrumah—the question was: how to advance "civilizational progress" while the bulk of the population, however eager to receive the benefits of modernity, was still tied to its origins, unwilling or unprepared to change. That was the question looming over Nkrumah's post-Bandung career, perhaps the most fascinating case of state-building in Africa.

Nkrumah's experience of the world was superficially more extensive than Nasser's, but he was impressed by the latter's rising influence. In 1956 he married an Egyptian woman (subsequently kept in the background)—the Bandung-inspired link to Egypt mattered, as Nasser considered sub-Saharan Africa within his sphere of influence. Compared with Egypt, the new Ghana was insignificant, without strategic importance or vital raw materials; it lacked the cosmopolitan elites of Alexandria or Cairo. Yet Nkrumah managed to raise it to global visibility.

Groomed to lead West Africa's first independent state, he was the beneficiary of a well-arranged transition to political sovereignty. Whatever his innermost convictions, until Independence Day he played his assigned part, a loyal Westernizer in the British tradition, endowed with unusual assets. The Gold Coast claimed the highest literacy rate and standard of living in any European colony in Africa. It possessed large financial reserves earned from the export of cocoa, a thriving business in the 1950s. In addition, Nkrumah himself was a credit to his country.

An articulate and intellectually ambitious state-builder claiming to be the apostle of African unity, he grandly played to the world audience which he had come to know in the United States and England. Admittedly, his platform was a tiny country, the bulk of its people still wedded to ancestral custom. But in these years of decolonization he travelled widely, his voice heeded around the world as the voice of emerging Africa.

Somehow his imagination and rhetoric—and even his own career—spanned the vast distances between soil-bound African tradition and the global universalism unfolding between the two world wars. He had benefited to an unusual extent from the opportunities for advancement offered by colonial rule, elevated far above his compatriots. Yet as a colonial subject he was attracted to the Marxist counterrevolution as the fastest shortcut from African weakness to the high-flown goals set by the West. And, like the Westernizing counterrevolutionaries of the 1920s and 1930s, his models, he was determined the lead his people—all the Africans—to the political kingdom of wealth and power in his own lifetime. Such things to be! Such risks to be faced so innocently in utter ignorance of the incompatibilities of culture (or of the hubris of his ambition)!

Master in his own house after March 1957, Nkrumah cautiously proceeded according to the vision mapped out in his London days. The transition, however, was gradual. The civil service continued in the British tradition with the help of expatriate Englishmen; a British general commanded the Ghanaian army, an important policy instrument on the way to pan-Africanism; an English lawyer acted as his constitutional advisor. Economic and cultural ties to Britain, and to the West generally, also remained close. Nkrumah was not above drawing on experts from Israel (a Western outpost), to Nasser's annoyance. All the while, however, the political atmosphere changed as Nkrumah rose far above the constitutional position inherited from British rule.

By 1960 he combined the position of head of state, chief executive, and commander-in-chief under a republican constitution designed "to serve the people" and freed from allegiance to the British crown (though Ghana remained a member of the British Commonwealth). As life chairman of the Convention People's Party he was the ideological guide of the country, the Man of Destiny, the source of Truth, *Osagyefo* (the Redeemer), high priest and paramount chief of all Ghanaians. He was likened to Confucius, Mohammad, Shakespeare, Napoleon, Saint Francis, Buddha, and Christ—indigenous culture furnished no measurements by which to assess the global galaxy of stars; any superlative seemed justified. His image appeared on coins and money bills; streets and squares were named after him; bronze busts represented him in all public centers. The new Ghana was built around his person, a target of two assassination attempts and increasingly protected by Soviet-trained security guards amidst a residential splendor befitting his station. He too sought to magnify himself through the grandeur of his surroundings. For ceremonial occasions he built the Black Star Square in Accra, his capital, to resemble the Red Square in Moscow. At Nkrumah Circle his name illuminated the night in flaming letters.

His vehicle for building a new state and eventually the Union of African Socialist Republics was his Convention People's Party (CPP). Originally designed as a mass party for achieving independence, it now was recast as an ideological vanguard in the manner of a communist party, complete with auxiliary organizations reaching into the daily lives of the population, such as the Trade Union Congress, the Farmers' Council, the Council of Ghana Women, or for a time the National Association of Socialist Student Organizations. It controlled the media and the police; it silenced opposition. For ideological guidance and indoctrination the Nkrumah Ideological Institute was established; Leonid Brezhnev attended its opening. Outside its gate loomed a statue of Nkrumah nearly forty feet high.

In the fall of 1961 Nkrumah toured a number of communist states, favorably impressed by what he observed. Subsequently, he shifted his political orientation toward the Soviet bloc in the hope of speeding progress towards his goals. Three years later, in 1964 (the year he received the Lenin Peace Prize), another new constitution declared Ghana a one-party state; it was ratified by a popular vote of 96% in its favor. Nkrumah pretended to consult the masses; "democracy" was an indispensable catchword in his vocabulary; "where control of the state is not vested in the people," he said in 1965, "imperialism takes over."[52] And Ghana indeed enjoyed a measure of freedom. Although controlled by his party, the Ghana National Assembly retained a modicum of independence; back benchers occasionally, though cautiously, criticized the regime. And talk ran freely among ordinary folk, without fear of a secret police.

The party was also responsible for building the economic foundations of the new state, with a socialist emphasis on planning. In preparation for independence, economic planning had started under colonial rule in 1951, promoting agricultural prosperity as a precondition to industrial development, much to Nkrumah's chagrin. After independence, planning took an ambitious turn to coincide with the establishment of the Republic in 1960. The Five-Year Plan of 1959, however, was soon scrapped and a Seven-Year Plan prepared to take effect in 1964, with the avowed purpose "to eradicate completely the colonial structure of our economy."[53] State-owned industries and state farms proliferated at the expense of private enterprise, absorbing the country's resources. Transport and communication were advanced, schools and hospitals built; a new harbor was constructed to the east of Accra.

The showpiece of Nkrumah's industrialization was the completion of the Akosombo Dam across the lower Volta River, Ghana's counterpart to the Aswan Dam. It was to create the world's largest man-made lake and provide a source of electric power for much of West Africa. Nkrumah succeeded in persuading two American companies (Kaiser Industries and Reynolds Metals) to take advantage of the Akosombo Dam's cheap electricity and build an aluminum smelter at the new port of Tema, with the approval of the American government. The terms of the contract, it has been subsequently argued, were unnecessarily favorable to the Americans, but Nkrumah was in a hurry. He hoped that the new smelter would draw on Ghana's bauxite deposits—instead the Volta Aluminium Company imported alumina from the West Indies. In any case, attracting American capital was not meant as a bow to the neo-colonialists. Nkrumah's newspaper greeted President Kennedy's approval in February 1962 with the haughty headline: "Dollar Boss! Better Late than Never."[54] Given the pro-Soviet orientation after 1961, continued reliance on Western capital had to be offset by anti-imperialist rhetoric, a common practice among the nonaligned.

While concerned with building an effective domestic power base, Nkrumah never lost sight of his main interest—and the source of his charisma: the promotion of pan-Africanism; he was resolved to give his people—all the Africans—a respected place in the suddenly widened-out world. Soon after independence his African policy went into high gear. In April 1958 the heads of the then eight independent states of Africa (Egypt, Ethiopia, Ghana, Liberia, Libya, Morocco, Sudan, and Tunisia) conferred in a demonstration of unity; subsequently, Nkrumah called on his guests at their own capitals. In November, inspired by the union of Egypt and Syria as a prelude to pan-Arab unity, he concluded a political union of Ghana and Guinea (the latter about to become independent under the leadership of Sekou Touré). In December an "All-African People's Conference" brought to Accra the representatives of political parties, trade unions, cooperatives, and youth organizations from the entire continent. The featured delegates included Frantz Fanon, Julius Nyerere, Patrice Lumumba; all listened as Nkrumah, attired in African garb, attacked the "balkanization" of Africa.

The 1958 conferences were followed by further initiatives. In 1960, the "year of African unity," Nkrumah pressed his claims at a conference in Addis Ababa. Yet gradually he encountered resistance. As other independent African states emerged, many larger than Ghana, rival ambitions for African leadership surfaced. By 1961 it became clear that the rise of independent African states led to disunity rather than pan-Africanism.

Meanwhile, Nkrumah pressed his claim for African leadership in a different way, by sending Ghanaian troops to the Congo and taking Lumumba's part in the civil war. That he by no means excluded military action on behalf of African unification was clear from his casual observation: "It is a pity we are in the second half of the twentieth century. If we were in another century, the Ghana Army would march into Togo, Upper Volta, Dahomey and so on, and before long we would have African Unity"[55]—by the sword. The Ghanaian intervention in the civil war in Congo, however, proved futile, as did Nkrumah's threat, in 1964–65, of armed intervention by African states to prevent Ian Smith from setting up a white regime in Rhodesia. His pan-Africanism fared no better than Nasser's pan-Arab unity.

When in 1963 the Organization of Africa Unity (OAU) was formed, Nkrumah's dream had been stalled and his ambition thwarted. The OAU established no all-African governing body, no joint High Command, as he had advocated; nor did it recognize his presumption to represent all of Africa. When in 1965 the OAU summit conference convened in Accra in order to vote for the formation of an all-African Union Government (as Nkrumah hoped), a sizable group of members boycotted the meeting; among those present, Nkrumah's motion lost.

The hard fact was that with the end of colonial rule the traditional tensions within that variegated and splintered continent reasserted themselves. It was not European imperialism, but the cultural conditioning of centuries, if not millennia, which had balkanized the continent. Liberation from colonial rule, combined with the heightened Westernization that came with statehood, aggravated and intensified the preexisting disunity, culminating all too often in unprecedented inhumanity. Nkrumah's dream of unity hung over Africa as a black star, a source of darkness rather than light. Considering the realities, the continent could have been brought under a common roof (if at all) only by brutal compulsions far exceeding Stalin's terror.

Yet even in Ghana forcible reculturation was out of the question. Despite his extreme ambition, Nkrumah—to draw a quick sketch of him at his peak—was an unusually benign dictator, by temperament and by necessity; for better or worse, he lacked the organizational and psychological resources for drastic measures evolved in Europe. By all appearances, the *Osagyefo* was a charming and simple man, dressed in the later years of his rule preferably in a plain tunic like Stalin or Mao, hardworking and ascetic in his habits (though in his testament of 1965 he acknowledged that he had children "begotten not in wedlock but in accordance with native customary law"),[56] a loyal and generous friend, unusually gentle and opposed to violence. With time, however, he suffered from the discrepancy between his ever larger responsibilities and his limited human capacity. Unable to live up to the demands of his vision, he became more aloof, more withdrawn, a megalomaniac without sense of the possible, and at the same time more traditional, surrounding himself with a bodyguard drawn from his own and a neighboring tribe, and seeking advice from fetish priests. Modern and traditional ways imperceptibly wove their nets around him, as he pushed forward toward his goals, the champion of the African masses and, by his own estimate, the embodiment of Africa. As he boasted in 1961: "all Africans know that I represent Africa and that I speak in her name. Therefore no African can have an opinion that differs from mine."[57]

As a Western-educated non-Western intellectual and admirer of Lenin, he buttressed his ascendancy with revolutionary theory, producing a spate of books spelling out the message of Nkrumahism as the Ghanaian—or even pan-African—embodiment of "scientific socialism." In 1958 he issued a volume, *I Speak of Freedom;* after 1963 followed three more books in yearly intervals, *Africa Must Unite* (1963, published to coincide with the founding of the Organization of African Unity), *Consciencism* (1964), and *Neo-Colonialism: The Last Stage of Imperialism* (1965).[58] Among these volumes, all ghostwritten or ghost-assisted, *Consciencism* stood out for its preposterous pretentiousness. It covered Western philosophy from Thales to Wittgenstein in an unsuccessful effort to link African tradition with modern socialism. All his books, obviously, were beyond the comprehension of the African masses; Nkrumah wanted to be counted among the elite of the world's political theorists.

Yet by 1965 his days as *Osagyefo* were numbered. From the start his vision of state-building had been founded on an illusory reality, perhaps more so than was the case with other leaders of the Bandung generation, all of them compelled to strike a synthesis between their indigenous roots and the new globalism. Nkrumah's fatuous reality was derived from an envious outsider's view of admired models in distant parts of the world unrelated to the ground-floor conditions with which he had to work.

The flaws of that disjointed image of the world had been unconsciously revealed in 1962 by Nkrumah's friend and intimate of long standing, Kofi Baako. "In an ex-colonial territory which has emerged successfully into independence, and which is anxious to repair a hundred years of damage done to its social, political and economic systems," Baako had said, "we can find no better solution to our problem than the adoption of *socialism, which in fact is our own traditional way of life.* We have to adopt a system which places social and economic power in the hands of the state"[59] (italics added). Put bluntly, all blame for the shortcomings of the regime was cast upon the imperialists (without appreciating their contribution to Nkrumah's or Baako's education);

all hope was concentrated on a socialist state which was identified, in defiance of all available evidence, with the traditional way of life. How in the face of such incomprehension could the extravagant vision be realized?

Up to independence local conditions and collective aspirations had been joined in relative harmony. Thereafter, through Nkrumah's leadership, the disorienting new globalism entered in full force. The discrepancy widened between the officially imposed superstructure and the attitudes and promptings of the people, even those at the very top (including Nkrumah himself). Prominent party members used their privileges to enrich themselves in traditional ways, despite Nkrumah's efforts to stop "corruption." He himself was not above reproach, having lost all sense of distinction between his private and public fortune. Nothing was more alien to traditional ways than the austere discipline required of model cadre. Why join the party if not to prosper? In addition, who possessed the managerial skills, technical competence, and trans-tribal overview needed for effective administration, whether in the party, the government, or the proliferating state enterprises? Who in a society so riveted to the immediacy of the present possessed the patience needed for long-range investment, the extended time scale required for planning? The ex-colonised man, like Nkrumah himself, wanted quick results—without knowing how to achieve them properly.

In pursuit of quick results, Nkrumah's government wasted the financial reserves accumulated under British rule and went deeply into debt, financially overburdening the country far into the future. Despite its promises to the masses, to the workers and farmers, it exploited their labor. In 1961, while Nkrumah was visiting the Soviet bloc, a strike of dockworkers was ruthlessly suppressed, in an action subsequently endorsed by him. The cocoa farmers fared worse. The country's foremost earners of foreign currency, its source of wealth, they found their profits drained into the public sector without receiving much benefit in return. All the state industries operated at a loss: so did the ever more numerous state farms. Meanwhile, private enterprise stagnated or contracted. Worst of all, the worldwide cocoa boom of the 1950s ended just as Ghana became a republic; but government planning remained blind to that fact, inclined to blame the neo-colonialists for the downturn. It became obvious by 1965 that the *Osagyefo* was not living up to his promise.

Soon after independence he had accepted a bet from his neighbor, the leader of the Ivory Coast, Felix Houphouët-Boigny, confident that socialism—a simpler system than capitalism, as he had argued in 1956—would produce more abundance in Ghana than the free-enterprise approach applied next door.[60] He proved utterly wrong. Freely admitting the neo-colonialists into the country and its economy, the Ivory Coast prospered. What mattered in francophone Africa was not cultural self-assertion or racial pride—Houphouët-Boigny had served in the French National Assembly, treated as an equal—but access to the West's material boons. Being able to deliver the goods, Houphouët-Boigny happily stayed in power indefinitely, while Nkrumah was rudely ousted in 1966.

Apprehension over the decline of Ghana's economy had become widespread; discontent was rife in the party itself. In addition, prominent army officers, trained in the British tradition, were dissatisfied over the process of "CiPiPification," the party's growing interference in military affairs; high-ranking police officers shared their feelings. Aware of popular disillusionment with the regime, they planned Nkrumah's fall. As it happened, in February 1966 Nkrumah was out of the country. Viewing himself as a potential peacemaker in the Vietnam War and eager to recoup his declining prestige, he had set out for Hanoi, against the better judgment of his advisors. Stopping in Beijing on the way, he was informed that he had been deposed by a military coup cheered by the market women of Accra, most Ghanaians, and the governments of England and the United States (the latter had played an active part in the event).

Undaunted, Nkrumah adhered to his mission even in exile. A Soviet pensioner residing in Sekou Touré's capital, he continued as the advocate of African unity under socialist auspices, writing books and pamphlets pleasing to his paymasters. He now wrote of class struggle in Africa and attacked the term "the Third World." For him there existed only two worlds: the evil one of the imperialists and the good one of the anti-imperialist freedom fighters. He died, age sixty-three, in a Rumanian hospital in 1972, the last survivor of the Bandung quartet of Westernizing anti-Westerners here considered.

By this time opinion in Ghana had again turned in his favor. The economy had fared worse under his successors, both military and civilian. He had, after all, made Ghana more modern, equipped with a better infrastructure, with better schools and hospitals, and with a university in the Oxbridge style at Legon, near Accra. Above all, he had supplied a much appreciated élan to life, the excitement of being important in the world. His body was returned to Ghana, lying in state in Accra while multitudes filed past in awed silence, before being ceremoniously interred in his hometown. Nkrumah has been worshipped as a national hero ever since—in a country still trapped in the agonies of state-building and dropped from global visibility.

XI

How, then, are we to judge these men and, through them, the Bandung generation of high-flying Third World leaders whose ambitions carried them far beyond their individual resources and those of their societies? They grandly employed the liberal ideals of the Western tradition for enlisting Western goodwill and magnifying their importance in the world, while they showed no capacity for carrying out these ideals in building up their own states—they rather followed the practice of the fascist or communist dictators of the interwar years. Did they merit the compassion and partisan support they received, most prominently from Dag Hammarskjöld, secretary-general of the United Nations at that time? Or were they selfish and deluded hypocrites, no better than the imperialists whom they denounced and far worse in their domestic politics?

In the perspectives here employed they stand out as immensely tragic figures. Men of extraordinary ability and scope, they were saddled with superhuman tasks. Yanked from their moorings by the new globalism, they found themselves suspended between incompatible compulsions. On the one side they were subject to the traditional ways still rumbling in their depths and guiding most of their followers. On the other side beckoned the multitudes of stimuli and incentives abroad in the world. They were exposed to all the multifaceted Western experiences; to all the social and political experiments, all ideologies, all scientific and technological innovations descending upon them at an overrapid clip. They suffered, in their smallish settings, from the pangs of a Western-induced global ambition that lacked the necessary cultural resources for living up to expectations. How were they going to fashion the inchoate ingredients in their own experience and that of their societies into a harmonious and meaningful philosophy of life or into an effective political ideology? Was it a wonder that they were headed toward catastrophe—for themselves or their peoples, or both?

Commentators in Western countries secure in their power and cultural continuity were tempted to mock the newcomers by advising patience and humility, or by recommending deference to the superior skills of the West as under colonialism. What, then, of freedom and self-determination, of the example of the American revolution (much touted at the time by American presidents)? The difference, of course, lay in the fact that the American revolutionaries were part of the West, possessing the basic skills for setting out on their own. The newcomers of the mid-20th century were not so lucky. The Western example had enticed their imagination and ambition, but then left them distraught and helpless to cope with conditions totally unknown both in Western and non-Western experience.

Under the cover of the idealistic rhetoric of the Bandung Conference, utterly unprepared and disoriented peoples began to face the heightened global competition for wealth and power. By any moral standards they surely deserved the compassion of a Dag Hammarskjöld.[61] They are also entitled to the sympathy of all who make the effort to understand their condition.

20

Chairman Mao

While the pacesetters of the Bandung generation coped with success and failure, another experiment of revolutionary Westernization through state-building was under way, of far grander dimension in the number of people affected and in the scope of leadership. That experiment was the rise of the People's Republic of China under Mao Zedong, who for a quarter-century was master of almost a quarter of the world's population.

I

On September 21, 1949, Mao celebrated victory in the civil war against the Guomindang that had begun after the defeat of Japan. "In the course of little more than three years," Mao proudly announced,

> the heroic Chinese People's Liberation Army, an army such as the world has seldom seen, crushed the offensive of several millions troops of the American-supported Kuomintang reactionary government. . . . We have a common feeling that our work will be recorded in the history of mankind and that it will clearly demonstrate that the Chinese . . . have begun to stand up. . . . Our nation will never again be an insulted nation. We have stood up. . . . The era in which the Chinese were regarded as uncivilized is now over. We will emerge in the world as a nation with a high culture.[1]

While apparently uncertain whether China had stood up or had merely begun to stand up, he had reason to be proud. His soldiers had defeated an enemy army four times as numerous and equipped with superior American weapons.

A few days later he restated his claim for a China that had come up in the world: "We have now entered into the community of peace-loving and freedom-loving nations of the world. We shall work with courage and industry to create our own civilisation and happiness and, at the same time, to promote world peace and freedom.[2] And on October 1, 1949, elected chairman of the new state, he officially proclaimed the foundation of the People's Republic of China. Having placed communist China on the political map of the world, Chairman Mao now faced the task of creating "a nation with a high culture" that could measure up to other great nations.

Mao came to that task with unusual assets. His political organization, the Chinese Communist Party, and his armed forces, the People's Liberation Army, possessed undisputed control over the Chinese mainland; Jiang Jieshi and the Guomindang had withdrawn to Taiwan. All foreign concessions and limitations on Chinese sovereignty were ended; after decades of civil war and political division national unity was restored (except for the offshore islands). The communists arrived as liberators, rescuing a once proud and much admired country from a century of defeat and humiliation inflicted by more powerful outsiders. Mao was the symbol of that yearned-for liberation, a symbol nourished by traditions of the imperial past. There was another asset for the future: in contrast with the Third World newcomers, China was already a nation, a nation, moreover, with an impressive cultural record and a well-defined identity, even though much of its past now stood as an obstacle to the adjustments required by the times.

Yet even in regard to those obstacles, Mao possessed a unique body of sociopolitical skills. In the Yanan years he had learned how to mobilize traditional society for untraditional tasks, for reculturation according to an alien model. From the Chinese countryside and the peasant masses he had raised a political force endowed with both military power and administrative organization, the most effective yet in all of Chinese history. His model for these skills had been Marxism-Leninism and the Soviet state as evolved under Stalin. But his success was based not on outright imitation but on his ability to translate the essentials of the Soviet model into Chinese routines.

He had applied the techniques of Soviet substitutionism to China with a Maoist twist that allowed him to draw political power from greater backwardness than had prevailed in Russia at the start of Soviet rule. Lenin had replaced the Western skills of spontaneous social cooperation with centralized organization, with planning and command by a political vanguard supported by a small industrial working class. In the Yanan years Mao had considerably broadened that prescription by insisting on perpetual interaction between the vanguard and the masses—peasant masses—as a precondition for building an up-to-date state and society. The "mass line," inspired by Mao's faith in the creative potential of plain people (like himself), projected a learning process of reculturation reaching deeper and more subtly into the recesses of traditional culture than any Stalinist command for change. The most dangerous enemy of the mass line, Mao well knew, was bureaucratic routine, especially rigid in its traditional Chinese mold. It stifled the spontaneous voluntary cooperation of men and machines in large-scale organization so essential in the vision of communism as well as in "capitalist" practice.

The Yanan skills had been evolved in a poor, out-of-the-way region under threat of wartime annihilation and subsequently affirmed by victory over the corrupt and demoralized Guomindang. Were they sufficient to repeat a similar transformation for all of mainland China? Now Mao dealt with far more complex conditions: the vastness and diversity of the country, its huge and rapidly increasing population, which included the inhabitants of towns and cities on the seaboard long tied to the outside, the "capitalist" world. That China was still profoundly traditional, "a sheet of sand," its people persisting in their granular, tightly knit petty loyalties under the capacious umbrella of a dimly remembered semi-divine imperial authority. Far more than Lenin's Russia, China persisted in inward-oriented seclusion from the outside world, resentful of all foreign inroads as a source of cultural—or even metaphysical—chaos. The new security tended to strengthen that traditional isolationism; gone now was the fear of Japan or of any comparable aggressor. What remained was the threat of backwardness, clearly perceived by only a few. Could this remote and abstract threat pry the bulk of the people from their accustomed ways?

The initiative for that feat now was to come mostly from within, from the Communist Party, and above all from Mao himself. The product of China's long years of troubles, he was steeped in Chinese history and culture yet also profoundly influenced by the Western impact, either directly or through the Soviet model. Strong-willed, articulate, and restlessly energetic, he was both a gifted theorist with unusual insights into the processes of reculturation and a ruthless political manipulator endowed with charisma. From his own experience he knew the power of ideas, their capacity to sway people uprooted by revolution, war, and civil war toward a vision of a new China radiating peace and glory. Endowed with an intuitive sense of the temper of the times, he considered himself the dynamo of revolutionary change in the tradition of Lenin and Stalin. On that memorable October 1, 1949, the fifty-five-year-old Mao confidently looked toward the future.

II

Already the previous July he had laid out his blueprint[3] for action, setting forth the ideological framework—and the metaphysical source of certitude needed for leadership—into which future policy was to be set. Marxism-Leninism, he stated, constituted the universal truth. Bourgeois democracy had failed in China; communists alone "understand the law governing the existence and development of things." Marxism-Leninism as embodied in the Chinese Communist Party also furnished basic perspectives for future development. China would move from democracy to socialism and hence to communism; to the abolition of class, state, and party; to universal fraternity.

As for the immediate future, Mao's message was somewhat blurred, hinting at unsolved problems. He argued that "the Chinese people" had learned a basic lesson from their recent history. "We must awaken the masses in the country. This is to unite the working class, the peasant class, the petty bourgeoisie, and national bourgeoisie into a national united front under the lead-

ership of the working class, and develop it into a state of the people's democratic dictatorship led by the working class with the alliance of workers and
peasants as its basis." The implicit circularity of the argument—the masses
are to be awakened by the masses—left open the question of leadership. In
addition, the list of classes united in the national front did not sound convincing. The "petty bourgeoisie" of craftsmen and small traders, or the
"national bourgeoisie" of big business and established professionals who had
not fled to Taiwan, conceivably responded to identifiable groups in Chinese
life. But what of a "working class with the alliance of workers and peasants
as its basis"?

China had not developed an industrial working class worth mentioning.
The industrial workers of the seaboard cities, a small segment of the population, had not shared the Yanan experience; they were not allied with the
peasants; they did not even possess a unifying class consciousness. Yet,
according to Mao, the basis of the "people's democratic dictatorship" was
"mainly the alliance of the working class and the peasant class, because they
constitute eighty to ninety percent of the Chinese population." The impre-
cision of political terminology employed by Mao testified to the continuing
prestige of the democratic ideal—there was no escape from the Western
model—to the strength of Soviet influence, and to the fact that the Chinese
communists possessed no class-based mass following. At any rate, the term
"people's democratic dictatorship" was more honest than that of "people's
democracy" applied at that time to the Soviet satellites in Eastern Europe.

As for the nature of that democratic dictatorship, Mao was more explicit.
Dictatorial measures were to be used only against the reactionaries, the partisans of the Guomindang. Otherwise the function of the people's state was
to protect the people, helping them by "democratic methods on a nationwide
and all-round scale to educate and reform themselves, to free themselves
from the influence of reactionaries at home and abroad (this influence is at
present still very great and will exist for a long time and cannot be eliminated
quickly), to unlearn the bad habits and ideas acquired from the old society."
Significantly, the people were to be helped by the "people's state" to reedu-
cate themselves; liberation from the reactionary influence of the past was not
an entirely spontaneous process. Yet Mao made clear that "the methods we
use . . . are democratic; that is, methods of persuasion, not coercion."

Regarding economic development, priority was to be given to industrialization Soviet-style with Soviet assistance. Yet the Soviet model also applied
to the management of agriculture. In this respect, Mao argued, "the grave
problem is that of educating the peasants; the peasants' economy is scattered.
Judging by the experience of the Soviet Union, it requires a very long time
and careful work to attain the socialization of agriculture." Yet if the peasants
had to be educated Soviet-fashion by nonpeasant outsiders, what about peasant participation as mandated by the mass line? In any case, the new China
had to rely on Soviet assistance, or, as Mao put it (subtly indicating China's
independence): "you lean to one side." He chided the idea of assistance by
the British or American governments as childish "at the moment," leaving

open the possibility of leaning to the other side (in March 1945 he had told an American diplomat: "America is not only the most suitable country to assist the economic development of China; she is also the only country fully able to participate").[4]

While plotting a broad-gauged and occasionally vague course for the future, Mao left no doubt about the many problems lying ahead. "There is plenty of work before us, and what work has been done in the past is like the first step on a ten-thousand-mile long march." That march called for continuing education. "We must learn economic work from all who know the ropes (no matter who they are). We must acknowledge them as our teachers, and learn from them respectfully and earnestly. We must acknowledge our ignorance and not pretend to know what we do not know, nor put on bureaucratic airs. Stick to it, and eventually it will be mastered in a few months, one or two years, or three or five years." Mao was both an optimist and a realist. He acknowledged that mistakes had been made in the past; they would be made again in the future: "Mistakes are unavoidable for any party or person, but we ask that fewer mistakes be committed."[5] These were wise precautions, considering the uncharted contexts of Chinese state-building.

The course of action anticipated in the summer of 1949 was ostensibly set into the (seemingly) self-sufficient framework of Marxism-Leninism: "The Communist Party of the USSR is our best teacher from whom we must learn." Yet the repeated reminders of Soviet and Chinese backwardness, the call for rapid industrialization, and the statement of September about China's entrance into "the community of peace-loving and freedom-loving nations of the world" indicated Mao's recognition of the wider arena of global power. Despite his claim "to understand the law governing the existence and development of things," his Marxism-Leninism was but an improvised experimental ideology of mobilization for matching the Western model; his vision of communism was but an adjustment to the pressures of the continuing world revolution of Westernization, a tool for catching up to the sociocultural preconditions of Western superiority.

Yet in contrast to the Soviet experiment—or to the four others discussed above (let alone the fascist experiments after World War I)—Mao's state-building did not throb with a sense of mission extending beyond his country's borders. There was no need to boost his legitimacy by expansionist visions like world revolution, leadership of the Third World, Arab unity, or pan-Africanism. Never having left China yet, knowing the outside world only from his reading, and carrying deep within himself the age-old sense of China's superiority, he took it for granted that China's progress toward communism possessed universal significance. Foreign policy as an instrument for achieving national greatness loomed far behind domestic development. Governing one-quarter of the world's population was challenge enough.

III

Under these assumptions the march toward communism under the people's democratic dictatorship began in late 1949, at a slow pace geared to "three years of reconstruction and ten years of development." Reconstruction started from a clean slate; after the collapse of the Guomindang the Chinese communists were free to impose their own government, following a familiar pattern. The new state was formed by a constituent assembly called the Chinese People's Political Consultative Conference; it was attended by noncommunist representatives, who for a time held high positions in the government. For evidence of the new regime's democratic nature, the existence of noncommunist democratic parties could be cited, or the marriage law of 1950, which provided equality between men and women. Further proof lay in the guarantee of basic democratic freedoms in the Constitution of the People's Republic issued in 1954.

Behind this fashionable facade Chairman Mao and the Chinese communists began to build their new China. Communist rule for the next four years essentially meant military rule (reaffirmed by the Chinese participation in the Korean War, which added to patriotic self-confidence). Through land reform the army and the party destroyed the power of the landed gentry, the stronghold of tradition in the countryside; for the first time in history the peasants in the diversity of rural life were free to run their own affairs. In the seaboard cities foreigners were expelled; factories and banks tied to the Guomindang and to foreign interests were nationalized, putting the bulk of China's industry under state control. Lesser enterprises were encouraged to maximize production capitalist fashion, with some regulation of wages and profits. Economic reconstruction was a joint effort, under party guidance, of the national and the petty bourgeoisie with the help of peasants in a relatively open market economy. By all accounts it succeeded within a short time, restoring a measure of normalcy without, however, remedying the country's backwardness or diminishing its poverty (especially noticeable in the cities).

Recovery was aided by the suppression of the drug trade and of crime, both rampant in the years before 1949; its puritan creed committed the party to the moral cleansing of Chinese society. It also introduced a simplified utilitarian dress code, making all Chinese look alike in Western eyes. With even greater determination it enforced political conformism, unleashing terror against all partisans of the old regime; two million "traitors" were executed in the first three years. Many more people were subjected to "thought reform" through "struggle sessions" in the manner evolved during the Yanan years. Even more effective in the long run were the urban residence committees introduced in 1952; each one of them put several hundred households under close supervision by the ever present Communist Party. And in order to keep people from falling back into their old ways once recovery was under way, the party, starting in late 1951, launched "rectification" campaigns enforcing its ideals. One was aimed at politically unreliable intellectuals, another at bureaucratic corruption and inefficiency, and a third at "counter-revolutionaries" generally—political deviance always loomed in the wings.

With the help of these measures the communists, consummating the revolutionary turmoil of the past hundred years, accomplished more changes in Chinese life than had been wrought in the previous two thousand years. The changes hardly originated in democratic consensus, but they could claim a large measure of popular support. Mao's prestige stood high. Stability had been restored; the new China was on the move toward the next stage, toward socialism and industrialization.

Emphasis in these years rested on the development of heavy industry after the Soviet model. In the winter of 1949–50 Mao spent several weeks in the Soviet Union (his only time abroad), wresting a promise of assistance from a Stalin both eager to rebuild his own war-ravaged country and suspicious of an ascendant China (the negotiations, as Mao observed, were like "tearing a piece of meat from the mouth of a tiger").[6] Further agreements were signed in the following years. By 1952 a Chinese State Planning Committee had begun to work; the first Five-Year Plan started in 1953 with the help of several thousand Soviet experts (to be assisted and replaced in the future by thousands of Soviet-trained Chinese). Under the plan private enterprise was gradually eliminated; the transition to socialism was getting under way.

IV

Soon urgent questions arose regarding party policies leading toward socialism. How, for instance, could the government, the party, and the masses be made "to unlearn the bad habits and ideals acquired from the old society"? A warning example was set in 1954 by the party boss of Manchuria, who was accused of running an "independent kingdom" in the manner of the former warlords; a similar case was reported from the Shanghai area.

Backsliding was also apparent in the countryside; "rich" peasants and former landlords resumed "capitalist" practices. And everywhere bureaucratism raised its ugly head, promoted by rapid industrialization and political regimentation and made more rigid by ingrained Chinese tradition. New rectification campaigns launched against callous bureaucrats and uncooperative intellectuals had little effect. All along, the rapid pace of enforced innovation set off uncertainty and tension, undermining goodwill for constructive action. External events added further complications. Khrushchev's attack on Stalin's personality cult in 1956 was bound to diminish Mao's ascendance, while in the following year *sputnik* lifted communist expectations (including Mao's) to extravagant heights. Everywhere problems piled up, adding to his impatience. How could he revive the vital revolutionary momentum, the essence of his leadership and the sole guarantee of rapid progress toward his goals?

Mao's impatience first surfaced in the summer of 1955 over the slow progress of rural collectivization. In July he attacked some comrades for "tottering along like a woman with bound feet" at a time when, as he saw it, "a new upsurge in the socialist mass movement is in sight throughout the Chinese countryside."[7] Reversing earlier policy, he now argued that "the formation of cooperatives must precede the use of big machinery." Claiming that Chinese peasants were "even better than English or American workers,"[8] he called on them "to organize, in accordance with the principles of voluntariness and mutual benefit, agricultural producers' mutual aid teams in preparation for small and eventually large cooperatives." In this manner he hoped "to raise steadily the socialist consciousness of the peasants . . . step by step. . . . "[9]

One may doubt the spontaneity of the "new upsurge in the socialist mass movement"—implementation, Mao indicated, would follow Soviet practice (though on the basis of voluntariness and mutual benefit)—but he proved his magic in unleashing what he called "a raging tidal wave" for rapid collectivization, backed by "immense enthusiasm for socialism." In early 1956 progress had been so fast that the completion of collectivization could be scheduled for 1958. Thanks to the fact that the Chinese Communist Party had risen from peasants, collectivization was carried out without drastic violence (though it drove many peasants to the cities).

Meanwhile, major controversies had arisen regarding industrialization, the role of the intellectuals, and the management of the emergent socialist society generally. Planned industrialization favored specialization, hierarchy, and the rise of a technical intelligentsia, which in the Chinese setting led to the reemergence of the divisive old ways. What was needed, Mao realized, were managers endowed with an un-Chinese adaptability, generalists with an infectious, outgoing revolutionary commitment, technically competent more in the manner of a "jack-of-all-trades" than of a specialist; a cadre, in short, both "red and expert." Socialist regimentation Soviet-style, whether in industry or life generally, did not promote such types, certainly not in the granular Chinese society. Traditional attitudes caused resentments in the ever expanding network of organizations, blocking the expansive fluency of socialization so crucial in Mao's political philosophy. How was he to deal with these tensions (or "contradictions," as he called them)?

His answer, set forth in the aftermath of Khrushchev's denunciation of Stalin and in defiance of leading comrades, was to bring the contradictions into the open and given them a positive meaning. "Some naive ideas seem to suggest that contradictions no longer exist in a socialist society. To deny the existence of contradictions is to deny dialectics."[10] Subsequently, in a major statement, he spelled out his assessment of the problems confronting the party.[11] Against prevailing consensus in favor of heavy industry and dependence on the Soviet Union, he argued that developing heavy industry called for developing light industry and agriculture as well. As for the even more crucial subject of the relations between the party and the population, he attacked "the inflated establishment" of bureaucracy, suggesting that two-thirds of the existing organs of party and government "be scrapped." He also advocated an un-Stalinist leniency in dealing with counterrevolutionaries: "no executions and few arrests." Tolerance should be shown also to those who had committed mistakes. "Within the party we have controversy, criticism, struggle. These are necessary" because "for 10,000 years to come there will always be two sides." According to Mao, the dialectics never ceased to operate; nobody had a monopoly on truth.

For that reason, Mao concluded, "China still had much to learn from other countries." "Our theory," he argued, "is made up of the universal truth of Marxism-Leninism combined with the concrete reality of China. We must be able to think independently." And thinking independently meant "that we should study all the good points of foreign countries, their politics, their economics, their science and technology, and their literature and art," learning foreign languages, "if possible several"—obviously, the Soviet Union was not the only teacher for communist China. In any case, while China's self-confidence needed to be raised, there was also need for humility. "Although our revolution is one step ahead of those of a number of colonial countries, we should resist the temptation to be proud of that [fact]."

Then he launched one of his most famous contentions: "We are very poor and have not much knowledge. We are first 'poor' and second 'blank'. By 'poor' I mean that we have not much industry and our agriculture is not so very advanced either. By 'blank' I mean that we are like a sheet of blank paper, since our cultural and scientific level is not high. Those who are poor want change; they want to have a revolution, want to burst their bonds, and seek to become strong. A blank sheet is good for writing on." In the next sentence, however, he pulled back a bit, admitting: "I am, of course, speaking in general terms. The laboring people of our country are rich in wisdom, and we also have a pretty good bunch of scientists."[12] But he left no doubt that "we must still learn from others. We must study for 10,000 years. What is wrong with that?" What other revolutionary state-builder in the 20th century has shown such humility when facing the problems of reculturation?

And what other political leader, at his risk, encouraged such open criticism as Mao did under the slogan, rebelliously set forth in May 1956, "let a hundred flowers blossom; let a hundred schools of thought contend"? He was determined to submit the flaws of socialist reconstruction to open discussion so that everybody, and the party foremost, could learn from their mistakes. Understandably, his plea was not welcomed by the party leadership wedded to the Soviet model and fearing a new cult of personality (after Khrushchev's speech denouncing Stalin). In the fall of 1956, at the first party congress held since 1945, Mao was relegated to the sidelines; the party, according to its new constitution, was no longer guided by "the thought of Mao Zedong."

Undaunted by this setback, Mao pressed his campaign to air the mounting contradictions in the new order, and especially those between the government and the masses. In early 1957, while insisting that the people "have to keep themselves within the bounds of socialist discipline," he stressed the fact that "contradictions in socialist society are not antagonistic and can be resolved . . . by the socialist system itself."[13] He therefore warned of unduly coercive measures against the critics. But his major thrust lay in putting the responsibility for the widespread criticism on "the bureaucratism of those in positions of leadership." It almost seemed that he was pitting the people against the party.

As a result of his agitation—he still enjoyed great prestige in the party as a charismatic leader—a mighty outburst of criticism developed through the spring of 1957, particularly among students and intellectuals, reminiscent of the May Fourth Movement in 1919. Citing the democratic freedoms promised in the 1954 constitution, the critics vigorously attacked the abuses among the new privileged class of bureaucrats, often contrasting Marxist ideals with the realities of the new socialism. Had not Mao himself deplored the unwillingness of many party bosses "to share the joys and hardships of the masses" and their search for "personal position and gain"?[14] Yet, given the anarchic nature of Chinese society and its penchant for intolerance, where would this outburst of protest end? How far could "socialist discipline" be stretched in the name of democracy or of the dialectic?

In early June 1957 Mao and the party, alarmed by the hostile torrent, sharply reversed themselves, denouncing the critics as "right-wingers" and their views as "poisonous weeds." The leaders of the democratic parties (then still in existence) were subjected to public humiliation, while the radical Marxists suffered worse. Yet repression was not to Mao's taste. To him the unexpected volume of criticism rather indicated the need for a massive new initiative recapturing the revolutionary momentum and rallying both the critics and the criticized. After *sputnik,* the time seemed ripe for a big communist advance to match the dynamic changes of the times; decolonization was under way, not to mention the worldwide economic upswing sparked by the United States. Thus from the late fall of 1957 into the next spring the sixty-four-year-old Mao prepared a spectacular demonstration of revolutionary impatience.

He expressed the spirit that moved him in a speech late January 1958. "When I review the past seven or eight years, I see that this nation of ours has a great future. Especially in the past year you can see how the national spirit of our 600 million people has been raised to a level surpassing that of the past eight years. After the great airing of views, blooming and debating, our problems and tasks have been clarified: we shall catch up with Britain in about fifteen years."[15] In May, racing ahead of Khrushchev's extravagant claims for the Soviet economy, he anticipated overtaking American steel output by 1966.

There was obvious need for a fresh start. The first Five-Year Plan had failed to reduce unemployment, especially burdensome in the cities; the second plan, then on the drawing boards (and soon discarded), promised no improvement. The basic problem was how to put China's biggest asset, its increasing number of human hands and brains, to productive use (Mao rejected population control as an undesirable restraint). He therefore began to oppose industrialization after the Soviet model. China did not have the resources for capital-intensive industries; it had to find its own prescription for economic development, counting on the superior size of its labor force.

V

Mao's revolutionary offensive was dramatically launched as the Great Leap Forward (an image copied from Sun Yatsen); it declared socialism accomplished and heralded the advance into communism. Its key concept was "simultaneous development" for all branches of the economy, heavy industry, light industry, and agriculture in dynamic interaction. The new drive was to stimulate overall productivity, especially in the countryside. It aimed to decentralize the economy, transfer initiative to local agencies, and promote rural industry for the production of consumer goods for local markets. It set out no blueprint, stressing diversity and local action according to local conditions. But it also decreed specific projects. The most publicized was the establishment of backyard iron and steel furnaces, designed to bring industry into rural areas; much was to be done also for irrigation and water conservation. In addition, Mao commanded: "now we are going to exterminate the four pests"[16]—rats, mosquitoes, flies, and sparrows (the latter soon taken off the list, because they were found to be useful after all, replaced by bedbugs). China's masses were set to hustle, bursting with constructive activity.

Increased productivity required that science and technology be introduced into the villages with the help of the new industries. New educational facilities were designed for combining work and study, while urban workers and intellectuals were dispatched to the villages to help raise the cultural level of peasants. Rather than create a privileged caste of technocrats even more exclusive in China than in the Soviet Union, Mao, an admirer of modern science and technology, looked forward to a nation of "socialist-conscious, cultured laborers." Countrywide material progress, he realized, required above all a mechanically minded popular culture (as was the case in England and the United States). Progress toward communism, like his own charisma, depended on radio and loudspeakers in every village, as on the most advanced means of communication generally, soon including television.

Such advance, however, was not possible without a new groundswell of revolutionary enthusiasm generating its own momentum toward more knowledgeable and expansive social cooperation. In these months Mao again stepped forward as the charismatic mover of the masses. "Now our enthusiasm has been aroused," he said in his January 1958 speech. "Ours is an ardent nation, now swept by a burning tide. . . . Our nation is like an atom. . . . When this atom's nucleus is smashed, the thermal energy released will have really tremendous power. We shall be able to do things which we could not do before." He praised impatience for quick results in the name of permanent revolution. "In making revolution one must strike while the iron is hot—one revolution must follow another, the revolution must continually advance. . . . Our strength must be aroused and not dissipated. If we have shortcomings or make mistakes, they can be put right by the method of great airing of views. . . . We must crave greatness and success. . . . We must indeed keep up our fighting spirit."[17]

In March 1958, however, he added a note of caution: "Right now there is a gust of wind, amounting to the force 10 typhoon. We must not impede this publicly, but within our own ranks we must speak clearly, and damp down the atmosphere a little. . . . Some of the targets are high, and no measures have been taken to implement them; that is not good. . . . We must have concrete measures, we must deal in reality. We must deal in abstractions too—revolutionary romanticism is a good thing—but it is not good if there are no [practical] measures."[18] Yet he was not afraid of disagreements. "The appearance of disorder contains within it some favorable elements; we should not fear disorder." In this spirit he continued his offensive: "We must improve our style of work, speak with sincerity, take a firm hold of ourselves and possess the spirit to sweep all before us and climb to the highest peak. . . . Unless you have the conquering spirit, it is very dangerous to study Marxism-Leninism."[19] Where there was a will, he was convinced, there was a way, regardless of material or cultural obstacles.

In these months he especially appealed to the zeal of the young: "Ever since ancient times the people who founded new schools of thought were all young people without too much learning. They had the ability to recognize new things at a glance and, having grasped them, they opened fire on the old fogeys"[20]—he cited Marx, Luther, Darwin, Ben Franklin, and Gorki (not to mention Chinese notables) as models, warning that too much concern for established knowledge inhibits creativity. Like the ignorance of youth, backwardness was an asset, a stimulus for a heroic advance. Throughout he now attacked the blind faith in the superiority of the Soviet model so widespread among leading Chinese communists. The Great Leap Forward (while resulting in a higher-than-usual level of imports) brought a new assertion of Chinese independence: "We must overthrow the slave mentality and bury dogmatism, . . . raise our ideological level, absorb the lessons of experience. . . . " "In the past," Mao asserted, "others looked down on us mainly because we produced too little grain, steel, and machinery. Now let us do something for them to see!" And as before, he sensed the impatience of the masses. "We ought to be leading the masses, yet the masses nowadays are more advanced than us."[21]

Indeed, in June and July, prompted by good summer crops, there miraculously emerged the rural People's Commune, the much publicized centerpiece of the Great Leap Forward. Once more Mao had succeeded in arousing a popular wave of spontaneous radicalism, especially among rural communists and poor peasants. An experimental commune in Honan Province furnished the model. In praising it, Mao exulted in the prospects: "It will probably take less time than previously estimated for our industry and agriculture to catch up with that of the capitalist powers. In addition to the leadership of the Party, a decisive factor is our population of 600 million. More people means a great ferment of ideas, more enthusiasm and more energy."[22] Soon the drive for communalization gained a seemingly irresistible momentum, quickly involving virtually all of China's half-billion peasants.

Hardly aware of the practical implications, Mao encouraged the stampede, over the doubts of top party leaders. Viewed in the light of revolutionary romanticism, the new communes aimed at no less than the introduction of a communist form of social life. They were designated as self-contained, self-administered units in charge of all aspects of life: political, administrative, economic, social, technical, educational, cultural, and military (men and women were trained to serve in the militia in case of war). Their members were to practice the wholesome egalitarian versatility once advocated by Marx and now endorsed by Chairman Mao (looming larger in the public eye than ever).

Praised as an antidote to bureaucratism, the communes operated under strict rules, combining many collective farms and setting up production brigades numbering thousands of workers and smaller work teams for lesser projects. The men marched to work like soldiers, sometimes to martial music; the work was organized as in a factory. As the men headed for distant projects, the women took over the fieldwork nearby. Exhausted from their new routines, men and women were fed in communal mess halls, the children taken care of in communal nurseries. During communalization family property was often collectivized; the family household ceased to be the center of life. In this manner peasant perspectives, hitherto narrowly restricted, were forcibly widened out by drastic innovation. The commune introduced small-scale industries for the production of agricultural implements and fertilizer; local coal mines provided fuel; handicrafts produced consumer items—all for the long-run benefit of the countryside, but at an exorbitant price for the moment.

In the fall of 1958 fatigue, low morale, and food shortages took their toll; the original spontaneity had faded, regimentation had quickly become the rule. The backyard iron and steel furnaces had proved a disastrous failure, many irrigation schemes faulty or ill conceived; ignorant zealotry and haste often spoiled sound projects. Outsiders had been impressed by the busy diligence of China's "blue ants," its people clad in the mandatory uniform of simple blue cotton garments, chasing down the four pests or marching to work. But underneath the bustle, the organic unity of the antheap was missing; Mao's emphasis on local initiative indeed had encouraged the traditional anarchy. Alarmed by the disorder, the party quickly curtailed the scope of the communes, restored the village-based collective farms, returned private property, and reaffirmed the authority of the central party organs. Mao himself was forced on the defensive, to the point even of abdicating as chairman of the People's Republic.

In the summer and fall of 1959 he fought a rearguard action,[23] admitting that those who said "we are in a mess" were right. "The more they say we are in a mess the better, and the more we should listen"; he was ready to "toughen [his] scalp and bear it." He frankly admitted that in 1958 he had been concentrating on revolution. As a result, "many things have happened which we could not possibly predict beforehand." The backyard iron and steel furnaces particularly were a "great catastrophe" for which he was ready to take the blame. He readily confessed: "I am a complete outsider when it comes to economic construction, and I understand nothing about industrial planning." Or, even more forcibly: "There are many things I haven't studied. I am a person with many shortcomings. I am by no means perfect. Very often, there are times when I don't like myself. . . . " Yet had not Marx or even Confucius made mistakes?

The essential point, Mao urged, was to keep learning. The concept of the commune was not necessarily wrong; it merely needed more time. "I said that for the transition to be completed from collective ownership to communist ownership by the whole people, two five-year plans was too short a period. Maybe it will take twenty five-year plans" (one hundred years). At one point he defiantly deplored the adverse publicity given by the Beijing authorities to the mistakes of the Great Leap Forward. If that negative attitude persisted, he threatened "to go to the countryside to lead the peasants to overthrow the government"—if necessary with the help of a new Red Army (though he was sure the People's Liberation Army would still follow him). At another point he insisted on the Leninist principle of discipline. Yet on the whole he showed repentance. "The chaos caused was on a grand scale and I take responsibility. Comrades, you must all analyze your own responsibility. If you have to shit, shit! If you have to fart, fart! You will feel much better for it."[24]

Unfortunately, he himself did not. By late 1959 he realized the necessity of dismantling the Great Leap Forward; the economic crisis was still mounting; it had spread from agriculture into industry and transportation, causing chaos in the entire economy. As if to show its displeasure, Heaven itself visited floods and droughts on the country in 1960, making the winter of 1960–61 the worst in memory; people starved. To compound the damage, Khrushchev, after heaping scorn on the naiveté of the Great Leap Forward and antagonized by Mao's repudiation of the Soviet model, abruptly withdrew his experts from the country, deepening the economic crisis. Mao himself was discredited and, although formally still chairman of the Chinese Communist Party, withdrew in anger and frustration from active participation in the government. He had no use for the "bourgeois" elements in the party, for Liu Shaoqi (who had succeeded him as chairman) and Deng Xiaoping, Liu's deputy, who now tried to restore order and production by government regulation and expert knowledge; he even suspected them of having sabotaged the Great Leap Forward.

VI

Thus ended Mao's revolutionary leap into communism. However justified in his Olympian perspectives, it foundered on the complexities of Chinese life and the stubbornness of tradition. No matter how eager for improvement or how swayed by the nebulous promises of an alien ideology, the masses were unable to rise above their ignorance or their country's poverty. The gap between Mao's ambitious goals and reality was too great to be bridged by "revolutionary romanticism." Reculturation double-quick was utopian.

Even Mao himself, now sobered by adversity, admitted as much, repeating in 1962 more emphatically what he had said in 1959: "In our work of socialist construction, we are still to a very large extent acting blindly. For us the socialist economy is still in many respects a realm of necessity not yet understood. Take me as an example: there are many problems in the work of economic construction which I still don't understand. I haven't got much understanding of industry and commerce. I understand a bit about agriculture, but this is only relatively speaking—I still don't understand much."[25] While still holding out hope that "the economic development of our country may be much faster than that of capitalist countries," he admitted that "it will be impossible to develop our productive power so rapidly as to catch up with, and overtake, the most advanced capitalist countries in less than one hundred years." Feeling in a prophetic mood, he added: "the next fifty or hundred years from now will be an epic period of fundamental change in the social system of the world, an earth-shaking period, with which no past era can be compared."[26] That was part of his wisdom when thwarted: to view his troubles in the largest perspectives.

Meanwhile, the party under Liu Shaoqi restored order and productivity. The authority of the communes was reduced, allowing a modest return to private farming and a free local market; harvests improved, enlarging, with the help of imports, the food supply. Industry was streamlined, output increased, and labor efficiency advanced with the help of differential wage incentives; much redundant industrial manpower was sent back into the countryside. The economy quickly recovered, thanks to restored discipline and guidance by technically competent administrators. While still concerned over the backwardness of agriculture, the party leaders tried to catch up with Western science and technology, improving education, especially higher education, and providing better medical services. In 1964 China tested its first atomic bomb, no mean feat scientifically and technologically.

These improvements, benefiting above all the urban centers and, within them, the elite of technocrats, signified a return to the pre-1958 pattern of planned economic development and, at the same time, an involuntary lapse into tradition. Under pressure for quick results, the experts tended to retreat into their mandarin aloofness, the administrators into bureaucratic arrogance, the peasants into "capitalist" practices and corruption; the gap between city and country widened. Amidst a mood of political apathy and public lassitude, people yearned for "bourgeois" comforts, especially for watches, radios, bicycles, and washing machines—a trend repugnant to the aging Mao watching from the sidelines.

Treated like a "dead ancestor" (so he complained), he was not inactive. Assuming responsibility for the failure of the Great Leap Forward, he still pleaded for his cause: social transformation—or reculturation—from below, with the help of the masses. Like Liu, he followed the Leninist principle of "democratic centralism," but with a decided accent on "democratic," aware, however, that under present conditions the consciousness of the masses still needed to be raised. For the double purpose of educating the masses and fostering their active participation in the unfinished revolution, he launched in 1962 a "Social Education Movement," justifying his initiative by arguing that the class struggle had by no means been ended in China; indeed there could be counterrevolution, a return to feudalism and capitalism. Obviously, he still yearned for a spontaneous upsurge from below as an antidote to the rigidities implicit in the party's current policies; large-scale cooperation and technical progress required social fluidity and individual flexibility. For the same reason, he continued to encourage dissent and criticism, stressing the beneficial dynamics of dialectics, the inevitability and ceaselessness of contradictions.

In these polemics he was aided by the People's Liberation Army, which, under its commander, Lin Biao, still cherished the egalitarian comradeship of the Yanan days. The army conducted training courses for teachers and party members, urging civilians to learn from the revolutionary spirit of the army. More significantly, the army now propagated a new cult of Chairman Mao as a political inspiration, a lodestar above the party and the government, a unifying force and a symbol of revolutionary certitude. In 1964 the army published a little book bound in red, *Quotations from Chairman Mao*, which soon was in everybody's hands.[27] The foreword hailed "Comrade Mao" as "the greatest Marxist-Leninist of our era"; he had raised Marxism-Leninism to "a higher and completely new stage." Readers were admonished to "study Chairman Mao's writings, follow his teachings, act according to his instructions and be his good fighters," driven by a spiritual strength as potent as an "atom bomb of infinite power." The cult of Chairman Mao rose like a red sun all over China; his picture loomed huge over city squares and public buildings; his effigy in bronze busts, photographs, and books abounded in stores and private quarters, giving him a semi-divine omnipresence. The "cult of personality" waxed more intense than in Stalin's Soviet Union and even more transparently as a political device.

In part the new cult originated in a concern for training Mao's "revolutionary successors." Aging, suffering from Parkinson's disease, and repelled by the "revisionist" or outright "capitalist" tendencies pervading the party and the country, Mao himself worried: who would carry on his revolutionary zeal after his death? In 1964 he again turned to China's young people, this time warning them of lapsing into Confucian scholasticism; he specifically cautioned against excessive book-learning. "At present there is too much studying going on, and this is exceedingly harmful. There are too many subjects at present, and the burden is too heavy, it puts middle-school and university students in a constant state of tension. . . . The present method of education ruins talent and ruins youth. I do not approve of reading so many books." He was particularly concerned with intellectuals in the humanities, the most prone to lapse into mandarin ways. They should know life firsthand: "We must drive actors, poets, dramatists and writers out of the cities, and pack them all off to the countryside. They should all go down in batches to the villages and to the factories."[28] How else could they be integrated into the mainstream of Chinese life? As for the practical benefits, it was not literary folk, however, but medical students and paramedics, the "barefoot doctors," who in the long run contributed most to village welfare.

Forever concerned with creating the social cement necessary to transform China's multitudes (now approaching eight hundred million people) into a true national community, Mao urged closer links between city and country. He tried his best to counteract the city-country polarization so common in non-Western societies under the Western impact, whether as a result of development policies or deliberate national mobilization. His command "to

go down to the people" as a method of thought reform or political reedu-
cation was designed to achieve the bonding effects of the relatively free social
mobility and openness found in the "capitalist" West. Under the circumstances, Mao's pleas made sense. Book-learning (especially in its Chinese
form) isolates or even desocializes individuals. And especially in times of rapid
change there is need for travelling light, without burdening the mind with the
accumulated (and often irrelevant) knowledge of the past. Mao wanted open
minds filled with a constructive vision of the future and ready for social and
political action. These were admirable prescriptions for a country long ruled
by an unusually rigid, book-oriented, and tradition-ridden elite and now compelled to compete with the outside world; in this respect, too, Mao was an
agent of the revolutionary West.

Yet, himself the product of backwardness, he was still caught in ignorance.
Escape from backwardness is impossible even by vigorously stirring and mixing up a backward society; such agitation might indeed make matters worse.
The country still had to learn foreign skills. Given China's isolation, how else
could one catch up to the West's superior ways except by books? How had
Mao learned about Marxism-Leninism, about the outside world, about the
Western thinkers who, as he repeatedly asserted, made their most significant
discoveries at a youthful age? And, yet, could he be expected to surrender his
revolutionary vision, his trust in the unceasing dialectics of learning from
contradictions, and his conviction that the best learning lies in the doing? As
the Mao cult grew more potent and the party leaders remained indifferent or
even hostile, he became agitated with the desperate impatience caused by
advancing old age. In this mood he planned his biggest revolutionary offensive ever.

VII

In late 1964 he talked of yet another Great Leap Forward. "We cannot follow
the old paths of technical development of every other country in the world,
and crawl step by step behind the others. We must smash conventions, do our
utmost to adopt advanced techniques, and within not too long a period of
history, build our country up into a powerful modern socialist state. . . . This
is an inevitable trend which cannot be stopped by any reactionary force."[29]
Early next year he urged the members of the Politburo "to criticize bourgeois
reactionary thinking." In November he casually called for a "full-scale revolution to establish a working class culture" while attacking a popular play,
written after the failure of the Great Leap Forward, which had questioned
his leadership. To Mao that play reflected the "revisionist" and "bourgeois"
attitudes that had permeated the Beijing cultural and political establishment.

From literary criticism Mao quickly advanced to a challenge of the party
leadership, calling on the People's Liberation Army for assistance. While the
party leaders began to strike back, Mao, in February of 1966, set up the Cultural Revolution Directorate to press for the full-scale revolution soon called
(in English) the "Great Proletarian Cultural Revolution." With the help of

his wife, Jiang Qing (his fourth), an ambitious if mediocre arbiter of the arts pursuing her private feuds, Mao directed his arrows against the arts faculties in the universities, especially in Beijing with its refined tastes. In May radical students from nonprivileged backgrounds at Beijing University responded. One of them put up a big-character poster attacking the university authorities, provoking conservative students encouraged by the party to oppose the radicals. Demonstrations led to counterdemonstrations and violence; hostility escalated. "You say we are going too far?" one radical student shouted. "To put it bluntly, your 'avoid going too far' is reformism; it is 'peaceful transition.' You are daydreaming! We are going to strike you down to the dust and keep you there!"[30] Out of these clashes rose the Red Guards, militant students committed to Mao's cultural revolution and soon enthusiastically endorsed by him. They quickly sprang up also at other centers of learning throughout the country.

Mao meanwhile moved himself into the foreground of public attention. At the age of seventy-two he took a much-publicized swim in the Yangtze River, testifying to his vigor; more significant, he talked bluntly to the party's Central Committee: "All conventions must be smashed. We should trust the masses and be their pupils before we can be their teachers. The current great cultural revolution is an earth-shaking event. Can we or do we dare undergo the test of socialism?"[31] Simultaneously he put fire under the radical students: "Youth is the great army of the Great Cultural Revolution. It must be mobilized to the full. . . . What are you afraid of? . . . You should replace the word 'fear' by the word 'dare.' . . . You must put politics in command, go among the masses and be at one with them, and carry on the Great Proletarian Cultural Revolution even better."[32] In early August he fanned the flames, demanding in a big-character poster designed by himself: "BOMBARD THE HEADQUARTERS"[33]—a blunt invitation for a violent onslaught on the enemies of the Great Proletarian Cultural Revolution. He himself took the drastic step of dismissing, contrary to party statutes, the head of state, Liu Shaoqi, and installing Lin Biao as his successor.

On August 8, the party's Central Committee signalled Mao's triumph by publishing its *Decision Concerning the Great Proletarian Cultural Revolution.*[34] It set out the program for "a great revolution that touches people to their very souls and constitutes a new stage in the development of the socialist revolution in our country. . . . " Continuing the inflammatory rhetoric of the big-character posters, it stated: "Large numbers of revolutionary young people, previously unknown, have become courageous and daring pathbreakers . . . they launch resolute attacks on the open and hidden representatives of the bourgeoisie. . . . " The youngsters were admonished to "trust the masses, rely on them and respect their initiative. . . . Don't be afraid of disturbances . . . revolution cannot be so very refined, so gentle, so temperate, kind, courteous, restrained and magnanimous. Let the masses educate themselves in this great revolutionary movement. . . . "

In the same document, however, the party advised restraint in case of disagreements among the people: "When there is a debate, it should be conducted by reasoning, not by coercion or force." Moreover, the Central Committee warned: "any idea of counterposing the Great Proletarian Cultural Revolution to the development of production is incorrect." The revolution should not interfere with the country's economic development—and certainly not with the work of "those scientists and scientific and technical personnel who have made contributions"; progress in atomic weapons and space rockets was not to be endangered. In any case, in order to forestall unwarranted expectations, the embattled revolutionaries were told that "the struggle of the proletariat against the old ideas, culture, customs and habits left over by all the exploiting classes over thousands of years will necessarily take a very, very long time." In conclusion, the "masses of workers, peasants and soldiers, the cadres and intellectuals" were admonished to take "Mao Zedong Thought" as their guide to action in whatever they did.

Ten days later, the Great Proletarian Cultural Revolution was consecrated in Beijing on Tian'anmen Square in front of the Gate of Heavenly Peace. Over a million young Chinese sporting the red armbands of the Red Guards waited at dawn for Mao's appearance, waving their copies of *Quotations from Chairman Mao* and chanting passages from it. As the sun rose in the east, Mao, clad in a soldier's fatigues, stepped out onto the gallery of the famous gate, greeted by a thunderous welcome rising from the vast crowd. Before him stood a youthful force setting out for revolution to the tune of "The East Is Red," their battle hymn. No other country could mount the impressive displays of massed humanity organized during the cultural revolution; every human atom seemed fused with all the other human atoms in an exalted common dedication to Chairman Mao.

Indeed, his *Quotations,* though occasionally contradictory, contained much good advice. Consider the following items:

> Place problems on the table. This should be done not only by the "squad leader" but by the committee members too. Do not talk behind people's backs. Whenever problems arise, call a meeting, place the problems on the table for discussion, take some decisions and the problems will be solved. . . . Nothing is more important than mutual understanding, support and friendship between the secretary and the committee members, between the Central Committee and its regional bureaus and between the regional bureaus and the area Party committees.
>
> "Have a head for figures." That is to say, we must attend to the quantitative aspect of a situation or problem and make a basic quantitative analysis. . . . To this day many of our comrades still do not understand that they must attend to the quantitative aspect of things—the basic statistics, the main percentages and the quantitative limits that determine the qualities of things. . . .
>
> Natural science is one of man's weapons in his fight for freedom. For the purpose of attaining freedom in society, man must use social science to understand and change society and carry out social revolution. For the purpose of attaining freedom in the world of nature, man must use natural science to understand, conquer and change nature and thus attain freedom from nature.

Things develop ceaselessly. It is only forty-five years since the Revolution of 1911, but the face of China has completely changed. In another forty-five years, that is, in the year 2001, or the beginning of the 21st century, China will have undergone an even greater change. She will have become a powerful socialist industrial country. And that is as it should be. . . .

The socialist countries are states of an entirely new type in which the exploiting classes have been overthrown and the working people are in power. The principle of integrating internationalism with patriotism is practiced in the relations between these countries. We are closely bound by common interests and common ideals.[35]

Regarding this last quotation, one wonders how the Red Guards—if they used their heads critically—reconciled it with the fact that in the spring of 1966 the Chinese Communist Party had broken all relations with the "fraternal" Communist Party of the Soviet Union.

The integration of patriotism with internationalism under the Great Proletarian Cultural Revolution as advocated by Mao, posed indeed a problem. Mao rejected any intervention in the Vietnam War as a distraction from urgent domestic tasks. Yet during the years of the cultural revolution he let Lin Biao take an ideologically more radical line. The "revolutionary countryside" around the world, Lin Biao argued from Chinese communist experience, would engulf and eventually overpower the "capitalist" cities of the West; he stressed solidarity with national liberation movements around the world as well as the dangers of imperialist aggression (the imperialists now included the Soviet Union as well). But, in fact, during the years of the cultural revolution Chinese foreign relations were virtually dormant.

VIII

As the Great Proletarian Cultural Revolution proceeded, all efforts were concentrated on domestic affairs. Although aimed especially at the "revisionist" party leadership and the cultural establishment, the mandarin guardians of academic learning, the cultural revolution was intended to promote *all* of Mao's revolutionary aims, especially the cultural advance of the peasants with the help of industry brought into the countryside and the promotion of greater equality between city folk and peasants.

Within that framework of emotional appeals, heady and vague ideological visions, national ambition, common sense, and ignorance of social dynamics in the Chinese setting, the Great Proletarian Cultural Revolution took its tumultuous course. Already in the late summer and early fall of 1966 the violence caused by the Red Guards grew to disturbing proportions. What else would one expect after hearing the official advice that revolutions cannot be gentle, kind, or magnanimous or, as Mao put it, that "destruction precedes construction"? By late October even he was worried: "I had no idea that one big-character poster, the Red Guards and big exchange of revolutionary experience would have stirred up such a big affair."[36] The next day he even admitted: "The Great Cultural Revolution wrought havoc. . . . The time was so short and the events so violent. I myself had not foreseen that . . . the whole country would be thrown into turmoil. . . . We were not mentally pre- pared for new problems."[37] It would take five years, he argued, to acquire the necessary experience to solve these problems. In the next two years Mao waffled, torn between inciting revolutionary radicalism (reviving even the Marxist vision of government by communes) and calling for order and unity.

And thus, in the name of Mao's lofty ideals, the Great Proletarian Cultural Revolution proceeded for almost three years, an elemental rampage of destruction propelled by its own unsuspected momentum. The protest against the "capitalist-roaders" in the party, government, and society was not only directed against abuse of power, official arrogance, or mandarin exclu- siveness, but also against the individual and collective discipline required for creating an effective modern polity. The changes brought about by the com- munist regime had built up their own hidden destructive dynamics.

Mixed with the traditional (and often appallingly cruel) fractiousness of Chinese society, with the heady Maoist visions of national power and glory, and with the intoxicating ideological incitements to drastic action, the ten- sions released an anarchic fury laced with refined brutality both psychological and physical. Mao's teachings about discipline had no effect (if they were understood at all), as wave after wave of Red Guard terror and counterterror rolled over the country. In 1967 the People's Liberation Army was called into action as a peacemaker, but, divided in its own ranks, it often added to the discord. Rural China, on the whole, was spared the clash of hostile factions. The cities, the industrial centers, and the institutions of higher learning suf- fered most.

The protest was directed against any kind of authority, whether of the party, Confucian teachings, parental authority, the weight of tradition, or the prestige of foreign manners, dress, or even musical instruments. One Red Guard suggested that the order of traffic lights (another indication of West- ernization) be changed: red as the sign to go, green to stop. Far worse, chil- dren were set to denounce—or even to condemn to death—their parents; families were torn apart. Priceless relics of China's past were destroyed, museums vandalized, libraries ransacked. Untold millions of human beings were annihilated, many under torture; the victims included Liu Shaoqi, once Mao's designated successor. Many more who survived found their careers and lives ruined. Invaluable human talent was wasted at a time when China sup- posedly was racing to catch up to the West. But the frenzy of the self- appointed young guards of revolutionary rectitude knew no limits; their capacity for worshipping Chairman Mao bordered on ecstasy—a response utterly incompatible with his expectations.

As the anarchy of the cultural revolution increased, even Mao became disillusioned. Already in the fall of 1967, while on an inspection tour, he admitted: "I think this is a civil war."[38] In July 1968 he was ready to call a halt, turning, reportedly with tears in his eyes, against the Red Guards: "You have let me down and, moreover, you have disappointed the workers, peasants, and soldiers of China"[39] (a response, incidentally, which showed how much he had lost touch with reality). It was high time. By late 1967 the party and the mass organizations under its wings were in shambles; as the anarchy grew, the economy began to collapse. It was no wonder that in 1968 Mao and Lin Biao began preaching discipline and unity, relying on armed force as the last resort of governmental authority. In April 1969, at the first party congress since 1955, they and their closest associates resumed control, trying to restore, Leninist fashion, from the top down, order in the party and the country.

Yet the Great Proletarian Cultural Revolution was never officially ended; Mao, now again at the center of power, would not admit defeat. Indeed, even in 1967, at the height of the cultural revolution, when he had feared civil war, he argued that "All Party members, and the population at large, must refrain from thinking that all will be smooth after one, two, three, or four cultural revolutions."[40] He stuck to his convictions at that party congress: "the socialist revolution must still be continued. There are still things in this revolution which have not been completed. . . . After a few years maybe we shall have to carry out another revolution."[41] And, for a rallying point after the chaos of the cultural revolution, the cult of Mao was raised to its highest pitch. The continued necessity of cultural revolutions—the Chinese version of the revolution of Westernization—remained part of "Marxism-Leninism-Mao Zedong Thought" until his death.

IX

After 1969 the ideology of the cultural revolution survived as the party line, but its practical application was drastically curtailed. Considering the exhaustion, apathy, and disillusionment that followed the appalling destruction of human lives, what other course was open? The Red Guards, now accused of "ultra-leftism," were banished to the countryside; the role of the People's Liberation Army was scaled down. Factory discipline was restored, worker participation in management reduced, while wages were scaled according to output as before 1966; productivity was trump. As for institutions of higher learning, Mao's directives for the cultural revolution continued in force. Enrollments ran far below 1966 levels; the most prestigious institutes remained closed. In line with Mao's preference, urban resources were directed to the countryside, industries and skilled manpower together with improved medical care and schools geared to rural routines. In the aftermath of the cultural revolution rural China was the chief beneficiary.

Indicative also of the new accent on efficiency in all aspects of government was the gradual reappearance of officials ousted by the cultural revolution. Even more remarkable was the new soberness in regard to Mao himself: the cult of Mao was considerably scaled down; in 1971 Mao himself publicly admitted that he was no genius. In the next year, as if to drive home that point, he circulated a scathing condemnation of himself written by Lin Biao and culminating in the question: "Is there anyone whom [Mao] had supported initially who had not finally been handed a political death sentence?" Rather than deny the charges, Mao added that Lin had not gone far enough.[42]

Indeed, after the failure of the cultural revolution Mao was never his former self; his China too lacked the confidence and zest of its early years. As he slowly faded from power, he collided with Lin Biao, his recent ally, in a deadly fight. In 1972 Lin, accused of plotting Mao's overthrow, was eliminated under obscure and sordid circumstances. His fall caused some embarrassment; he had been, after Mao, the most prominent supporter of the cultural revolution. At first accused of "ultra-leftism," he eventually found himself denounced as an "ultra-rightist"—one of the cultural revolution's legacies was a greater-than-usual disarray in ideological consistency. His demise was followed by a new rectification drive aiming at more rigid conformity. Given the need for order and stability, the enforcement of social and political discipline at the place of employment or through the urban residence committees was tightened, now also for the sake of limiting populating growth. In the 1970s China, especially urban China, was a strictly regimented country.

The hero of Mao's last years was his longtime associate Zhou Enlai, who, though a moderate, had retained his confidence throughout all crises. With Zhou's help, some of the pre-1966 leaders, including Deng Xiaoping (who had barely survived the cultural revolution), were restored to power and influence, in constant battle with unrepentant Maoists led by the "Gang of Four," headed by Mao's wife. Under Zhou Enlai's guidance a chastened communist China also made an about-face in foreign policy. After the break of 1966 Sino-Soviet relations had deteriorated further; border incidents had led to talk of war. As a result, communist China began to veer toward the United States, then at the turning point in the Vietnam War, as a counterweight to the Soviet threat. President Nixon's visit to Beijing in 1972, a year after the People's Republic had been admitted to the United Nations, heralded an opening to the West unthinkable during the cultural revolution.

Mao spent his last years in increasing isolation, debilitated by the advanced stages of Parkinson's disease; he last appeared in public in 1973. Responsibility thus fell on Zhou Enlai to sum up, at the National People's Congress in 1975, the results of a quarter-century of communist rule. They were indeed impressive. Total industrial output in 1974 was nearly 200% above that of 1964, with more impressive results scored in some branches of production, such as petroleum, coal, electricity, and tractors. Above all, on the way toward growing into a modern industrial power, China had also become self-sufficient in foodstuffs, despite its rapidly growing population.[43] However traumatic the upheavals, the country had made remarkable progress in its material welfare; life expectancy, counted at thirty-nine years before 1949, had risen to sixty-four in 1980.[44]

In 1976 Zhou Enlai died, followed a few months later by the eighty-two-year-old Mao. It was a subdued end of a turbulent era in Chinese history. Were the advances in these years sufficient to satisfy the expectations expressed at the beginning of communist rule? Had China emerged as a "powerful modern socialist state" able to hold its own in the world? At the time of Mao's death, his charisma had faded. His achievements were beginning to be questioned; he had hardly lived up to his promise. But, then, had not he himself given frequent warnings that it would be a long march to the promised goal?

X

Considered in the context of the 20th century, Mao was another tragic hero, the greatest and most thoughtful of the human dynamos of enforced and manipulated revolutionary change. He dealt with the largest human aggregate in the world, one most set in its ways through several millennia and hardly touched in its depths despite a century of bitter crisis. How was he to recast such a huge collection of human atoms into a more cohesive social and political mold? How, in addition, was it possible with the available human raw material to form an effective government reaching into the depths of that granular society? How could he, a single human being no matter how capable and energetic, dominate that far-flung government? And more: how could he lift his country to the level of the West, the ultimate model recognized essentially only by its visible forms of power? How indeed could he gain insight into the dynamics of reculturation after an alien model? Clearly, the task was over his head, the challenge too overwhelming. But, then, did anybody else possess a deeper knowledge of these matters?

The Yanan experience, evolved under simpler conditions and wartime necessities, obviously was inadequate for dealing with all of China in more secure times. In addition, his reliance on Marxism-Leninism in its Stalinist version posed problems: its concepts and methods, artificial in the Russian setting, fitted Chinese conditions even less. Mao realized that sociopolitical change in China had to aim at a much deeper level of popular culture; there had been no autocracy to prepare the way. He therefore wisely insisted on the "mass line," on a perpetual interaction between leaders and the led, between the party and the masses. At the same time, however, he struck to the Leninist principle of regimentation through the Communist Party as the ultimate authority. How was he to reconcile the contradiction?

The contradiction pursued him throughout his quarter-century of ruling China. At times he indulged in extravagant praise of initiative from below. And indeed, in 1958, as in 1966, he met with surprising response to his appeals for drastic change. Since the May Fourth Movement in 1919 Chinese society, especially in the cities, had shown a remarkable capacity for sudden upwellings of patriotic (though uninformed) desire for change. His own experience had taught him to draw on this political force as an agent of social and political transformation.

Yet, as proven by the consequences, he could not afford to let these promptings take their spontaneous course; they needed guidance in the right direction. Distrusting the masses, he again limited their spontaneity, for good reasons. Left to their own devices, the peasants, the students, or any group of Chinese activists only produced more Chinese disorganization. From what sources within their own experience could they learn the alien ways of social cooperation transcending all traditional barriers? Thus there was need for Mao's preaching: "Place problems on the table. . . . Do not talk behind people's backs. Whenever problems arise, call a meeting, place the problems on the table for discussion" in the hope that a new transcendent consensus would emerge. Imagine the time needed for several hundred millions of people to reach a national consensus by open discussion (even assuming that they were capable of open discussion).

In addition, could they learn from each other how to adjust to the alien world outside? The hard fact was that in terms of cultural evolution China, even after 1949, was not an independent country. Unlike the West, it could not rely on the dynamics of internal adjustment for gaining national wealth and power; it had to follow the Western model, whose nature was known, if known at all, only to a small minority. Only Mao and, through him, the Communist Party knew the course to be steered in the wake of that model. And they knew it but partially, from books or from the surface tokens of power.

Under these adversities Mao's work and thought appear all the more remarkable. The most impressive feature of his work was his willingness to learn dialectically, that is, to analyze the inevitable mistakes in a humble and nonviolent mood, no matter how tense the situation. Of all the major counterrevolutionaries pitting their countries against the Western impact, he was the one most open to the difficulties he faced, the most modest in the estimate of his own ability, the least militant. Even in questions of raw power he argued that it was not weapons but human beings that counted. Freedom to him meant "the recognition of necessity" and thereby a tacit submission to the demands made by that necessity through trial and error.[45]

He was humble also in his patriotism, admonishing his people to keep learning from all foreigners, admittedly not in slavish imitation but in adapting foreign skills to indigenous conditions (and thereby subtly changing tradition). With equal humility he insisted that the process of adaptation required constant interaction with the masses. Reculturation meant cultural deep plowing and deep planting, which, he knew, could not be achieved by command and the breaking of wills. In this respect he was wiser than Lenin and Stalin—and less panicky (he did not preside over a traditionally insecure multinational empire at the edge of the European state system threatened, toward the end of World War I, in its very existence).

Admittedly, his sense of time was subject to extreme fluctuation. Had China stood up or was it still in the process of standing up? In essentials of industrial production England could be overtaken within fifteen years, he argued at one point; at another he envisaged China's surpassing the advanced capitalist countries within a century. But, then, he also repeatedly warned that it would take many cultural revolutions or ten thousand years to accomplish that feat. Learning from his big mistakes, he always extended far into the future the time limits for China's regeneration.

Were, then, the Great Leap Forward and the Great Proletarian Cultural Revolution big mistakes? The initiative in each case certainly was his; he incited drastic action. He therefore also carried the responsibility for the consequences. One might perhaps plead, however, several mitigating circumstances. In the first place, his Olympian aloofness after 1949 had made him idealize the Yanan experience; he had grown apart from the rough ground floor of Chinese life, from his followers' ignorance and raw temperaments. When his disciples acted destructively, as was most obvious in the cultural revolution, he felt betrayed. Second, his impulsive impatience, his eagerness to achieve his goals—his country's greatness—in the shortest time possible was hardly a personal failing; he shared it with all the other great counter-revolutionaries of the age. Given the wide disparity of resources and the overpowering force of the Western model, such impatience was a symptom of the times for which, in the last analysis, the West itself bore some responsibility. Was he—or his country—not entitled to make every conceivable effort to achieve equality as soon as possible? And, finally, who in the world understood the problems he was trying to solve? His ignorance was another universal symptom; the West certainly did not offer any helpful analysis. That fact made him a tragic figure, but, to repeat the point, one of exceptional insight. Who else among his peers pleaded such sensible truths: learn humbly from the inevitable mistakes; avoid violent solutions; involve the people themselves in the learning process; and most important: when in trouble, take a long view?

It is not, however, in this sympathetic light that his successors view him. The Chinese are no longer an "ardent nation," having come to distrust grand ideals and the enthusiasm of mass demonstrations; he has taken the ardor out of them. People keep coming to catch a glimpse of him at his grand mausoleum, where he lies, like Lenin, entombed in a glass case. Yet the immense gains of the past quarter of a century, the remarkable unanalyzed and unconscious progress in the collective disciplining of the population, are taken for granted. Considering China's continued poverty and backwardness, it is natural that people sobered by his failures ask: was the high price of Maoism justified? Would communist China not have fared better under the guidance of Liu Shaoqi? Is it not better off under Deng Xiaoping, who opened China to the Western world, provided the material incentives of an open market for consumer goods, and even allowed talk about a pluralist society in China?

All we can say is that under Deng Xiaoping and his like-minded successors another experiment, another cultural revolution, is under way, more peaceably but with its own problems. The new access to the outside world and the new freedoms do not automatically promote the sense of national unity and common purpose, nor the individual and collective discipline, required for wealth and power. All too frequently the gains disappear into the petty channels of traditionalism; they lead to "corruption," to tensions in the body politic and practices not in the country's overall interest. The problem does not lie in a lack of individual capacities—the Chinese are a gifted people—but in the insufficient socialization of the individual ego. For that reason the Communist Party must still stand guard over China's destiny; it allows only limited spontaneity of social, economic, or political development. Formally still committed to "Marxism-Leninism and Mao Zedong Thought," it remains the ultimate arbiter of right and wrong, trying to adjust over one billion Chinese people still set in millennia-old tradition to the demands of national and global interdependence.

In that world the outsiders, the barbarians, continue to hold the keys to the future, thereby indefinitely prolonging China's externally imposed cultural revolution.

XI

Now a look, for contrast's sake, across the China Sea to Japan—very briefly, because its success has been well analyzed. The rise of Japan from defeat and foreign occupation to the preeminence of an economic giant—some call it an economic superpower—stands out as a singular historic miracle. It represents the most startling triumph of the world revolution of Westernization: a non-Western country signally successful in terms of Western achievements.

In 1945 Japan was utterly defeated, its Co-Prosperity Sphere destroyed, its cities devastated (two of them by atomic bombs), its economy shattered. With demobilization unemployment soared; famine stalked the land. The country's disorientation was aggravated by the American conquerors' aim to recast the Japanese polity, with only the emperor's symbolic authority spared, into a peaceful democracy American-style. Under these circumstances the defeat of 1945 constituted a far more drastic humiliation than the opening of Japan to the West in the mid-19th century.

Stunned by unprecedented catastrophe, the country stagnated in its misery for two years, quietly reassessing its fate. Then its spirit revived. By 1952, when the American occupation ended, Japanese industrial production had almost reached prewar levels. Twenty-five years later Japan had emerged as a major industrial power; its genius for technical efficiency and innovation threatened its major competitors in Western Europe and the United States; it outproduced even the Soviet Union. It was a breathtaking advance, despite discouraging obstacles.

American occupation had foisted major changes on Japanese state and society. Japan had lost its control over its destiny, relying now on the American nuclear umbrella for its national security, and thereby dependent on American policy and power generally. While its own martial traditions were discredited, it had to accept American military bases on its soil. After 1947 it also had to accommodate itself to a foreign-imposed democratic constitution, establishing a parliamentary government controlled by political parties (a plethora of them at the outset). The American occupation authorities also conducted an extensive land reform, destroying the tradition-oriented big landowners and creating a new class of landowning farmers. They encouraged the rise of trade unions, reorganized the school system, the police, the press, the welfare system, and Japanese business practice. By all outward appearances Japan was launched on a thoroughly Westernized course, though—miraculously—without break in the texture of civic cooperation, without disruption through cultural disorientation.

The Japanese people faced still other adversities, for instance, in the extreme crowding of their islands; nowhere else in the world lived so many people (110 million and more) in such small quarters—mostly urban quarters, as one-third of the population had moved into cities. To prevent further congestion strict population controls went into effect—by the traditional method of abortion. Other disadvantages lay in the absence of basic raw materials like fuel or minerals; nor could the country grow enough food. Japan's cramped people were singularly dependent on trade with the rest of the world for their survival—a hostile world in East Asia after years of Japanese occupation.

Yet the country also possessed vital assets. Recognizing their responsibility and self-interest, Americans assisted in the country's economic reconstruction. Subsequently, the Japanese, by their own initiative, took advantage of the postwar expansion of the world economy, soon becoming its chief beneficiaries. They also gained cheap access to the new technologies developed in the United States and Western Europe, quickly improving them by their own ingenuity. In addition, they were spared the outlays for national defense; all available resources could be invested in industrial development.

More broadly, barred now from the costly race for world power, they could concentrate on making their mark, within their shrunken area of self-assertion, in constructive competition, if possible outpacing the West—even the United States. The country's collective ambition had not been broken by the recent humiliation; sensitivity about their country's standing in the world was keener than ever. It was now combined with a dogged determination to wipe out the shame of defeat by prevailing in the peaceful achievements of fine arts, high tech, international trade, and even in sports. It was not a matter of imposing Japan's culture on the rest of the world—Japan's policy in this respect was like China's—but of showing the world how successfully the Japanese could prevail in the new globalism; they would be thoroughly Western, although still Japanese in spirit.

Their biggest assets, of course, lay in their cultural and ethnic homogeneity bred through endless centuries of insular isolation, in their sense of collective discipline and the effective coordination of individual wills within their crowded islands. The imposition of a democratic constitution stressing individual self-determination did not lead to conflict. On the contrary, the new emphasis on the rights of individuals and of low-status groups reinforced the traditional practice of consensus-building within a hierarchically ordered process of decision-making; it cemented the traditional quest for solidarity.

After the war, "democracy" was a popular concept, testifying to the profound impact of Western superiority. Seemingly without effort or protest, the Japanese adapted the alien institution to their own heritage of civic discipline, the only non-Western people among whom democratic government did not augment social and political discord. The number of political parties quickly shrunk; communists and socialists never enjoyed widespread support; the conservative-minded majority parties ruled alone or in coalition; no strong man was needed at the helm. Politics never disrupted the economic advance; effective government indeed accelerated it.

The economic advance, at the core of Japanese postwar nationalism, arose from the firm habits of self-denying and self-affirming teamwork in the factory, laboratory, and the office. There was no protest against placing industrial investment ahead of popular welfare; even the poor put the future prosperity and prestige of their country first. All worked exceedingly hard to learn the superior skills of foreigners, always incorporating them into their own routines, with admirable humility and yet also with growing pride in their Japanese methods of operation. Despite rapid technical innovation and social change, the collective discipline prevailed.

Even the marked discrepancy between traditional culture and the Western impact did not undermine national morale. Internally generated tensions over individual and collective achievements rose to high levels; they were at their worst among teenagers who all too early, by Western standards, must prove their capacity for success (thereby escaping the dissipations demoralizing their Western peers). But the country escaped the social evils associated with Western urban-industrial societies. The Japanese people were dedicated to cultivating public order and social harmony.

All told, in the unique case of Japan—the exception to the rule—there was no need of reculturation for the sake of matching Western standards of living and industrial productivity. Out of its own tradition the country furnished the necessary civic commitment and social discipline, so conspicuously as to impress envious Western observers. Japan was eulogized as Number One, as the First Twenty-first Century Country, as the Coming Superstate, causing concern or even apprehension among Americans hitherto safe from humiliation in invidious comparison. Yet, in terms of raw power, the ultimate arbiter of global politics, Japan constituted no threat. Its challenge was—and still is—limited to competition in peaceful pursuits.

And in one respect crucial in an interdependent world, Japan still suffers from a marked disadvantage compared with its Western rivals. Japanese culture lacks the cosmopolitan openness of the Western model. Its creativity is contained within its ethnic—some observers have termed it racial—homogeneity. While prodigious in their capacity for assimilating foreign skills, the Japanese have shown little propensity for admitting foreigners into their midst or mingling freely with them abroad. Their admirable qualities can only be affirmed within the group, among fellow Japanese; they cannot be shared or universalized. Western cultural truths are anchored in the individual, derived from a religious belief in a universal God supplying personal guidance anywhere in the world.[46] The Japanese, by contrast, live by a social morality socially enforced; outside the group they feel ill-at-ease and ineffectual. For that reason it seems unlikely that in the foreseeable future Japan will attain that leadership in an interdependent world which its admirers assign to it. It will rather continue to be a self-contained stimulant for material progress in a world that craves transcendent universality.

21

The Burden of Development

Rather than examine additional case histories of responses to the pressures for reculturation in the years after 1945, we now turn to essential overall effects of the revolution of Westernization on the newcomers to statehood. The focus here will shift from Asian countries, which, on the whole, possessed more suitable cultural assets, to their African counterparts, which suffered the worst.

I

One of the most striking and fateful effects of the unification of the world achieved by that revolution was the emergence of an unofficial but generally recognized hierarchy of cultural standing which put the newcomers near or at the bottom of the scale. This hierarchy was based on the Western-derived universals expressed through the capacity to prevail in the comparisons of political power, standards of living, technological progress, and cultural creativity. In these terms the "developed" countries, essentially the West plus Japan as a newcomer—the countries united in the Organization for Economic Cooperation and Development (OECD)—occupied the top rank. Below them were arranged the Soviet bloc countries of the "Second World" and the newcomers classified as the "less developed" countries (LDCs) of the "Third World." At the bottom lingered the "least developed," sometimes grouped together as the "Fourth World," found mostly (though not exclusively) in central Africa. The newcomers, ushered into a Westernized world, were locked into a position of inferiority recognized, however reluctantly, even by themselves.

There could be disputes about a country's precise location in the intermediate range of "developing" countries. The terminology for the hierarchy, too, has been subject to change, so as to protect the most vulnerable peoples from the stigma of categorized inferiority. Gradually the differences were simplistically subsumed in the contrasts between North and South, or between rich and poor as measured by various indices. According to the Brandt report of 1979,

> the North including Eastern Europe has a quarter of the world's population and four-fifths of its income; the South including China has three billion people—three-quarters of the world's population but living on one-fifth of the world income. In the North, the average person can expect to live for more than seventy years; he or she will rarely be hungry, and will be educated at least up to secondary level. In the countries of the South the great majority of people have a life expectancy of closer to fifty years; in the poorest countries one out of every four children dies before the age of five; one-fifth or more of all the people in the South suffer from hunger and malnutrition; fifty percent have no chance to become literate.[1]

Whatever the bases for evaluation, agreement prevailed about a universal scale for measuring a people's or country's standing in the global hierarchy of power and prestige. The newcomers were trapped.

The disparities measured by this scale were not, however, of political power, standards of living, or economic productivity—these were but surface phenomena. What counted were the cultural resources required for living up to the models that set the pace, the skills of individual and collective conduct needed for meeting the demands of the times. What was the nature of these skills?

Put simply: they encompass the human capacity for managing the ever more refined complexity of large-scale societies organized in nation-states now linked in global interdependence. The central factor is the human ability to cope with an ever more minute division of labor in dealing with the environment, with machines, and with other human beings, while at the same time evolving a corresponding community of cooperation among millions and hundreds of millions of human beings. That capacity has two interrelated dimensions, one concerning the individual, the other the coordination of individual wills in society.

As for the individual, the skill of effective cooperation with society and its technical tools requires a refined ability for coping with complexity through cerebral functions, through abstract thought in intellectual pursuits, through meaningful generalizations covering an ever larger volume of knowledge. It means shifting vital energy from physical action to cogitation, sublimating bodily impulses into wide-ranging intellectual awareness. It means restraining the individual will and adjusting it to constructive cooperation with ever larger numbers of fellow citizens. The larger the group, the more disciplined and pliant individuals must be in their innermost selves, obedient toward all the human institutions and the artifacts of technology that uphold their lives. A society dedicated to an ascetic discipline of life, to a practice of controlling the libidinous body for the sake of heightened mental and spiritual outreach, possesses a major advantage in achieving the large-scale cooperation required for success in the rising global competition.

Yet individual discipline is not enough. Highly developed in many of the world's cultures (especially in Asia), it may sometimes reach extraordinary intensity (as among Asian immigrants in the United States). What is still missing is the civic component, the coordination of wills, the individual's subliminal integration into the body politic; individual discipline has to be socialized. The social discipline which cements the "developed" countries into effective consensus-based polities, however, is an elusive achievement, the product of history—of chance as much as of individual effort—among socially alert and open-willed individuals under favorable circumstances; it cannot be legislated or mandated; it cannot be forced.

Social cooperation, in the West as elsewhere, started with small-scale societies based on family and lineage enlarged eventually to tribal status woven out of lineage relations. Then, over two thousand years ago, a quantum jump in the social learning process occurred with the rise of the Greek *polis*, an institution transcending all blood relations and creating a superior "political" dimension of collective life. The Roman Empire and subsequently the universal monotheist Christian church added a further bond of abstract unity, reinforced by a strongly ascetic ideal of life. Elsewhere collective unity was still guaranteed by a human being, a semi-divine paramount chief or emperor (in imperial China or even contemporary Japan) as the embodiment of unity.

As an immensely abstract entity above all blood relations, above all local and regional interests, the modern nation-state owes its cohesion to these transcendence-promoting antecedents; it also owes much to the European hothouse combination of unity and diversity, to nationalism, to a politically motivated sense of spiritual mission, and to success as a model for other peoples. Good communications, an unusually high level of material security, and the industrial revolution certainly contributed their share to building national solidarity. But like all other human achievements, the material assets were human creations, depending on the mental and spiritual skills of their creators, on the pliancy and openness of their wills, both in regard to other human beings and to their ever more complex mechanical tools.

Now compare these assets and the corresponding cultural skills found in the West (and to some extent in Japan) with the cultural resources of other peoples. Where else do we find a similar blend of capacities? Taking a hard look around the world, one might argue that in this respect conditions have not changed much since the days before Western colonialism; the basic disparities have not disappeared despite the Western outreach. One can even argue that they have become aggravated. The gap between North and South, between rich and poor, has widened. Among the "developed" countries the hothouse pace of innovation has increased, thanks to the enlarged scale and opportunities of interdependence. The others suffered not only from the comparative "backwardness" of their cultural tradition, but also from the disorientation and subversion caused by the Western impact. Deprived of the freedom of independent cultural evolution, they were reduced to imitating the powerful models; their natural creativity was stunted. It could well be argued that they were being *under*developed.

It can hardly be maintained, of course, that in absolute terms the poorest were worse off than they had been before the arrival of the Europeans; all evidence regarding life expectancy, survival rates, or access to opportunity points to the contrary; in these respects the revolution of Westernization has had a constructive effect. But in the psychologically more important relative terms—in the ever more intensive comparison of wealth, power, and human rights (including freedom from unmanageable external pressures)—their fortunes have declined. The powerful, the rich, did as they would, and the powerless poor obeyed as they must. Political and economic decisions affecting their welfare were taken in the capitals of the "developed" countries by governments or by the ubiquitous multinational corporations far richer than any number of "developing" countries combined. As regards shipping and insurance fees, import regulations, or the terms of trade generally, the rich helped themselves; the poor lacked clout to redress the imbalances. The call for a New International Economic Order, loudly heard in the 1970s, went unheeded. However strident their protests against neo-colonialism, the poor became more heavily indebted to the rich (directly, or indirectly through the World Bank) and thus firmly dependent on them; any repudiation of their debts would set back their economies still further. Even playing the super-powers against each other brought no relief.

The hard facts stands out: the newcomers began their political careers as the underdogs in a world of accentuated and deeply rooted inequality. That inequality has become part of the human condition in the contemporary age, at a time when the dominant ideal universalized by the Western model is equality—equality applied not only to individuals but also to their collective identity in cultures and states, the indispensable instruments of individual advancement. As argued before, the world revolution of Westernization has not created those elemental cultural disparities. Its contribution has been to reveal and to aggravate them.

The disparities now have become public knowledge, viewed from a single perspective, classified, and made subject to intense and often impassioned debate over justice and morality as well as over "development." The anti-Western protest, long simmering, has been shared among the victims of that revolution; it was the common bond at the Bandung Conference. It has been aired in the General Assembly of the United Nations, which declared the 1960s the "first development decade" and subsequently backed the New International Economic Order; it troubled the conscience of the rich, whose prospects are beclouded by the helplessness of the poor. In this way the stage was set for constructive action. But the intense agitation never altered the foregone conclusion: for the newcomers the only alternative to ignominy was going up, up the steep ladder of "development" as defined by the Western model.

The only comfort in the ascent stemmed from the fact that in the West condescending terms like "the natives" or "tribalism" inherited from colonial rule were dropped from learned discussion; the "people without history" were treated with greater respect. More important, the terms "modernization" and "development" have provided a culturally neutral conceptual framework for liberation from dependence and ignominy.

II

Whatever the terminology, liberation from Western domination, attempted in many variants, implied or openly promoted further Westernization. Elemental resistance through rigid adherence to custom, sometimes carried to violence, ended in defeat for lack of modern weapons and grasp of basic realities; defeat taught submission to change, however unwanted. More effective was the literary, artistic, or scholarly glorification of tradition (a romanticized tradition)—but only thanks to the Western skills of literacy combined with typewriters, scholarly libraries and research techniques, telephones, publishers, banks, not to mention the media, especially television. The plea for cultural pluralism, for cultural diversity as a prerequisite for worldwide human progress, often associated with the highbrow defense of indigenous culture, was similarly broadcast through the uniform universals of modern communication located in the Western metropoles. In order to be heard, the anti-Western protestors travelled back and forth by the most modern means of transport.

Other non-Western observers trained in Western religion, philosophy, and social science protested the injustice and immorality of the outrageous inequalities found in the world—thereby universalizing their Western standards of humaneness (which were not part of traditional practices). Critics deplored the imposition of imported Western schooling on preliterate people who rather needed education suited to their still limited grasp of the world. It was equally inhuman, so they argued, to make these people into thoughtless consumers of refined—or not so-refined—Western consumer items (like Coca-Cola); they rather required goods specifically related to their own needs and resources. But, considering the rapid advances of the metropolitan centers, would such slow apprenticeship in modern ways, however humane, not widen the cultural distance between the periphery and the center?

And did not the leaders of the newcomers set a public example to the contrary by enjoying the latest conveniences? Even Nyerere, the advocate of socialist austerity in Tanzania, admitted that "today the standard of living in the United States of America is part of Tanganyika"[2] (as his country was originally called). He used a Mercedes-Benz car and a British airplane piloted by a British pilot for his transport (Nkrumah was conveyed in a Rolls Royce). The ex-colonised man was an envious man; he wanted the latest, at least vicariously through his leaders, in order to be equal. Let romantics like Senghor plead: "Africans do not want to become consumers of civilization."[3] Given his choice, the ex-colonised man would rather drink Coca-Cola.

The Marxist socialists among the newcomers, of course, were determined modernizers. As Marx (unaware of the cultural dislocation caused by industrialization) had prophesied: "the country that is more developed industrially only shows to the less developed the image of its own future." Industrialization was one of the newcomers' primary goals, as they aimed for a social order superior to "capitalism." As Marxists, they had little use for the remnants of "feudalism" in their midst, although some of them, including Nkrumah or Nyerere, sought, like Russian slavophiles, to link what they called traditional communalism with the socialism of the future—unsuccessfully, as events were to show. The application of refined socialist or Marxist theory to the conditions of underdeveloped countries, especially in Africa, posed many problems. Was it a wonder that it abounded with the contradictions forever bedevilling the Westernized anti-Western intelligentsia?

Consider the contradictions in the writings of Frantz Fanon. A French-trained, Marxist-inspired black psychoanalyst from the French West Indies, he was decorated for bravery in the war against Hitler and eventually sided with the rebels in the Algerian war for liberation. His attack on colonialism led him to the affirmation of the essentials of Westernization, to the endorsement of the nation-state, a united Africa, and a transcendent humaneness.

The fury of the Algerian war still seethes in his analysis of colonialism and his cry for decolonization in his book *The Wretched of the Earth* (1961). "'The last shall be first and the first last.' Decolonisation is the putting into effect of this sentence,"[4] he declared with utter contempt for the imperialists. "When the native hears a speech about Western culture, he pulls out his knife—or at least makes sure it is within reach"[5]—in a gut reaction similar to that of the Nazi stormtrooper who at the mention of the word *Kultur* reached for his pistol. The Europeans in their wealth, Fanon fulminated, were war criminals: "This European opulence is literally scandalous, for it has been founded on slavery, it has been nourished with the blood of slaves and it comes directly from the soil and from the subsoil of that under-developed world. The well-being and the progress of Europe have been built up with the sweat and the dead bodies of Negroes, Arabs, Indians, and the yellow races."[6] Against that vicious force Africa had to be united, not under the leadership of the Europeanized African bourgeoisie, but "through the upward thrust of the people, under the leadership of the people,"[7] not the people of the new capitals but of the African bush.

Yet did Fanon trust "the people"? Obviously not. "Everything depends on the education of the masses, on the raising of the level of thought, and on what we . . . call 'political teaching.'"[8] In short, out of their native inheritance the "people" were not capable of moving forward. Hence arose the need for guidance: "The task of bringing the people to maturity will be made easier by the thoroughness of the organisation and by the high intellectual level of its leaders."[9] And for what ends? "In the colonial situation," Fanon explained, "culture . . . falls away and dies. The condition for its existence is therefore national liberation and the renaissance of the State. The nation is not only the condition of culture, its fruitfulness, its continuous renewal, and its deepening. It is also a necessity."[10] But it was even more. The nation-state reshaped the people's culture into a "new humanity." "The building of a nation," Fanon topped his argument, "is of necessity accompanied by the discovery and encouragement of universalising values. . . . National liberation leads the nation to play its part on the stage of history"[11]—global history now, eons removed from traditional culture.

Yet could the African peoples really accomplish the transition by themselves? "Let's be frank," Fanon advised; "we do not believe that the colossal effort which the under-developed peoples are called upon to make by their leaders will give the desired results. If conditions of work are not modified, centuries will be needed to humanise this world which has been forced down to animal level by imperial powers."[12]

How, then, was the change to be accomplished? Fanon's answer, not surprisingly, resembled Lenin's, who realized that socialism in Russia could be achieved only with the help of the workers of Europe. As Fanon wrote: "This huge task which consists of reintroducing mankind into this world, the whole of mankind, will be carried out with the indispensable help of the European peoples, who [however] . . . must first decide to wake up and shake themselves, use their brains and stop playing the stupid game of the Sleeping Beauty."[13] Europe—the West—still was the ultimate source of salvation. And the anti-Western hatred of the colonized, fanned to high heat, was merely a means of preparing them for the massive transformation that was to make them like humanized Europeans in a new global fellowship transcending all cultural barriers. Did not that vision, evolved from an attack on Western civilization, constitute the ultimate victory of Westernization?

III

Whatever Fanon's conclusions, statehood was an inescapable necessity in the global state system, and state-building the challenge of the age. What, then, were the conditions under which the newcomers were forced to build the states that were their instruments for rising to respectability and power, to the good life?

The hard fact is that the new states were built from the outside in, and from the top down. The nation-state was an alien institution derived from Western experience and imposed by Western-trained intellectuals upon uncomprehending and unprepared peoples—for their own good in an inescapably interdependent world. It set a new and greatly enlarged framework for all human activities which diminished the self-esteem of individuals long before it enlarged their opportunities. That state introduced an alien structure of government, an executive complete with ministries served by civil servants ideally committed to the welfare state, a standing army, a diplomatic corps, a new superstructure complete with national flag and national anthem. In the case of English or French ex-colonies, the new states were launched under a democratic constitution establishing an elected national assembly as the source of governmental authority—all exceedingly complex and unheard of in traditional society.

As an exotic import manipulated by a foreign-trained indigenous elite, the new state within its Western-imposed boundaries was extended downward into the mainstream of traditional life, often in marked continuity with colonial rule. The country's capital and its state offices remained the same; in most cases, the colonial master's language became the official, the only unifying, means of verbal communication, his culture the standard for collective and individual life—superficially, at least. The downward extension of the alien model, however, caused endless problems. The imported institutions would not take root among the bulk of the population; the great majority resisted or rejected the uncongenial innovations. For the run of people certainly the new nation-state was just another case (in Lord Lytton's words) of "the application of the most refined principles of European government, . . . of the most artificial institutions of European society, . . . scarcely, if at all, intelligible to the greater number of those for whose benefit it is maintained." No doubt, the nation-state was a manifestation of neo-colonialism, even if imposed by people in revolt against colonial rule. It tied everybody to a hostile and overpowering world that undermined or destroyed all familiar relationships. Now imperialism had been indigenized; its battles were fought internally.

Wherever one looked in the new states, one found a clash of incompatible cultural traits. Even the new ruling class, the foreign-trained intellectuals turned leaders, politicians, or professional experts, suffered from the strains of cultural disjointedness. They had acquired the manners and the knowledge of the metropolis while their ancestors still rumbled in their souls. On assuming power back home, they encountered these ancestors among the people whom they wanted to convert into citizens. Inevitably, the ancestors asserted themselves, perverting the new institutions to their own benefit and corrupting the elite. "The state" carried little inherent authority, while family, clan, clients, or ethnic groups remained powerful constituents. Society was still held together by the cement of tradition, at the expense of the state, which required loyalties far transcending the old bonds.

The new state perversely even affirmed these bonds. It was the link to the outside, to the Western metropolitan centers, to wealth and prestige measured by traditional standards. In order to be effective among its constituents—and even to buttress its own self-esteem in the comparison with Westerners—the elite had to bring home the material trappings of success. Living like Westerners supported the liberated colonials' claim to equality; it also raised their status by traditional rules, imparting a psychological boost to lesser folk who—strangers to the sentiments of the class struggle—took a vicarious pleasure in the glories of the rich. In this manner, valuable state resources were diverted from their intended purpose.

The pull of the metropolis asserted itself in contradictory ways. It served as the ultimate arbiter of the conduct of government and the progress of the country; it offered expertise and financial aid. Yet it also built up new obstacles. The most talented natives tended to leave for greener pastures abroad, depriving their compatriots of needed brainpower. Likewise, the income, legal or otherwise, of successful politicians was likely to be lost for domestic investment, drained to secret accounts in foreign banks (a rational response perhaps to the perennial instability of the new states, yet also a cause of it). In any case, the new institutions of government, barely started, carried little moral authority as compared both with the Western model and with the established ways of the past.

The ways of the past suffered likewise, subverted (as under colonialism) by the inroads of modernity with its many temptations. Modernity intruded through fancy consumer goods, new ways of self-indulgence, undigested information about the outside world, book-learning that bore little relevance to community needs or job opportunity, or raised expectations that could not be met. All caused distrust of traditional authority which, obviously, no longer applied. All told, modernity promoted not only statehood and opportunity for advancement, but also alienation, disorientation, hopelessness, or worse: violence, criminality, or outright brutishness. If the destruction of traditional restraints (in any case less extensive than in Western tradition) was by chance combined with political ambition, military power, and external incitement to terror, the result might be raw cruelty in civil war, or at least a string of political coups forestalling the continuity needed for state-building.

IV

The overriding need certainly was for unity within the often arbitrary boundaries inherited from colonial rule. The new states were riven from the start by sharp disagreements among their subjects; they were multiethnic multireligious polities lacking a sense of unity (apart from the desire to be rid of Western domination). Liberation provided no common bonds; on the contrary, it affirmed self-determination; it strengthened narrow traditional identities. What loyalties, then, could be invented for a transcendent nationalism, what symbols be devised? Could the people be left to their own devices? Did Western democracy as embodied in the new constitutions fit the indigenous conditions? Or, in the absence of any preexisting consensus, did unity have to be artificially created?

After the collapse of the constitutional regimes bequeathed by the departing colonial powers, the most suitable source of unity, as illustrated earlier, proved to be the charismatic leader, who, adorning modern institutions with indigenous symbols, continued the appeal of traditional rulers as guarantor of peace, order, and dignified collective survival. Yet how, in the absence of consensus, was he to overcome the deep-seated disunity? The most attractive prescription proved to be a military regime. It could count on a minimum of discipline, at least among the officers trained in the military academies of England or France; it commanded the force of arms for compelling submission. Yet, even so, it could not guarantee success. The military men lacked the political qualities and the professional skills needed to run an effective government; they could not elicit the voluntary cooperation of the civilians. Moreover, in a fragmented country they often disagreed among themselves, sometimes to the point of civil war.

Apart from military rule, government by a one-party dictatorship also had its appeal. Most commonly, that party's leaders tried to follow the Leninist model of deliberate substitutionism, engaging in sociocultural engineering with the help of "scientific socialism." That prescription promised quick results; it would make Africans run while the others walked (as Nyerere put it). The flaw in this prescription was the absence of any African equivalent to the Soviet Communist Party. As Nkrumah and Nyerere discovered, there could be no socialism without socialists; there could be no militant party without the long cultural conditioning of the Russian revolutionary movement. In every respect, there could be no catching up for lack of the required collective discipline.

It was no wonder, then, that everywhere heads of state clamored ever more urgently for "discipline," among soldiers, civil servants, professional experts, business people, students, workers, and citizens generally. And "discipline" meant above all preventing "corruption," preferably not through external restraints Soviet-style (however appealing under the circumstances), but through self-imposed austerity, through a more ascetic conduct, despite the temptations of easy self-indulgence broadcast by the rich metropolis. It called for transcending the traditional narrow loyalties for the sake of serving the all-encompassing abstract state, which, through endless discouraging crises, all too slowly assumed the human reality of a nation. But how to achieve that discipline in the face of the ideals of self-affirmation, spontaneity, freedom, and individualism fostered by the Western model as well as by tradition? How to achieve it despite poverty, hunger, and debility caused by malnutrition, and last, not least, despite the continued promptings of dark magical practices deeply embedded in the popular mind?

Meanwhile, there loomed the urgent task of building the necessary infrastructure of statehood, for instance, a school system topped by institutions of higher learning. Obviously, the country had to train its own professionals. But who was to teach them? Where were the necessary buildings, the suitable textbooks? Was there enough paper, even for the national press? And who would pay for these foreign innovations? Similar questions arose regarding surface transportation by car, truck, or railway (somehow the new states found it easier to establish their own airlines—for travel by the elite or foreigners). And what about adequate health services? Where was the money to come from?

Money caused the biggest headaches for the governments of the newcomers and for their peoples as well. The obvious answer was export of raw materials where these were available, whether of minerals or agricultural luxury items like cocoa, coffee, or tea. Such export put the country at the mercy of foreign firms and the world market. Yet the establishment of domestic processing industries, recommended under the circumstances, called for foreign investment, another source of dependence. The development of domestic productivity generally was hampered by the absence of trained experts or skilled workmen. Foreigners, mostly from Lebanon or India, ran the small shops providing modern consumer goods; for the bigger jobs, Western expatriates were hired—there were more whites in Africa after liberation than before. With their help the new master started big attention-catching projects involving heavy foreign investment to the neglect of simpler and more immediately productive enterprises which might have strengthened the economy and the national budget.

Even more than under colonial rule, people hankered for imported goods. Luxury items were channelled into private consumption, industrial equipment to the new enterprises; only imports could furnish the higher standard of living promised by liberation from colonialism. In the ensuing competition the privileged helped themselves, illegally if necessary, while the public services, and the economy generally, suffered. All along, people drifted from the villages into the cities, lured by hope for a better life and closer access to the excitements of modernity. As a result of the emphasis on export crops, of the rural experiments of "scientific socialism," or of the rural exodus generally, local food production suffered; people went hungry. Most African countries became food importers—and went more heavily into debt and dependence, despite all well-meant efforts to develop their economies.

V

The majority of development projects, so hopefully started in the first development decade under the auspices of the United Nations, Western governments, or private initiative, did not alter the adverse drift. Development was viewed as a transfer of methods of productivity—institutions, technologies, cultural skills—from the developed to the less developed countries, based on the assumption that the recipients were prepared (like the beneficiaries of the Marshall Plan) to continue on their own; their economies would take off. Western economists believed in the existence of a universal pattern of economic growth without taking into account two crucial factors. First, an economic system freely evolved out of its own inherent dynamics cannot be a model for economies disoriented by foreign intervention and governed by external forces. Second, economies function in the context of cultures—cultures which by their nature are mutually incompatible. Prevailing Western opinion, however, held that Western achievements in demand throughout the world could be lifted out of their invisible cultural wrappings and readily implanted elsewhere. What needed to be arranged were only the details of transfer.

There, unfortunately, lay the rub. The cultural contexts among the recipients distorted or even destroyed the potential of the development projects, especially the biggest and most prestigious. At best, they benefited the Westernized elite at the expense of less spectacular innovations or even popular welfare. They did not reach the common people, among whom change was needed most urgently (and yet resisted most forcibly). In the villages the disparity between traditional resources and the skills needed to make the imported plans work invariably caused disappointment. The local population certainly possessed superior knowledge of local conditions, but lacked any means to communicate that knowledge to Western experts promoting assistance schemes conceived under Western conditions.

Complex technologies proved of no use. For want of specialists, spare parts, and supporting social and political conditions, they commonly came to grief. Even simple technologies did not take hold. While well-meaning Westerners talked of the need for "intermediate technology," its prospective beneficiaries lacked the prerequisite mechanical aptitude, motivation, or money to buy it. The best chances lay in small and simple projects undertaken with a maximum of interaction between givers and takers. Yet as common Peace Corps experience indicated, even the rudimentary improvements frequently did not survive the departure of the volunteers; life soon ran again in the old grooves which had stood the test of time. Why engage in unsettling experiments that increased dependence on uncontrollable outside factors? The prevailing instability was a further factor hampering the transfer of productive skills. National liberation and independence brought uncertain times indeed. Who, anyway, knew their way in the Westernized world?

Growing disillusion with wasted effort and resources gradually led to a more informed and sober assessment of development. Aid from the outside encouraged passivity and lethargy among the recipients. What was needed was internal mobilization, a new willingness to change, and submission to the Western discipline of productivity, however uncongenial. As a result, expectations were lowered while attention shifted from industry to agriculture, to the needs of the majority so often neglected in the early and optimistic phase of development.

Meanwhile, problems emerged that lay beyond the scope of development projects. The population exploded, thanks to medical advances introduced from the outside (another result of Westernization). More people meant more demand for food, education, employment, and public services; it meant greater pressure on the existing natural resources. Tropical forests were cut down, savannah lands overgrazed; deserts advanced, abetted by natural catastrophes like drought. In addition, the new countries quarreled with their neighbors over ill-defined boundaries, migrant workers, or refugees. As their internal problems spilled across their borders, they were in danger of becoming entangled in the superpower rivalry; some fell victim to permanent civil war fostered from the outside. Nonalignment proved an elusive goal for weak states. They could air their grievances in the UN General Assembly, but for what gain? The ideal of African unity certainly was a mockery. Violent disunity escalated throughout the continent.

VI

In the light of these facts, statehood, that Western product so eagerly sought as an instrument of liberation from Western domination, turned out to be a dubious boon. On the one hand, it put the ex-colonial country on the world map; it established a claim to be respected. On the other hand, it imposed a new, if more subtle, form of subjection. It demanded submission to the novel discipline of statewide cooperation, of coping with Western technology, and of survival in the competition of the global state system. Some newcomers, like the Ivory Coast in West Africa or Kenya in East Africa, managed fairly well by letting their citizens learn on their own how to cope with the challenge of modernization as a creative opportunity. Other countries under more ambitious centralized planning fared worse; still others collapsed into anarchy or dire poverty. All have as yet failed to cope with excessive population growth.

The worst case perhaps—and one of the most tragic in Africa—was that of Uganda, under the *pax Britannica* a promising territory. After independence, however, tribal rivalry rose to the fore, and with it martial rituals sharply at odds with missionary teaching or Western standards of humaneness. In the escalating disorders violence begat more violence, marked by excessive brutality. It reached its culmination in the regime of Idi Amin. He was the son of a witchwoman, who had risen, without benefit of formal education, in the British army in East Africa until appointed, not long after independence, as commander-in-chief of the Uganda army. Soon he seized power for himself. He ruled from 1971 to 1979 by methods of repression copied from Soviet practice and spiced with self-indulgent delight in cannibalistic murder of the most barbaric variety.

Awakened in the turmoil of decolonization and stimulated by high office, Idi Amin's brutal instincts were refined by modern weapons and techniques of torture. His ambitions were encouraged by Yasser Arafat and his protector, Colonel Qaddafi, for their own ends; but he was also respected by the members of the Organization of African Unity. Only his neighbor Nyerere protested, deploring "this tendency in Africa that it does not matter if an African kills other Africans. . . . Being black is now becoming a certificate to kill fellow Africans.[14] But why should blacks be different from whites, who in their own way had shown worse license as recently as World War II? Idi Amin was yet another symptom of the times, a combination of deepest native tradition with the most modern methods of power in an ex-colony's moral no-man's-land in which everything was permitted. His Uganda was merely the most glaring case of conditions prevailing throughout much of Africa. Below the superstructure of modern statehood a primeval streak of raw brutality continues, ready to rise elementally to the surface in the all-too-frequent political or economic crises.

Such inhumanity was not limited to Africa. In Southeast Asia Idi Amin's ferocity was matched with a more pronounced ideological flair in Kampuchea, the former Cambodia revamped under communist rule that began immediately after the end of the Vietnam War. The Kampuchean communists, called Khmer Rouge, were determined to accomplish at one stroke the changes achieved in China under Mao Zedong. After seizing the capital, Phnom Penh, in April 1975, their peasant soldiers drove its approximately three million inhabitants into the countryside. The process of "ruralization" in Phnom Penh and lesser cities was carried out with the utmost ruthlessness. Untold thousands, including women and children, were killed, many after torture and mutilation. The soldiers went on a rampage of insensate brutality refined by modern techniques of destruction popularized in the Vietnam War.

The instigators of these appalling atrocities were a small group of intellectuals apprenticed to radical politics in Paris; the figure best known among them was Pol Pot, who in 1976 emerged as head of the government. Returning from Paris, these intellectuals applied their theories without the restraints invisibly attached to them in their Western European context. They took no part in the atrocities; as communist intellectuals eager to impose their puritanism on the population, they found no personal gratification in murder. Their satisfaction lay in social engineering on the grandest scale, combining Stalin's and Mao's terror. Inevitably, after such senseless large-scale destruction of human lives and valuable material assets, their experiment proved a dismal failure—a further symptom of the time. What else should one expect, given the explosive tensions between the ground-floor realities of life in Southeast Asia and the high-flown visions of power and social progress derived from the Western metropolis of radical thought?

VII

More broadly in conclusion, what else should one expect, given the universal ignorance of the revolutionary effects of Westernization now at its culmination, given also the pressures within the newly created global state system? There was no escape from elemental invidious comparison in the global competition for wealth and power. The instruments of power were derived from Western experience; in order to be powerful, one had to master these instruments, both the weapons and, more demandingly, the skills to produce them. Above all, one had to build a viable state, without which all the other accomplishments were unattainable. All these achievements combined constituted a complex cultural phenomenon that could not be readily transplanted.

Here lay the flaw in the valiant efforts of the impatient Third World statebuilders and imitators of Western power, or in the grand development projects advanced by Western countries. Westerners and non-Westerners alike did not know the basic fact: cultures evolved in different natural settings are, in essentials, incompatible with each other, like languages. External manifestations like weapons, machines, written constitutions, or political visions can be transferred, but not the aptitudes and the social habits responsible for their successful operation. Unless these already exist in some form in the receiving country—as in Japan—their acculturation cannot be forced by the will of a leader or a decision of government. The transfer of cultural achievements demands no less than a permanent revolution of reculturation, the re-creation of the original setting in a new and uncongenial environment, a feat never yet accomplished.

Societies or polities culturally conditioned over long stretches of time cannot readily transform themselves according to a different cultural pattern; non-Western cultures cannot follow the Western upward-bound route. As with economic development, so with cultural growth generally: the evolution of Western culture, freely shaped by its internal dynamics, can be no guide to the cultural development of others who, deprived of cultural sovereignty and operating individually and collectively at greatly lower levels of vital human energy, must adapt themselves to a ready-made all-inclusive alien model evolved under exceptionally privileged conditions. Moreover, there possibly exist biological limitations. Essential qualities of a culture may be anchored in corresponding physical—or even genetic—traits by a natural selection of the human qualities fittest for the given circumstances. Under favorable conditions isolated individuals may reculture themselves (up to a point), but not collective bodies, especially large ones with a long record of cultural continuity.

The foregoing assertions obviously lead to a rigid cultural determinism repulsive to all progressive thought. Yet unless we learn to understand the limitations imposed on human beings by their cultural conditioning, we have little chance to cope rationally and constructively with the differences compressed into the hierarchy of cultural standing created by the revolution of Westernization; we remain victims of unexpected calamities or barbaric catastrophes. The pessimism implied in this cultural determinism is justified only as long as we do not understand the nature of the intercultural contact enforced by the ever growing global interdependence. Accepting and understanding the force of cultural conditioning we have at last a chance of liberation from its most baneful consequences. According to these assumptions, we have to live with glaring inequalities in the world for a long time, whether of political power, the capacity to manage effective states, or all other forms of human creativity. What, then, should be done? What should be changed?

The answers have been suggested by the foregoing analysis. First of all, obviously, we must learn to recognize the dynamics of culture, of cultural conditioning, and of inter-cultural contact. Just as human beings could not fly until they had understood the law of gravity and a host of other necessities, they cannot coexist peaceably in ignorance of the causes of tension in a multicultural world united by one dominant culture. A better command of reality will also give rise to patience and humility—patience not to expect quick changes and to be satisfied with slow progress where it counts most, in individual consciousness and wills; and humility to submit willingly to the necessities caused by cultural conditioning. From humility also will grow the compassion needed for coping humanely with the existing inequalities. That compassion has to guide both the agents and the victims of cultural change, with special protection offered to the human dignity of the most helpless, so as to ease the hardships of inevitable prestructured cultural change. All must humbly strive to transcend their inherited cultural limitations and search for more inclusive human bonds.

In this change of heart, Westerners must accept the blame for the hardships and tragedies in the developing countries. The anti-imperialist radicals are right. Before the Western impact, traditional societies existed in reasonable harmony within the intellectual, spiritual, and material resources at their command, in precarious balance with their environment. It was the Western impact which forced them, against their will, into a complex world beyond their comprehension and resources, destroying into the bargain the former bonds of community. The anti-imperialists in turn should recognize the inevitability of inter-cultural contact in a shrinking world; let them also appreciate the goodwill and the opportunities that came with the West. In any case, there is no chance of returning to the precolonial era, nor comfort in the nostalgic yearning, sometimes heard, for the good old days of colonialism. Both sides have no choice but to look forward, recognizing that in the global confluence of innocently incompatible cultures they face a huge problem never encountered before in all human experience: how to overcome the obstacles to inter-cultural communication in a tightly compressed world, how to minimize the violence caused by cross-cultural incomprehension?

The initiative for such "humanization" (as Fanon called the liberating process) must obviously come from the West, from the most powerful competitor in the unpremeditated uneven cultural evolution around the world. The developed countries must recognize their ignorance of the consequences of their triumphs. Despite their much-vaunted technological advances, the most advantaged human beings still live in profound incomprehension of the complexity of human relations in their all-too-suddenly enlarged world. Fanon's "Sleeping Beauty"—the self-centered West so blindly convinced of its superiority—must wake up.

Yet as we look around the contemporary world, what evidence do we find for such down-scaling of vanity and pride, for a willingness to take a humbler view of our command over reality?

22

The United Nations as an Agency of Westernization

Let this section conclude with a comment on the United Nations as an additional engine of the world revolution of Westernization in its final phase. As an agency endowed after the horrors of World War II with a vision (inscribed in its Charter) of a more peaceful future for the world, it became in the years following the Bandung Conference the stage for the newcomers in world politics on which to manifest their presence and plead their needs. Guided by long-established precedent, they used the Western ideals embedded in the Charter—above all peace, freedom, social justice, and equality—not only to stake out their claims but also to justify their imitation of Western achievements needed for gaining a proper share of the world's riches and of political power. Thus the envious, critical, and occasionally angry newcomers, who soon formed the majority of the UN's membership, transformed that agency into yet another instrument of anti-Western Westernization operating uneasily amidst the global tensions.

I

The inspiration behind the United Nations unquestionably came from the depths of Western tradition. Prominent writers and public figures had long projected Christian pacifism into international politics, trying to overcome the inhumanity of Europe's perpetual wars. Drawing on that tradition as well as on American idealist universalism, Woodrow Wilson had launched the League of Nations (in which the United States did not participate). Learning from the League's failure, the framers of the United Nations shaped, in the name of the wartime coalition against Germany and Japan, a more ambitious and inclusive agency with global responsibilities, backed up by both superpowers.

317

In that agency the Security Council (composed of representatives from the four most powerful countries—the United States, the Soviet Union, England, and France, plus Jiang Jieshi's China, as permanent members—assisted by representatives from originally six and later ten other countries drawn, by rotation, from the general membership) was entrusted with the overall political responsibility. But the General Assembly, representing all members on an equal footing, also had the right to discuss any business "relating to the maintenance of international peace and security brought before it. . . ."[1] The administrative work of the United Nations was entrusted to the Secretariat, headed by the secretary-general, whose influence varied according to the quality of the person holding that office and political circumstance, yet who always acted in the spirit of the Charter.

Whatever their differences, the superpowers held out hopeful promises not only in regard to preserving peace in the future but also, more important in this context, to improving the life of all humanity as a precondition to peace. In ringing phrases reminiscent of the American political creed they, together with the forty-nine other signatories of the Charter, agreed to "reaffirm faith in fundamental human rights, in the dignity and worth of the human person, in the equal rights of men and women and of nations large and small, and . . . to promote social progress and better standards of life in larger freedom."[2] Admittedly, the superpowers disagreed on the precise meaning of these phrases—the term "democracy" is not mentioned in the Charter—but they also promised "to achieve international co-operation in solving international problems of an economic, social, cultural, or humanitarian character, and in promoting and encouraging respect for human rights and for fundamental freedoms for all without distinction as to race, sex, language, or religion." The United Nations was to be "a center for harmonizing the actions of nations in the attainment of these common ends."

Subsequent sections of the Charter contained specific promises of special interest to the newcomers. "The United Nations shall promote (a) higher standards of living, full employment, and conditions of economic and social progress and development; (b) solutions of international economic, social, health, and related problems; and international cultural and educational co-operation; and (c) universal respect for, and observance of, human rights and fundamental freedoms for all without distinction as to race, sex, language, or religion."[3] The Charter established a special agency, the Economic and Social Council, to take charge of the United Nation's work "with respect to international economic, social, cultural, educational, health, and related matters."

Even more important to the statesmen assembled at Bandung and to the leaders of national liberation movements in countries still under colonial rule was the promise by the founding members of the United Nations to end colonialism. The colonial masters were enjoined "to develop self-government, to take due account of the political aspirations of the peoples, and to assist them in the progressive development of their free political institutions. . . ." The UN Trusteeship Council was to supervise the advance toward self-government in the territories placed under its care, thereby setting an example for decolonization generally.

The sweeping guidelines set down in the Charter were underscored in 1948 by the Universal Declaration of Human Rights. Announcing that the "recognition of the inherent dignity and of the equal and inalienable rights of all members of the human family is the foundation of freedom, justice and peace in the world," the members of the United Nations (except for the Soviet bloc and the Union of South Africa) voted that this Declaration be proclaimed "as a common standard of achievement for all peoples and all nations. . . ." All were obliged "by progressive measures, national and international, to secure . . . universal and effective recognition and observance [of the human rights as defined by this Declaration] both among the peoples of Member States themselves and among the peoples of territories under their jurisdiction."[4]

Article 1, forever quoted in subsequent UN documents, solemnly opened the inventory of human rights: "All human beings are born free and equal in dignity and rights. They are endowed with reason and conscience and should act towards one another in a spirit of brotherhood." Other articles continued in the familiar vein: "Everyone has the right to life, liberty and the security of person. . . . All are equal before the law. . . . Everyone . . . has the right to be presumed innocent until proved guilty." Further clauses affirmed the right to private property, to social security, to work, to health, and to education. In this manner the Universal Declaration of Human Rights universalized the standards which Western society had set for itself since the late 19th century and which Lenin had grandly co-opted for Soviet communism. These were also the standards which appealed to the newcomers; that was how they wanted to be treated.

That inventory of human rights, one might argue from a Third World perspective, lacked one highly desired item, the precondition of all the others: the right to live in a respected, secure, and above all a powerful state. The liberal mind-set which had framed the definition of human rights was deficient in one basic respect: it disregarded the power factor. Taking their own fullness of power for granted, liberals were not ready to confer a similar boon on the others whom they dominated politically or culturally. In their eyes (although not in the eyes of the envious others) equality in political power was not a human right. Yet the Charter of the United Nations at least upheld the "sovereign equality of all nations,"[5] stipulating that there should be no interference in their domestic affairs. All states, in short, were to be treated legally alike, though not in their political ambition. The arena of power politics lay beyond the competence of the United Nations, although its wrangles constantly spilled over into its work. In fact, the United Nations was a microcosm of the global state system, fiercely engaged in the power politics of moral indignation. Viewed in the broadest contexts, its functions could never be separated from the overall contests of anti-Western Westernization.

II

During its first decade and a half the United Nations was patently an instrument of American policy counterbalancing its Cold War hard line. With the help of the United Nations the United States could magnify its peaceful postwar outreach and achieve a universal presence, preparing the way for reshaping the world after its own image. The United States at the outset paid 40% of the UN budget; it supported the largest delegation. It succeeded in establishing the UN headquarters in New York City, the metropolitan center of the world and the most desirable assignment, especially for Third World diplomats. From the UN headquarters American goods and styles of living radiated into the world, establishing the United States as a worldwide model of the good life—and further upsetting the cohesion of traditional societies.

The high point of American ascendancy over the United Nations was reached in 1950. Because the Soviet Union boycotted the Security Council over the nonadmission of communist China, the United States obtained UN backing for resisting the North Korean invasion of South Korea and carrying the war to the Chinese border. Although almost entirely an American campaign, the defense of South Korea was legitimized by the United Nations. How thoroughly the United States had its way in that world body may also be seen in the fact that it did not have to use its right to veto any undesirable Security Council decision until 1970.

It was the Soviet Union which, in a minority position from the start, was driven to that expedient safeguarding the sovereignty of the Security Council's permanent members. Yet, while handicapped, the Soviet Union too considered the United Nations as a tool of power politics. As the UN General Assembly turned into the "town meeting of the world," Soviet power with its Marxist-Leninist globalism was to be heard loudly—once even with the banging of Khrushchev's shoe. That gesture, however, made a bad impression because the decorum of Western parliamentary practice had become part of UN style. Western organizational procedures and formalities, incidentally, also prevailed in the Security Council, the UN Court of Justice, as well as in the Secretariat, absorbing endless time and energy. Western skills and perspectives likewise dominated the global network of the UN agencies, fashioning from people still tied to their governments and their cultures a reasonably effective international civil service somewhat protected from the direct impact of global power politics.

In short, whether considered as a tool of the superpowers or as an autonomous body, the United Nations was a factor in the subtle power game of cultural influence, an agency for the Westernization, if not the outright Americanization, of the world. It was the guardian, watchdog, and symbol of the new globalism created by the expansion of Western culture, yet offering something for all, especially for the newcomers to global politics. It was charged—to focus here on those functions of special interest to the latter— with the responsibility for helping its member states give reality to the Charter's vision. The newcomers were soon the UN majority.

III

In the late 1950s the balance of power in the global arena, and consequently the political orientation of the United Nations, began to shift as the newcomers to statehood—the nonaligned countries of the Bandung Conference—asserted themselves in international relations and took the lead in the UN General Assembly. In 1960 a large batch of newly independent African states was admitted: Cameroon, the Central African Republic, Chad, Dahomey, Gabon, the Ivory Coast, Malagasy, the Congo, Senegal, Somalia, Togo, Upper Volta, and Nigeria. In late 1961, after the death of Secretary-General Dag Hammarskjöld in the Congo, the two top UN officials came from Asia, the new secretary-general, U Thant from Burma, and the president of the General Assembly, Sir Zafrulla Khan from Pakistan; paired with Ralph Bunche from the United States as undersecretary for special political affairs was another Asian, C. V. Narasimhan. By that time, not surprisingly, the Soviet Union had seized the initiative in the General Assembly for a demonstration of anti-colonialism, urging the adoption of a Declaration on Independence for Colonial Countries and Peoples. The political climate in the United Nations was changing.

That Declaration, illustrating once again the dynamics of anti-Western Westernization, began by reciting the promise of the Charter "to reaffirm faith in fundamental human rights, in the dignity and worth of the human person, in the equal rights of men and women and of nations large and small . . . and to promote social progress and better standards of life in larger freedom." Calling attention to "the passionate yearning for freedom in all dependent peoples and the decisive role of such peoples in the attainment of their independence," it solemnly proclaimed "the necessity of bringing to a speedy and unconditional end colonialism in all its forms and manifestations."[6]

The discussion pitted the Soviet Union against the former colonial powers, restating the arguments already heard at Bandung. The Soviet speakers, including Khrushchev himself, pointed to the fact that more than 100 million people were still living in conditions of colonial oppression and exploitation. The speakers for the colonial powers countered by pointing to the success of decolonization and by citing the fate of the peoples placed under Soviet rule against their will. In the end, the Declaration was approved by a vote of 89–0. Nine countries abstained from voting for what they considered a one-sided document, including France, Belgium, Portugal, Spain, South Africa, Great Britain, and the United States—the traditional representatives of the West.

After this prelude, liberation from colonial rule was moved to the foreground in UN discussion, revealing the complexity of the issue. In 1961, when Indian troops liberated the small Portugese colony of Goa, passion flared high as the Afro-Asian bloc in the General Assembly voted against a resolution calling for a cease-fire in that petty war. If, as India's representative, Krishna Menon, argued, "colonialism is permanent aggression,"[7] counterviolence was justified whatever the ideals of the United Nations. In the same year the savagery of decolonization in the former Belgian colony of the

Congo revealed not only how difficult it was to reconcile, in an African setting, the passionate yearning for freedom with peace and human rights, but
also how little the UN could do to help.

IV

Deriving its name from its great river, the Congo was the largest sub-Saharan
African state, inhabited by a much fragmented population lacking a common
language (except the official one of French) and common political institutions. The Belgian government had done nothing to prepare the people for
independence. Self-determination after liberation from Belgian rule in mid-
1960 led to anarchy; soldiers mutinied and people turned violent. The political parties formed in anticipation of independence proved ineffectual, riven
by factions and personality clashes. Internal discord was fanned by powerful
competing outsiders. The country's wealth lay in its mineral deposits of
cobalt, industrial diamonds, and copper, all mined in the Katanga (Shaba)
area dominated by big business tied to the industrial countries; Katanga was
held by white mercenaries (some hailing from South Africa). At the opposite
end of the political scale, the Soviet Union sided with the most radical anti-
imperialist faction headed by Patrice Lumumba; Khrushchev was eager to
gain a Soviet foothold in central Africa. Was, then, the African heartland in
these years of proud decolonization and pan-Africanism to become a battlefield of civil war and foreign intervention?

Here beckoned an opportunity to carry out the mandate of the United
Nations. Dag Hammarskjöld, its secretary-general, known for his sympathies
with the emerging Asian and African countries and his eagerness to give reality to the UN's ideals, was ready to seize it. To supervise the establishment of
an independent Congolese government, he assembled a sizable UN expeditionary force (known as ONUC), recruited from African and Asian armies
plus Canadian, Irish, and Swedish units and counting at its peak twenty thousand men; it was to supervise the establishment of an independent Congolese
government. ONUC faced an exceedingly difficult task, in part because of
opposition within the Security Council. The Soviet Union opposed UN intervention, which weakened its own opportunities, while the Western European
countries wished to preserve their hold over the Congo economy. Both camps
were suspicious of Hammarskjöld's initiative.

Even greater difficulties confronted the UN operations within the Congo
itself. How could an ill-assorted and inexperienced multinational military
force operate in a tropical country in which all public service, minimal to start
with, had collapsed and where the accumulated tensions were ready to break
into open war at any moment, between Africans themselves and between
blacks and whites (with the missionaries sometimes the hapless victims).
Besides trying to impose order by armed force, the United Nations also
imported technical experts and schoolteachers to man essential public services, with little effect. Eager to assist in the work, Hammarskjöld went again
to the Congo in September 1961—to die in an airplane crash.

His successor, U Thant, continued the United Nation's mission, managing after further fighting to subdue the secessionists in Katanga yet unable to install a firm government. In the summer of 1964 the UN expeditionary force was dissolved, for lack of financial support rather than for success in its work. The Congo operation had been a costly venture, leaving the United Nations in financial distress and the Congo in turmoil, until in 1965 General Joseph-Désiré Mobutu, educated at mission schools and trained in the Belgian army, seized power. In the following year he established a presidential dictatorship lasting, with the help of massive foreign credit and drastic abuse of power, to the present. In 1971 he renamed his Congo river state "Zaire," after the Portugese term for "river"; in the following year he changed his name to Mobutu Sese Seko. No respecter of human rights, he was yet in his Congolese way a Westernizer promoting development—in pursuit of "national authenticity." The function of the UN secretary-general, meanwhile, was considerably curtailed. Hammarskjöld had acted too boldly and too expensively. As an autonomous agency for peaceful and unobtrusive Westernization, the United Nations possessed only a limited capability; it was not to enter the arena of power politics.

V

Yet the United Nations continued to press its civilizing—or Westernizing—mission. In the face of the inhumanities committed in the struggle of decolonization and in the experiments of state-building, it reiterated its humanitarian principles in two documents approved in 1966. The first, called the International Covenant on Economic, Social and Cultural Rights,[8] summed up the nonpolitical objectives of independence. After observing in passing that "the individual, having duties to other individuals and to the community to which he belongs, is under a responsibility to strive for the promotion and observance of the rights recognized in the present Covenant"—a generally much neglected appeal—it listed, in weighty UN legal language, the obligations assumed by all UN member states. They were to guarantee the right to work, to form trade unions (with a proviso that "lawful restrictions" could be imposed upon them by the police or armed forces), to enjoy an adequate standard of living (including food, clothing, and housing). As for food, more of it was to be provided by "making full use of technical and scientific knowledge." Education, too, was prominently included, "for the full development of the human personality and the sense of its dignity. . . ." For a touch of realism the covenant allowed that the state may restrict such rights, but "solely for the purpose of promoting the general welfare in a democratic society." Living up to these standards was a difficult task in the year of military coups in Ghana and Nigeria—coups that yet could be justified in just these terms, because the regimes just overthrown had not lived up to expectations.

The second document, the Covenant on Civil and Political Rights,[9] supposedly restrained all military regimes. It spelled out the conditions protecting individuals: the rule of law, constitutional procedures, limited use of the death penalty, no torture, no slavery, and, in case of a criminal charge, a "fair and public hearing by a competent, independent and impartial" legal tribunal. In addition, this covenant insisted that "every citizen shall have the right and the opportunity, . . . without unreasonable restrictions, to take part in the conduct of public affairs, directly or through freely chosen representatives; to vote and to be elected at genuine periodic elections" by secret vote under universal and equal suffrage. The right of peaceful assembly was also recognized, subject to restrictions "in conformity with the law" when necessary "in a democratic society in the interests of national security or public safety. . . ." In laying down this code of political conduct the United Nations affirmed the highest standards of Western statecraft as a global guideline that could not be formally denied no matter how much violated in practice by the new military regimes.

In some ways the United Nations was a pulpit from which to preach political morality to a global congregation of would-be believers too busy—and too cynical—to pay serious attention.

VI

More serious attention meanwhile was given to the practical ways of building the material base for the humane system of governance envisaged in those covenants. One of the major preoccupations of the United Nations since the 1950s was pushing development, assisting the developing peoples to match the comforts and resources of the developed ones. The simplified UN vocabulary, carefully avoiding all reference to "capitalism" and "socialism," listed merely three categories of societies in the international community: the developed, the developing, and the least developed, hierarchically arranged according to Western standards. In the name of development, the United Nations was an active agent of global Westernization, an agent with many arms.

To list only a few of its specialized agencies and autonomous organizations concerned with development (noting how many carry that term in their name): the International Atomic Energy Agency (of special interest at the Bandung Conference), the General Agreement on Tariffs and Trade, the Food and Agriculture Organization, the United Nations Educational, Scientific and Cultural Organization, the World Health Organization, the International Monetary Fund, the International Development Association, the International Bank for Reconstruction and Development (also known as the World Bank); the International Fund for Agricultural Development, the UN Conference on Trade and Development, the UN Institute for Training and Research. All of these agencies were tied to member governments, especially those, like the International Monetary Fund and the World Bank, concerned with the flow of money and credit, the ultimate fountainhead of development. Whatever their duties, all of these agencies universalized Western forms of organization and Western accomplishments, imprinting them, however imperfectly, upon the entire world, ostensibly in the service of development.

The 1960s were declared a "Development Decade" in which, in 1966, the UN Development Program was established, together with the Industrial Development Organization. Yet progress was disappointing, especially when, in the early 1970s, the world economy slowed down, in large part because of rapidly rising oil prices imposed by a triumphant OPEC—all the more reason to press the issue.

The 1970s, declared the Second Development Decade, introduced some novel factors into UN development politics. In 1972 Mao's China became a permanent member of the Security Council, competing with the Soviet Union in inciting the anti-imperialism of the newcomers, a minor complication. More important in UN business was the conspicuous rise of a new awareness regarding worldwide economic and ecological trends. Scientists asked: what were the consequences of the hothouse pace of the new globalism upon the world's rapidly increasing population and upon the common habitat? The new sense of worldwide responsibility for the earth's physical resources was expressed in a series of world conferences, the first held in Stockholm, "The United Nations Conference on the Human Environment"(1972). It established a UN Environmental Program, an earth-watch on pollution of air and water, based in Nairobi, Kenya. Two years later followed the "World Population Conference"(1974), exploring ways of coping with the rapid population growth, especially in Third World countries. Subsequent UN-sponsored world conferences were devoted to the problems of women, to human settlements, to renewable energy sources. Thus the West's critical rationality regarding human and environmental problems was popularized and universalized.

The centerpiece, however, of the Second Development Decade was the UN drive for a New International Economic Order (NIEO), launched in 1974.[10] As described in the usual long-winded UN rhetorical style, the new order was "based on equity, sovereign equality, interdependence, common interest and co-operation among all states irrespective of their economic and social systems, which shall correct inequalities and redress existing injustices, make it possible to eliminate the widening gap between the developed and the developing countries and to ensure steadily accelerating economic and social development and peace and justice for present and future generations. . . ." It addressed "the most important economic problems facing the world today." The declaration establishing the NIEO then listed the major changes of the recent past justifying the call for a new economic order: decolonization, which had made a large number of peoples and nations into "members of the community of free peoples"; and technological progress in all spheres of economic activities, which had provided "a solid potential for improving the well-being of all peoples." These were positive achievements.

But—to quote in full the ever present traditional anti-Western suspicions:

> the remaining vestiges of alien and colonial domination, foreign occupation, racial discrimination, apartheid and neo-colonialism in all its forms continue to be among the greatest obstacles to the full emancipation and progress of the developing countries. . . . The benefits of technological progress are not shared equitably by all members of the international community. The devel-

oping countries, which constitute 70% of the world's population, account for only 30% of the world's income. It has proved impossible to achieve an even and balanced development of the international community under the existing international economic order. The gap between the developed and the developing countries continues to widen in a system which was established at a time when most of the developing countries did not even exist . . . which perpetuates inequality."

The complaints, however, were not unduly pressed. The accent rested on the fact that "the political, economic and social well-being of present and future generations depends more than ever on co-operation between all the members of the international community on the basis of sovereign equality and the removal of the dis-equilibrium that exists between them." As for the last point, the developed countries were admonished to give "preferential and non-reciprocal treatment for developing countries wherever feasible, in all fields of international economic co-operation whenever possible" (the qualifications here underlined indicated the supplicatory character of the plea for the new economic order—what power did the UN majority have to enforce its wishes?). The developing countries, however, were advised to do their share by concentrating their resources on development and cooperating for this purpose among themselves.

Due emphasis was given to the need among developing countries for access "to the achievements of modern science and technology" and for promoting the transfer of technology for their benefit. The expansion of industrialism was urged even more strongly. All necessary efforts were "to be made by the international community to take measures to encourage the industrialization of developing countries" and to increase their share in the world's industrial production. No explanation was offered, however, how this might be accomplished, except indirectly. The declaration affirmed "the right of every country to adopt the economic and social system that it deems the most appropriate for its own development"; socialist planning was a legitimate method. The subsequent technical advice contained in this document dealt with the terms of trade and with the practices of international commerce and finance that were to be changed in favor of the developing countries. These technicalities as well as the policies advocated imposed further burdens both in expert knowledge and domestic policy on officials in the developing countries, contributing to their apprenticeship in Western skills.

The purpose of the New International Economic Order was further spelled out in the Charter of Economic Rights and Duties of States debated subsequently.[11] Faithfully restating the ideals of the United Nations, that document stressed the right of each state "to regulate and exercise authority over foreign investment within its national jurisdiction in accordance with its laws and regulations and in conformity with its national objectives and priorities." It further entitled each state "to regulate and supervise the activities of transnational corporations within its national jurisdiction," with an accusing finger pointing at transnational corporations: they "shall not intervene in the internal affairs of a host state." The host state was even empowered "to nationalize, expropriate or transfer ownership of foreign property" (with due compensation to the owners). To avoid possible discrimination on that account, each state was guaranteed "the right to engage in international trade and other forms of economic co-operation irrespective of any differences in political, economic or social systems."

The list of duties imposed upon the members of the United Nations by this document reads like a sermon. They were "to co-operate in the economic, social, cultural, scientific, and technological fields," especially for the benefit of the developing countries. The developed countries in particular were "to co-operate with the developing countries in the establishment, strengthening and development of their scientific and technological infrastructures and their scientific research and technological activities. . . ." All states had "the duty to cooperate in promoting a steady and increasing expansion and liberalization of world trade and an improvement in the welfare and living standards of all peoples," with the promotion of "general and complete disarmament" thrown in for good measure. Special attention was given to the protection of "the sea-bed and ocean floor and subsoil thereof"—these were "the common heritage of mankind" to be exploited in the interests of all, not of the rich and powerful countries alone.

The charter concluded in the customary UN pulpit style: "All states have the duty to contribute to the balanced expansion of the world economy, taking duly into account the close interrelationship between the well-being of the developed countries and the growth and development of the developing countries. . . . The prosperity of the international community as a whole depends upon the prosperity of its constituent parts." What could be more obvious?

Yet, given the bias in favor of the developing countries, was it surprising that leading developed countries, including West Germany, England, and the United States, voted against this charter, and that others, like France, Israel, Japan, and Canada, abstained from voting? Undaunted, the General Assembly in 1974 endorsed it with overwhelming support. Paradoxically, the anti-Western sentiment reflected in this vote, as in the sentiments of the UN General Assembly as a whole, arose from the fact that the developed countries— the old West—would not help the newcomers to Westernize themselves rapidly enough; it seemed as if the developed countries wanted to keep the developing countries under their thumb. In response, the latter turned anti-Western because of their impatience to be Westernized.

VII

But development was more than a power struggle. It entailed a profound recasting of traditional life, a subject not examined in the United Nations. Examining in the larger contexts the UN commitment to development, continued after 1980 with increased vigor into the Third Development Decade, one cannot help being distressed by its unreality. One finds no consideration how the recommendations sent down from the lofty heights of the UN headquarters could be reconciled with the tensions prevailing on the ground floor of life. Admittedly, in November 1981 the General Assembly adopted a resolution "calling for further UN work on the national experience of developed and developing countries in achieving far-reaching social and economic changes for social progress." But it merely "invited Member States to give special attention to the social aspects of development so as to increase the well-being of the population based on its full participation in development and a fair distribution of benefits." These were pious generalities without relevance to the real problem of development: how in the process of nation-building to jump over centuries or even millennia of cultural evolution; how to reculture a people entirely ignorant of the road to an externally prescribed future.

By its very structure, one might argue, the United Nations was incapable of coping with this central aspect of development. The UN resolutions and declarations were based on agreement among the sovereign member states, many of them in the midst of uncomprehended cultural change and rather eager to keep their domestic turmoil out of sight to avoid embarrassment. The forever restated principle of noninterference in the domestic affairs of the member states prevented open discussion and analysis of the most pressing and baffling tasks within the purvue of the United Nations. UN staff members could offer no relevant insight. And would the most vulnerable member states, promised equal rights and consideration, lay bare the insufficiencies of their cultural resources? Would their audience—especially in the metropolitan centers—be able to respond with sufficient compassion and understanding?

One can hardly avoid the conclusion, then, that the improvements attempted by the New International Economic Order were sadly superficial. They offered a package of skills and techniques based on Western experience; they raised popular expectations to the height of Western standards, all within a global perspective derived from Western premises. But how were the developing countries to put that package to effective use on the ground floor of life in their distraught societies? And, more disturbing, if they succeeded even minimally, would the improvements contribute to peace, as was assumed in the United Nations? Or would the enhanced resources of developing countries merely stimulate their political ambition following the example set by the superpowers? Events since the 1960s certainly proved how little the developing countries were capable of, or interested in, preserving peace among themselves.

VIII

One wonders, then, in conclusion, to what extent the United Nations, with its high-flown ideals and incomprehension of the dynamics of the age, could make a positive contribution to the elemental needs affecting all humanity. It has not reduced the gap between the rich and the poor. Its achievements as a peacekeeper are debatable. And despite unending protests about the exorbitant costs of the arms race, it did not stop the escalation of hostility between the superpowers. All told, it offered only incomplete answers to the big problem: how in the age of nuclear weapons could the precipitous global confluence of the world's disparate cultural and political traditions be channeled into orderly cooperation, reconciling competitive human wants with the limited resources of Planet Earth. Did the United Nations advance humanity's political skills to match its scientific and technological skills? Or did it merely preach a well-meant but unrealistic sermon based on the limited human insights available?

All told, the United Nations has been only a minor force in the world revolution of Westernization. Its budget of approximately $1.5 billion at its peak was very much smaller than that of the great multinational corporations; its staff of eleven thousand could not match these multinationals' armies of employees. Its significance, and consequently its finances, depended on the goodwill of the superpowers, especially the United States; and that goodwill in turn depended on public awareness of its functions and appreciation of their usefulness. Its future is uncertain, subject to factors discussed in the final section of this book.

One may approvingly argue, however, that in the decades after World War II the United Nations has helped to raise human awareness to global perspectives, thereby tightening the cords binding all mankind together, for better or worse. It has advanced the constructive skills of Western culture as well as its political combativeness. Yet it also has rung more loudly than had been done before in all history the bells for humaneness, peace, and worldwide cooperation in the human family, undeterred by widespread indifference and the incomprehension of the forces at work in the contemporary age.

After nearly ten years as secretary-general Kurt Waldheim in 1981 summed up the motives supporting the vision of the United Nations:

> Our experience in this century has shown beyond the shadow of a doubt that a world organization must be developed without delay and with the widest possible participation to enable us not only to deal with the effective maintenance of international peace and security, but to bring order into many other aspects of human activity, which, owing to the technological revolution, are now closely and vitally intertwined. In other words, we are living in one world whether we like it or not, and we have to develop institutions capable of regulating and guiding that world.[12]

The basic question in his mind was: "whether we shall be able to take advantage of our awareness and knowledge to act together and in time before our problems overwhelm our capacity for dealing with them in an orderly and peaceful manner."[13]

Which takes us to an independent assessment of the world in which we live.

VII

The Contemporary Age:
Reflections on the
Global Present

The characteristic danger of great nations, like the Roman,
or the English, which have a long history of continuous
creation, is that they may at last fail from not
comprehending the great institutions they have created.

Walter Bagehot

23

The Human Condition at the
End of the 20th Century

To return now to the overall perspectives, focusing on the present moment that shapes—or should shape—our understanding of the past by our assessment of "what is expected and what is hoped for" (as Hegel said).

The present moment covers a flexible and many-layered span of awareness. We are caught in our daily routines of petty decision-making. Yet we also think ahead, compiling schedules for future obligations, choosing roads for travel of long duration. While looking forward we also glance back over decades and centuries, celebrating birthdays and anniversaries. The size of the present moment expands with advancing age and responsibility as we rise to more inclusive perspectives above the ground floors of immediacy. Taking a global view especially requires a copious sense of time. In our most reflective moods we place ourselves into the largest time scales, backward as well as forward, enlarging our sense of selfhood and magnifying our identity *sub specie eternitatis*. In this expansive, detached, and inescapably impressionistic spirit we now take a look at the present age, compressing its immensity—an immensity even greater, it seems, than the immensity of the past—into a short summary flavored with a moral bias geared to "what is hoped for."

I

The world revolution of Westernization, "a gradual but gigantic revolution—the greatest and most momentous social, moral, and religious, as well as political revolution which . . . the world has ever witnessed," has run its course, virtually completing its work. "The most refined principles" of Western society have now been implanted in a non-Western population "in whose history, habits, and traditions they have had no previous existence." In the non-Western world the revolutionary impact from without has been replaced by the internal revolution of reculturation through "development" in conformity with contemporary requirements. At the same time, the anti-Western counterrevolutions have been absorbed into the domestic politics of developing countries and into the routine wrangles of the global state system; the Western tools of cultural and political domination are now the property of all humanity. The West has been proportionately scaled down, although it still retains basic assets in its cultural continuities. As an analytic concept for application to the present age, however, it is becoming increasingly meaningless.

Most crucially, in completing its course the world revolution of Westernization has recreated on a global scale the preconditions for its own advance; the Western hothouse combination of unity and competitive diversity has been extended from Europe and North America to cover the entire earth. The contemporary world, far more than traditional Europe, constitutes a close-knit network of almost instantaneous interaction based on an inescapable uniformity of assumptions about essentials, above all the essentials of power. And even more dangerously than traditional Europe in its heyday, the contemporary world abounds with diversity—diversity of languages, religions, standards of living, historical experiences, and culture. The uniformities do not extend into these inward worlds, where the source of war and catastrophe lies.

Unity and diversity now combine in generating an extraordinary competitive creativity. The competitors can draw on the cultural skills and material resources of all parts of the world; they also benefit from the vastly accelerated tempo of interaction. The techniques of communication, in word and even more importantly in image, are producing an intensely competitive self-consciousness among all peoples of the earth, whatever their differences. Individual and collective egos in heightened sensitivity are incited to invidious comparison across all cultural boundaries. The accumulated tensions explode in small-scale terror and war, while the superpowers pile up weapons capable of destroying all civilized life. The hothouse pace of Western evolution in war and peace has been superheated into an uncontrolled global race along a narrowing track between good and evil, between human progress and utter devastation. The global community has become both more uniform and more anarchic.

This unprecedented and still uncomprehended condition of human existence deserves a closer look.

II

The world revolution of Westernization has covered the world and all its diversity with a thick layer of separate but interrelated uniformities. The first and outermost layer is the hardest, concerned with power and statehood. It stems from the universal urge of individual and collective life to prevail through the arts of peace or war, to impose change on others rather than be changed by them. In the absence of a universal culture, conflict is bound to take the form of violence as the ultimate means of communication. Peace and mutual service presume the existence of common bonds; violence is no respecter of cultural differences. The perennial human assertiveness has been collectivized in statehood. The power that counts in the contemporary world is that of states, the primary agents in the dynamics of the anarchic global community.

Statehood is now the universal framework for human existence around the world. In the West it has long been established; elsewhere non-Western peoples have recently been recultured to comply with its requirements, however superficially and reluctantly in many cases. Statehood has universalized Western institutions like government bureaucracy, diplomatic service, and armed forces. It has transformed nonpolitical people into citizens, often against their will, always in an effort to mold their capacities for the global competition of power. It has thereby made into universal requirements all Western accomplishments that contribute to power, including science and technology, literacy and education, industrialism, effective communications, and large-scale organization, as well as, more subtly, the cultural skills that sustain the visible foundations of power. There is no escape from the pressures of the global power competition and from the necessity of keeping up with its strenuous pace.

The global state system is nervously sensitive to the slightest seismic disturbances whether caused by discord in its member states or by the tensions between them. The shock waves are recorded everywhere and apprehensively assessed for their effects on the overall balance of forces as well as on regional frameworks of conflict. The common apprehension has caused a Machiavellian alertness among the competitors, a readiness to relegate domestic politics to second place behind foreign policy. The correlation between the dynamics of domestic affairs and the pressure of foreign relations all too often inclines toward the latter. In this manner the universals of power are ever more deeply implanted in global society, whatever the divisive cultural differences.

Enforcing a continuous mobilization for competition, these universals are also likely to preserve the existing distribution of power in the world. Given the pervasive spirit of competition, none of the contenders can afford to slacken their competitiveness, whether in the arms race or in political cohesion and industrial productivity. While the preeminence of the West is being levelled down, its direct descendants, including the United States, will not disintegrate like the Roman Empire; "capitalism" will not collapse. The bearers of the Western tradition cannot be defeated on military grounds—except by a nuclear holocaust destroying all human accomplishments.

There will no doubt occur minor ups and downs, and the relative position of each state will be more accurately defined by its basic resources. The majority of states are so small and weak as to carry virtually no weight in the global scale; yet even they play their part in regional subdivisions within the overall balance of power. The competitive urge is too strongly entrenched among states (as well as among their citizens) to allow a collapse of their political will and their desire for a respected global presence. By necessity, every state, every government, has to keep up with the global competition and adjust—or reculture—its domestic arrangements accordingly. By the same necessity, foreign policy, in the present global system, as formerly in the European state system, tends ultimately to rely on the force of arms. The arms race is part of every competitive state system. Now its pace is set by the two superpowers; the others follow suit as best they can, often abetted by the big rivals. The United Nations certainly can prevent neither a major war nor, more generally, the aggravation of anarchy through the universalization of power-oriented statehood.

Admittedly, the polarization of global politics has led to a measure of international cooperation built around the superpowers as represented by military, economic, and political alliance systems (in the manner of NATO or the Warsaw Pact). These associations offer, at least on the American side, a modest sign of progress toward transnational association. The European Economic Community is another example—based on common origins in Western culture and safeguarding the sovereignty of their member states. In addition, the industrial democracies—the United States, Japan, the countries of Western Europe—constitute a political triangle of states with globally converging economic and political interests in mutual accommodation.

There is hope, then, for peaceful cooperation with the help of the second and somewhat softer layer of universals operating underneath the hard layer of state power and to some extent counteracting it: the layer of community-building trade, finance, scientific research, and humanitarian concern. These universals, growing out of the age-old practice of transcultural contacts reinforced in the mid-20th century by the peaceful outreach of American universalism, are expressed most clearly in the Charter of the United Nations and the guidelines of its various agencies, including the powerful World Bank and the International Monetary Fund, which cover the world even though they do not formally operate within the Soviet bloc. Associated with the United Nations is an expanding body of international law (some of it of long standing) facilitating interaction through the regulation of time zones, postal service, the global standardization of weights and measures, the interchangeability of money, and communication by sea, air, or airwaves, and by satellites in outer space. Health, safety, crime control, and even the conduct of war are likewise covered, although perhaps inadequately, by international agreements; weather forecasting too has gone global. Cooperation in this layer again promotes or reinforces other universals like science and technology, the techniques of large-scale organization, and even the use of English as the preferred international language.

With the help of these ground rules of global interaction, the multinational corporations, the big trading companies and banks, impose their own universal practices, whether of management, industrial production, or the sale of commodities with a universal appeal. Automobiles, radios, electronic equipment, computers, medicines, and soft drinks are the same around the world. These commodities, directly or indirectly through the exchange of personnel, in turn enforce new uniformities, as of styles in clothes and architecture, or in international sports. Every four years the Olympic Games serve as a minor flywheel of universalization in tastes, lifestyles, and popular expectations. Tourism links the wealthy and the poor, as does academic or business study of world conditions. Every day the media transmit news from all over the world, leaving a vague residue of common impressions and ambitions.

As in power politics, so in peaceful cooperation: keen competition prevails, among rivals from within a given country as from around the world. Scientists and engineers vie with each other as do bankers, businessmen, scholars, artists, musicians, marathon runners, and fashion designers. The worldwide competition encourages bigness of operation. In order to be counted, one has to be omnipresent, at least in image. The new globalism has raised the pressure for competitive universalization; anything of significance has to be put on display in the global supermarket. As a result, the traders in that market are highly sensitive to the slightest change in prevailing conditions, whether caused by financial or commercial transactions, by new inventions, by natural catastrophes, rumors of war, or war itself. This layer of universals contributes most effectively to building a true global community. It limits the competition among power-oriented states; but it also is at their mercy. The community-building forces in the global world unfortunately do not outpace the global anarchy, even with the help of yet another layer of universals.

The third layer of universals is the newest and softest, the most idealistic, yet also the least binding and most troublesome among the ties that bind together contemporary humanity. It can best be stated in terms of the Universal Declaration of Human Rights adopted by the United Nations in 1948. That document—its ideals deserve repeated mention—stipulates "recognition of the inherent dignity and of the equal and inalienable rights of all members of the human family" as "the foundation of freedom, justice and peace in the world." It further enjoins "all peoples and all nations . . . to strive by teaching and education to promote respect for these rights and freedoms . . . and to secure their universal and effective recognition and observance. . . ." While the declaration stresses the protection of individual rights in the manner of Western legal practice, it also includes "the right to a nationality" and to "a social and international order in which [these] rights and freedoms . . . can be fully realized." Obviously, the provisions of this Declaration are widely, if not universally, disregarded. Yet they cannot be repudiated in principle; they are used, however hypocritically, to justify policies and practices within countries and in international relations. They are an ideal, a lodestar for human aspirations as well as a measurement of humaneness in the international competition for prestige. Which country observes these ideals more diligently?

Paradoxically, this soft inner layer of idealistic universals closely interacts with its opposite, with the underlying cultural diversities which divide humanity. It stimulates them and thereby promotes counterforces hostile to its conciliatory intentions. Given the stark differences in the world, the Universal Declaration of Human Rights, rather than advancing peaceful cooperation, endangers it; trying to cure the global anarchy, it promotes it.

In the absence of a unifying common culture the universalizing appeals of freedom, justice, and peace escalate the anarchy of competitive interaction to a new and often dangerous intensity; they harden diversity. Freedom especially is a troublemaker. Liberty, as Paul Nitze had rightly observed in 1950, "is the most contagious idea in history"—but with upsetting effects. Freedom encourages spontaneity; it justifies assertiveness. It promotes a particularist self-consciousness in protest against the universals of the outer layers or against unfriendly neighbors. The right to be free not only legitimizes and further enhances the existing differences, it also arouses ambitions for giving indigenous cultures a worldwide significance. Everybody feels entitled to be counted, if necessary by shots heard around the world. In the sharply divided human family the quest for freedom, justice, and peace always raises the elemental issue of power: freedom for whom? justice for whose benefit? peace on whose terms? In the name of the highest ideals the very unity of the global system promotes disunity, within insufficiently cohesive states as well as between them.

III

The range of diversity around the world—to examine now the major causes of global anarchy—is obvious, but the depth of the differences is not commonly appreciated. In the global setting the differences are far greater than in traditional Europe. Humanity represents a disorderly mosaic of sharply contrasting cultural identities. People are set within their separate language communities, their separateness often reinforced by a common religion and historic experience. In human evolution, cultural and political evolution has varied profoundly. At one extreme we still find stateless nomads, at the other the sophisticated and resourceful metropolitan citizens of the United States or Japan, with all manner of gradations in between.

Within each of these cultural islands the deepest layers of psychic sensibility, of will and motivation—the sources of individual and collective identity—persist despite the disorienting inroads of universalizing reculturation. These tradition-bound mini-worlds allow scant communication with the outside. Their deities appear in different shapes, teach different creeds, and speak in different tongues (often written in different scripts, if written at all). Though adjusting to change, indigenous languages act as storehouses of collective memories and cultural uniqueness, even where a European language serves as the language of national unity, as in many African states. More generally, there exists no common grasp of an ultimate reality for all humanity, no common metaphysical truth, no shared psychic foundation for enduring economic or political cooperation, or even for survival. For the bulk of the world's peoples the outer layers of globalism and the inner worlds of the spirit do not harmonize. The unity persists only among the peoples of the West (and even here perhaps with lessening intensity).

Since it is not commonly realized, especially in the United States, how deep the barriers of incomprehension are between cultures and how much they aggravate global anarchy, it is necessary here to add a cautionary note about the problems of inter-cultural understanding on the ground floors of human existence. Westerners—and Americans foremost—are conditioned to think in terms of a culture-transcending common human nature. Continuing an ancient effort at transcultural communication embodied in Roman law, that notion has facilitated peaceful cooperation within the European system; it has lent moral zest to Western expansionism. Yet it has also obscured the differences separating peoples and cultures; it has not prevented war and genocide. Commodities indeed can easily cross cultural borders, yet not complex cultural accomplishments, and least of all sentient human beings, as the experience of non-Western intellectuals or immigrants anywhere shows. For the purpose of counteracting the global anarchy by peaceful understanding at the most sensitive center of collective human awareness, it is wisest to assume that we have inherited no common human nature from the past; it has to be created in the future, under immense difficulties. A conditional cultural relativism is a necessary step toward transcultural understanding.

IV

In the light of these reflections on the interaction of unity and diversity in the supercharged furnace of globalism, let us next explore—tentatively, as viewed from a high vantage point—the human condition at the end of the 20th century.

In the global confluence of all cultures, religions, and historic experiences evolved over millennia, all of humanity's cultural heritage has now come into full view. Trying to preserve their traditional identity, people of different creeds and lifestyles are driven to advertise their insights and skills, taking advantage of the new facilities of globalism. In the great metropolitan centers, especially in their intellectual circles (or their peripheries), the world's great religions vie with each other; lifestyles from different parts of the world are on display. The world has become a shopping mart crammed full with humanity's riches, ranging from items for the crassest sensual self-indulgence to compendia of practical knowledge, labor-saving devices, industrial machinery, and psychological advice and spiritual verities. The present generation is born to shop—or at least window-shop—in the world's supermarket, challenged but also bewildered by the choices offered and increasingly overtaxed by the decisions to be made.

There is no doubt: the heated interchange of the global confluence has ominously complicated human life. All people suffer from the overload of detail to be managed, of information to be digested, of stimuli to be sorted out, and of civic duties to be performed. The contemporary world, adding a global story to the already overtowering structures of the nation-state, has grown over everybody's head. Even in the most privileged societies, with their abundance of labor-saving devices and energy-preserving services, there is never enough time in the day nor enough physical vitality to cope with the obligations imposed by professional responsibility or conscience. The details of technical knowledge required in the management of modern society become daily more complex, enforcing yet greater specialization at the price of a corresponding shrinkage of overall awareness; more attention is given to machines and organizational technicalities than to human need. At the same time, human relations have become more complex, as have public issues locally, nationally, and internationally. Ever more information becomes available, much of it dubious, biased, or contradictory, calling for independent assessment for which there is neither time nor capacity.

That overload has engendered a mood of frustration, helplessness, and fatalistic escapism, which—especially in affluent societies—provides opportunities for the ever more numerous purveyors of hedonistic distraction and self-indulgence, some of them drawn from easygoing preindustrial cultures. They add to the burden of choosing and decision-making, sapping vital energies from constructive attention to the individual psyche, to family, to local and national community, and to the world at large. The range of intellectual and spiritual outreach as well as the capacity for mutual adjustment shrink under the overload of detail, important or trivial, to be managed in daily life. At a time when enlargement of perspectives and flexibility are needed more urgently than ever, people become more immorally self- or even body-centered. The globalization of life is encouraging a counteruniversalist contraction of human awareness, an aversion to globalism.

If the overload of globalism has such constricting effects on the most privileged and affluent people, their poorer and more inexperienced neighbors can bring even less energy and attention to the management of global affairs. Hard-pressed by the necessity to conform to the requirements of an unfamiliar urban-industrial society or of reculturation generally, and, even more frequently, barely surviving even by backbreaking toil, the great majority of humanity can hardly raise its perspectives above the ground floor of subsistence. The global world, with its many demands for cooperation, lies beyond their comprehension. How can they realize that they live in One World?

Under these conditions, the tasks of government everywhere are made more difficult. Governments are caught between conflicting pressures, suspended between the needs of domestic politics and the urgencies of international relations, between private individuals on the ground floor of life and the outer universals of power politics where the fate of humanity in war and peace is determined. Peaceful cooperation between states is made easier if the public in each state can rise to the necessary overall perspectives. The more restricted the public's perspectives, the more self-centered public opinion, the more difficult, obviously, international cooperation will be, especially between governments proud of their power.

In any case, governments have to stay close to the ground floor majority, and not only under democratic constitutions. To obtain the consensus necessary for the conduct of foreign policy in the common interest, rulers of uninformed or apathetic people are compelled to become manipulative or authoritarian. They will cultivate baser instincts of pride and power rather than disseminate mind-taxing and disconcerting information about alien realities abroad. Their conception of the national interest will likewise tend to be narrowly nation-centered rather than overall world-oriented. Public sentiment favors resistance to outside pressures demanding domestic changes no matter how beneficial in the long run for the country and the global community combined. Or, given a chance, it may prefer foisting its narrow perspectives on the world by armed might.

Like their subjects, governments suffer from the overload of information as well as from the common ignorance; for them too the world has grown beyond their comprehension and sense of control. Expanding with the scale of the circumstances with which they must deal, they are handicapped by the complexities of large bureaucracies, divided in themselves and guided by partisan counsel. Ruling over ignorant constituencies, they attract officials and agents equally ignorant but representative of the prevailing state of mind; more farsighted civil servants drop out of government service.

Considering the general reliance on goods and services from around the world, public ignorance has perhaps less divisive effects on the universals of material interdependence; but even they are affected by the shrinkage of public awareness. While, for instance, enlightened opinion may favor worldwide free trade, the ground-floor view of short-run advantage and resistance to the required adjustments obstructs all too often the common long-run benefits. By popular judgment, freedom and power stand for resistance to change under external pressures, even in the matter of worldwide uniformity of weights and measures. Freedom and power guarantee the continuity of custom and accepted beliefs that gives meaning to life.

The search for meaning further aggravates anarchy. What is true in the burdensome overabundance of the global supermarket? What has become of the absolute truths that in the preglobal past have held the individual and the community together at their most sensitive center of inward awareness? Sacred beliefs are reduced and relativized by submersion among all the other hallowed truths on global display. Perhaps the most ominous consequence of the emergence of the global hothouse was the rise of an insidious Babylonian confusion. The more open—or exposed—to the world's diversities the inhabitants of the many-storied global labyrinth become, the more vulnerable they are to metaphysical uncertainty. What in the competing array of truths and convictions is really true? That question is not primarily addressed to philosophers; it stirs most powerfully below the threshold of consciousness, troubling especially the common folk with the least intellectual awareness. In the past, they managed by faithfully structuring their ego according to accepted religious beliefs. These beliefs are now in danger of being outdated or disproved. And the individual spiritual discipline which they supported disintegrates or turns stale (to be replaced, most likely, by external compulsions).

Is it surprising then that, in the absence of any overarching and universal Truth or of a culturally integrated common human nature, people are divided; "the best lack all conviction, while the worst are full of passionate intensity." The majority are inclined to retreat to their roots, to the tried old ways which in the past have guaranteed metaphysical certitude. At present, the Babylonian confusion within global interdependence encourages a shrinkage of moral sensibility through a relapse into a divisive and self-righteous fundamentalism, whether in religion or politics—the ultimate and most powerful cause of anarchy and violence. No common bonds exist between impassioned fundamentalists of incompatible convictions. In the absence of any capacity for transcendence of cultural boundaries, the universalized appeals to freedom and self-determination lead not to peace but to war—and to the most unforgiving kind of war at that. The contraction of perspectives engendered by the global overload, by the deep sense of helplessness, and by the backward drift toward roots adds an especially ominous note to the global anarchy.

V

In the light of the tendencies just outlined, an elemental fact emerges about the source of hostility in the global furnace of competitive diversity: a vital collective otherness in any form, whether religious, political, or cultural, tends to constitute by its very presence an act of defiance against the established order. Forever searching for certitude, all people unconsciously universalize their communal way of life, judging by their limited experience of the world the actions of people living under entirely different circumstances, waxing morally indignant over events of which they know nothing. Such universalization is an intrinsic part of human assertiveness; it lies at the root of cognitive imperialism.

Human beings need the psychic reinforcement derived from the conviction that life as they know it is universally the best life; any suggestion to the contrary is subversive. That need for practical as well as metaphysical reinforcement applies not only to small ethnic and religious communities but also to superpowers driven for their cosmic sense of security to create a world conforming to their domestic arrangements. In the preglobal past the subversive impact of otherness was generally limited by distance and slow communications. In the age of the Great Confluence it is relentless, immediate, and intense. Cultural pluralism in peaceful association is possible only where there exist strong transcendent bonds (as within established polities, e.g., the United States, as argued in an earlier chapter).

The most vulnerable people on the whole are those who live in the developing countries, subject to reculturation. They are confronted not only by the alien institutions of statehood and modernization, but also by ethnic and religious diversity among themselves. Rendered doubly insecure, they often incline toward a fundamentalist militancy. Westerners, on the other hand, protected by their past superiority, generally feel less threatened in the face of otherness; they still look at non-Westerners with a measure of self-assurance. But even they have reacted with force when challenged, and their less enlightened fellow citizens have often discriminated against aliens.

Americans, exceptionally secure territorially and accustomed to transforming immigrants from diverse cultures into citizens, have had perhaps least difficulty in coping with otherness—a supplicant otherness; yet their response was—and still is—somewhat different toward their black, Hispanic, or Oriental fellow citizens. And the moment their country was challenged at the very core of its guiding ideology, as by the Bolshevik revolution, it reacted according to the general rule: a defiant otherness constitutes an existential threat (and any plea for cultural relativism is an act of subversion). In the Soviet Union, a country long suffering from the invasion of otherness, culturally, politically, and militarily, that defensive fear has not only taken deep roots but also long sponsored appropriate countermeasures.

Wherever we look, in short, we observe an intensified sense of insecurity aroused by daily confrontation with a subversive disproof of the verities that weld the community together. The defensive reaction of one polity stimulates further insecurity among others, combined with a heightened power-oriented political alertness. In the absence of a common ability for building bridges between cultures, cross-cultural communication takes the form of the only common language left, which is threats of violence or violence itself. Thus the global confluence has led to an incessant increase in the level of violence, the unfailing gauge of the prevailing sense of insecurity.

Violence in turn has become more effective, thanks to the unceasing perfection of weapons and their universal availability. The progress of science and technology has left its mark on the design and production of arms; they are more easily concealed, more deadly, and cheaper to purchase. More weapons than ever are on sale in the global supermarket, supplied by the United States, the countries of Western Europe, the Soviet Union, and Israel as the major exporters, for commercial as well as for political gain. Denounced after World War I as "merchants of death," the arms manufacturers now enjoy widespread moral support.

The use of violence also has become more sophisticated, benefiting from the perfection of communications and the expansion of government intelligence networks. Terrorism as an instrument of politics has become a fine art, practiced by states large and small under the protective cover of secrecy, national security, and moral righteousness. The superpowers have taken the lead in the escalation of violence, setting a challenging model for all others of security through armed power, inciting them to a matching response and in turn escalating the level and infamy of violence. The more tightly people around the world are compressed into interdependence, the more the intensity of violence is bound to rise throughout the global system, with minor opportunities for some, added dangers for others, and with a high cost of human lives for all. Pity especially those newly united countries in Africa that lack the social cohesion and individual self-restraint to contain the drift into an uncontrollable civic violence.

Pity, however, is in short supply. The tripling of the world's population within this century, the continued rapid population growth in the Third World, and the moral fatigue caused by the global overload have made people into exceptionally expendable cannon or bomb fodder. What does the anonymous individual matter among five and more billion fellow human beings? What, in the ultimate-solution scenarios of nuclear war strategists, do millions of faceless individuals count?

The material benefits of global interdependence unfortunately do not offer a sufficient counterweight to these tendencies. The circulation of goods and services does not create a countervailing capacity for transcending the violence-producing cultural incomprehension. The sense of insecurity even attaches to crucial—or strategic—raw materials, to industrial productivity, and national prosperity; in the showdown, national security takes precedence over the universals of global interdependence. By contrast, the forces of goodwill in search of transcendent truths are splintered and ineffectual. There is no evidence of a ground swell rallying all humanity, or even a significant minority, to a common creed (like supporting the United Nations); the differences are still far too deep, even between states closely associated economically and politically.

VI

And that is the human condition at the end of the 20th century: we doggedly continue life's ground-floor routines when for survival's sake we should realize that we live in an intensely interdependent world in which all the earth's peoples, with their immense differences of culture and historical experience, are compressed together in instant communication. The confluence of all human skills and energies in the global furnace of unity and diversity has produced a staggering advance in all fields of human endeavor, but especially in science and technology; it has provided the potential for a massive advance in controlling human destiny. Yet that intense interdependence is laced with hostility and terror-minded hatred. As Perez de Cuéllar, secretary-general of the United Nations in the 1980s, observed forty years after the founding of that organization: "we face today a world of almost infinite promise which is also a world of potentially terminal danger."[1] The chief contestants, the superpowers, are fatally locked into their separate worlds, equipped with weapons that can destroy human civilization and much nonhuman life as well. Never since the beginning of human life millions of years ago have human beings existed in such precarious balance between life and death.

In the competitive hothouse of the European state system the scales of destructive and constructive, dehumanizing and civilizing ingenuity have been weighted in favor of the latter, at least until the world wars of the 20th century. Where will the explosive and morality-undermining pace of the global state system, with its utterly irrational arsenal of nuclear weapons, take the present generation? As Perez de Cuéllar concluded: "the question is whether the Governments and peoples of the world are capable, without the spur of further disasters, of together making the right choice; for the choice and its implementation will, in many important ways, have to be collective."

Are the governments and peoples of the world making the right choice?

24

Explosive Confrontations

With that question poised in our minds, we now look at a few sources of escalating violence which illustrate in specific contexts the tendencies just outlined. The mood here, as throughout, is one of cathartic detachment. We deal not with right and wrong, but with tragedy, tragedy pregnant with unprecedented inhumanity unless—an unlikely outcome—the contestants manage to rise to a reconciling overview that provides common ground. What counts here is not detailed historical analysis but conciliatory realistic perspective.

I

First, a relatively minor case in the global context, illustrating how ethnic conflicts, subdued for centuries, were fatally sharpened by the new intensity of inter-cultural contact, by the ideas of nationalism, and by the demand for unity within a newly independent state. In the relatively small island of Sri Lanka off the southeast coast of India, with less than twenty million inhabitants, two rival ethnic groups live side by side, the Sinhalese, who comprise about three-quarters of the total population, and the minority Tamils, numbering about one-seventh. The light-skinned Sinhalese originally came from northern India; their religion is Buddhism. The more dark-skinned Tamils, adhering to Hinduism, hail from southern India; they maintain close ties with fellow Tamils on the mainland. Sinhalese and Tamils are also divided by language (even in regard to their alphabet) as well as by their historic sense of cultural identity. Engaged in open conflict to the 16th century, they settled down to coexistence until the end of the colonial era.

Not long after independence in 1948 the conflict revived, politicized under the impact of the heady ideals of self-determination and political freedom. The Sinhalese naturally dominated the new state of Sri Lanka, making theirs the official language; the Tamils complained of discrimination, backed by their kinsmen on the mainland. Some Tamil patriots even demanded an independent homeland to be carved out of the island. The conflict escalated as the island's population nearly doubled between 1950 and 1970 and economic conditions failed to meet rising expectations; cautious concessions to the Tamils over the language issue aroused Sinhalese extremism.

By the mid-1980s, as Tamil demands were not met and the Sinhalese government suppressed the secessionist agitation, hostility had turned into guerilla warfare and terrorism; in the fighting the casualties ran into thousands of lives. More bloodshed loomed as the government, posturing as a democratic regime confronted by a communist threat, appealed for Western help. If its offer of a limited federalism were not met by Tamil leaders, so the head of the government announced, the government would have to declare open war upon them. Although divided among themselves, the Tamil leaders were not ready to submit, leaving the conflict mired in more terrorism and guerilla action. The threat of civil war casts a shadow over tourism (a lucrative source of income), troubles relations with the Indian state next door, and adds to the world's sense of insecurity.

India meanwhile faces a similar escalation of conflict with its Sikh minority in the northwest. The Sikhs trace their religious and cultural identity to a 16th-century guru who combined Hinduism with Islamic religious inspiration, abandoning the caste system and allowing women to participate in public affairs. Threatened by both Muslims and Hindus from the start, Sikhs cultivated the military virtues, which eventually endeared them to the British *raj*. After independence their status was in doubt, their freedom of worship limited, their area of settlement divided in the partition of the subcontinent and the allocation of state boundaries within India. Their resentments led to demands for an independent state, setting a dangerous example for other ethnic and religious minorities in India. Repression and resistance escalated, leading to terrorism and, in 1984, even the assassination of Indira Gandhi, the prime minister. Violence continues, with no end in sight.

II

Take now a worse case, the conflict between black and white in South Africa. There a group of Dutch colonists had settled on the Cape of Good Hope during the 17th century and, in conflict and cooperation with encroaching English colonialists, had trekked into thinly inhabited territories to the north, displacing or subjecting the powerless African population. Clinging to their Calvinist creed in the face of African customs that grated on their puritan sensibilities, they built, eventually with British help, an outpost of European culture and economy endowed with vital raw materials at the strategic juncture between the Atlantic and Indian oceans. Their prosperity was not without benefit to the neighboring African states emerging into independence in the 1960s.

But, by contrast with decolonization elsewhere in Africa, the local African population, though by far the majority, remained in its subject status. Indeed, in the face of the rising anti-colonial agitation around the world, white rule hardened, surrounding itself with legal barriers that kept black and white strictly apart. Inevitably, "apartheid" practice antagonized and outraged Africans trained under Western influence. In the name of freedom and human dignity they enlisted moral and political support in Africa and around the world, while treated with relentless and often bloody repression at home, without hope of peaceful accommodation.

Nowhere in the world do such stark cultural contrasts clash. The Afrikaner minority proudly points to its accomplishments unique in Africa: a modern economy tied to the industrial and financial centers in Europe, the United States, and Asia, assisting the economies of neighboring countries, and contributing (if minimally) to the welfare of its oppressed African majority. Afrikaners ask: do the Africans possess adequate cultural skills for maintaining, under the hoped-for majority rule, the complex economy of South Africa? They point to the many divisions among the African population, to intertribal hostility and riots against the Indian minority, to illiteracy and lack of technical training, as well as to the unsolved problems of state-building elsewhere on the continent. Can they be expected to entrust their achievements, their accustomed way of life, to their African subjects? Yet in their arrogance of cultural superiority they defy the moral precepts of their own creed and of contemporary world opinion, sometimes in the extreme terms of racism.

On the opposite side, the representatives of the African majority cite the Universal Declaration of Human Rights and the ideals of Western democracy, while downplaying the Africans' pervasive lack of experience, leadership, and cultural skill in large-scale organization which in the West have given substance to those lofty ideals. In any case, how under prevailing repression and discrimination could Africans prepare themselves for equality? Yet opinion is divided: what consequences would their participation in managing state and society have for their country's economic buoyancy and political stability? Conversely, what of stability if they are not allowed to participate?

Among Western humanitarians (who have never faced any threat to their proud ways of life) moral indignation over apartheid practice runs high, in characteristic disregard of the dynamics of cultural incompatibility. Downplaying these differences presumably promotes cross-cultural understanding; it counteracts racism. Yet ignorance of basic realities also begets intolerance and violence, especially where the contrasts are so extreme. In the rising agitation the moderates on either side of the cultural divide stand little chance. Reculturing the black population for the required responsibilities, like alleviating the apprehension of the white population, will take a long time, too long considering the impatience engendered among Africans and their partisans by the pressures of globalism; all African states, together with the overwhelming majority of the UN members, oppose the apartheid regime. Meanwhile, the superpower competition assures, in a pinch, the white regime of at least minimal Western support, which might attract the Africans toward the

Soviet side, adding to the unending and widening confrontation. There spreads an ominous cloud over the future.

III

An even darker cloud hovers over the Mideast, in the Israel-Arab conflict. Its roots date back to antiquity, to the clash between Hebrew monotheism and Hellenistic polytheism. Why would the ancient Hebrews, the chosen people of an all-encompassing supreme God, refuse to worship the local gods as other people did? Why subsequently did they persecute the heretic Jesus and his followers, who tried to adapt the best of Hebrew monotheism to universal use, de-ethnicing their God and recruiting followers among the varied peoples under the Roman Empire? Why did they stick to their old creed and ethnic identity against the new transethnic religious universalism, even to the point of being driven from the Temple in Jerusalem and dispersed throughout the Mediterranean basin and its hinterlands?

A peculiar people insisting self-consciously on their own ways as the best, they inescapably incurred suspicion where they exercised no power, or fear where they did. Among Christians they posed a persistent challenge to basic truths, part of it drawn from their own heritage. Persecuted for that reason, especially at times of metaphysical uncertainty and change, they precariously held their own, building admirable moral and intellectual strength around their faith and bequeathing it, by intermarrying among their own kind, to their progeny. That strength enabled the dispersed and increasingly diversified Jewish communities to face the challenge of alien or hostile environments. They resisted assimilation, a trying challenge of otherness to equally self-righteous neighbors. By their faith they always won out to their own satisfaction in invidious comparison: their God was superior and they with him. All told, Jewish survival was perhaps the most impressive collective experiment in the crucible of universal history.

When in the age of the Enlightenment and secularism they were released in western Europe from their religious and cultural seclusion, their cohesion weakened as their opportunities in economic and cultural creativity broadened. Their cultural skills superbly fitted the liberated, secularized, and de-ethnicized descendants of the ghetto for the creative competitiveness in Western society then reaching its culmination. The open society of the United States offered especially congenial advantages. It justly provided people of Jewish descent with influence and power far beyond the weight of their numbers. By contrast, among the less competitive peoples of central and eastern Europe they aroused bitter hostility in the subtle power struggles of cultural traits, personality types, and social status.

All along, the consciousness of Jewishness persisted, vaguely among the majority and firmly among the orthodox guardians of the faith. In the areas of pronounced anti-Semitism, as in central and eastern Europe, Jewish identity remained strongest, adjusted by the end of the 19th century to contemporary political ambition. In the age of nation-states and culture-centered national ideologies, why should Jews not have a state of their own, centered on Jerusalem, the holy city of Zion? Theodore Herzl's dream of Zionism projected a new Israel living in prosperity and harmony with the surrounding Arabs. Yet as the dream approached reality, it became corrupted by the bitterness of power politics.

World War I advanced the cause of Zionism as part of the dismantling of the Ottoman Empire; in 1917 the British government expressed its support for "the establishment in Palestine of a national home for the Jewish people" without prejudice to "the civil and religious rights of existing non-Jewish communities." The resident Palestinians, feeling threatened by the influx of Jewish settlers, protested. Violence began in 1921, escalating apace through World War II, after which the shadow of the Holocaust descended on a land where Jews and Palestine Arabs had for years built up rival military gangs. Jews who had escaped the Final Solution streamed to it, eager to establish at last a secure Jewish homeland, if necessary with the help of terrorism.

In 1948 the state of Israel was proclaimed, an outpost of Western cultural skills reinforced by Jewish ability and determination; Palestinians and Arabs possessed no matching resources. Their growing hostility transformed the peaceful Zion of Herzl's dream into a first-rate military power, its 3.5 million people holding the surrounding states at bay, admittedly with the help of American political and economic aid. What counted most, however, was their will to self-determination derived from two thousand years of dispersal and persecution, and reinforced by occasional ruthlessness born of the horrors of World War II.

As for the displaced Palestinians and their Arab—or Muslim—sympathizers, their state of mind was shaped in response to the challenge. The promise of 1917 for peaceful coexistence was not kept. What mattered in their eyes was the intrusion of self-righteous aliens with the capacity to enforce their superiority, agents of a long-hated Western imperialism. In response they developed a matching resolve, drawing on a profound religious commitment of their own—Jerusalem is sacred also to Muslims. Their resources, however, did not equal their conviction. The Arab population of the Mideast is backward by comparison. It is divided in itself, never more fiercely than in the sectarian fights between Shiites and Sunni (the source of a protracted war between Iran and Iraq). Through its oil resources it is also economically—and even politically—interdependent to varying degrees with the United States and the industrial countries allied with it. Opposing Israel by war led to several humiliating defeats.

How, then, could the mounting Arab hostility to Israel and the frustration over past and present weakness, especially strong among the younger generation caught in the disorientation of globalism, be expressed? The answer, long prepared by the violence accompanying the arrival of Jewish settlers and the Islamic tradition of holy war, was terrorism, terrorism carried to extremes, directed not only against Israel but against its American ally as well. The wave of terrorism lapping in the 1980s from the Mideast over the Mediterranean into Western Europe predictably escalated the scale of violence, intensifying political tension and the sense of insecurity around the world.

Peaceful coexistence between the opponents obviously is impossible, given the enemies' self-righteous resolve to prevail on their own terms. Here too the moderates are at a disadvantage. Compromise, the search for transcendence, is a sign of weakness; any recognition of justice on the other side undermines the conviction of one's own moral superiority. The conciliation-minded left in the Israeli political spectrum loses out to the "Judeo-Nazis" (as the right-wing hard-liners are sometimes called in Israel).[1] The conflict festers and deepens, spreading through the Palestine Liberation Organization into neighboring countries, destroying the fragile Lebanese polity, arousing Muslim communities in Africa and Asia.

Inevitably, the Israel-Arab conflict also aggravates the hostility between the superpowers. Because of Israel's association with the United States, Israel's enemies look toward the Soviet Union for support—admittedly with caution, because Marxism-Leninism and Islam scarcely harmonize. In addition, the plight of Jews in the Soviet Union and the Soviet opposition to Jewish emigration is a perpetual complaint in Israel; it also incites anti-Soviet agitation in the United States. More than any other hot spot of political violence in the world the Arab-Israel conflict feeds emotional intensity into the confrontation of the superpowers, the overriding cause of tension and violence in the contemporary world.

IV

For understanding the contemporary dynamics of the superpower confrontation, the most alarming in all human affairs, we must first consider the contestants separately before looking at their escalating hostility, beginning with an update of the earlier assessment of America's adjustment to the new globalism.

In the United States, the driving force in superpower rivalry, we observe during the post-Vietnam years a profound cultural and political change, barely recognized in the hectic ground floor bustle of life yet amounting to a significant recasting of the American identity. Deeply immersed in the affairs and the economies of other countries around the world, Americans have moved beyond the foundations of their earlier exceptionality. They are becoming more apprehensive about national security, like other peoples long exposed to unsettled conditions at home and abroad.

The new sense of insecurity in foreign affairs is not caused by a lack of conventional sources of power. The United States still stands out by its strength; although heavily in debt to other countries, it impresses its creditors by its might. What lies at the root of that fear is ignorance of the non-American world. Viewed superficially, that ignorance does not entirely stem from a lack of information; ever more data are available. By an outside perspective, however, that information is Americanized in the transmission and thereby contributes to ignorance. Any evidence of a genuine otherness is unpopular; it deepens the sense of insecurity and evokes a protective counterassertion of patriotism. For the average American the balance between foreign and domestic news, in any case, inclines in favor of the latter. Foreign realities enter American life very superficially, if at all. What counts in the headlines is the country's might in the world—a world that has grown over people's heads.

An unknown world is a hostile world, most fiercely so in the case of the uncomprehended Soviet ambition backed by nuclear weapons and Third World terror directed against Americans and their allies. Americans no longer feel safe to roam the world as they formerly did. They are even becoming apprehensive about their neighbors in Central America, weak as these are. Their only security seems to lie in their own country—Fortress America. Rather than bequeath their best qualities to the rest of the world—as still was the case in the early decades after 1945—Americans are increasingly tempted to rely on force in trying to reshape the world after their own image. Only a world composed of democratic, or at least pro-American, states is a truly safe world.

But insecurity is at large even within the citadel of superiority. Life has grown over people's heads. What has become of the self-assured American outgoing sociability? The ever increasing overload of detail to be handled in everyday routines has encouraged widespread withdrawal into civic passivity and individual self-absorption. Although still protecting patriotic consensus, affluence has had its proverbial corrupting and desocializing effect. But so has population pressure: more people have to be dealt with, actively or passively (as in traffic jams). In addition, the influx of alien ideas, alien attitudes, and alien people insisting on speaking their own language overtaxes the traditional capacity for assimilating outsiders on the socially more disciplined American terms (which in any case are weakening). Meanwhile, for the run of Americans the pace of professional work has speeded up, in the volume of data to be handled, information to be processed, innovation to be digested. The pacesetting workaholics, absorbed in their specialties, make poor citizens in communities where public well-being used to depend on civic participation.

What time and mental attention is given to human relations near and far in the popular fascination with technological progress? However complex they may be, machines are simple as compared with human beings and social relationships. In an age of disorientation in human affairs, machines become attractive because they are impersonal and reasonably predictable; they make limited demands on the management of emotions. A preoccupation with machines for work or play—especially for play—represents a form of escapism into a safe, nonhuman world. One may even suspect that the obsession with technology in the arms race stems from an urge to avoid the human aspects of superpower hostility; the national-security mind-set of the weapons planners certainly is devoid of any touch of humaneness. In one way or another, however, the human contexts of technology always assert themselves—for worse if they are not duly taken care of. Technology is set into society; it can serve its purpose only if properly adjusted over the long haul to human need in all its diversity.

Whatever the reasons, American society as a whole has become more variegated, more brittle, more litigious, its members more self-righteously withdrawn into themselves, more quarrelsome, more inclined to selfishness or even corruption (notice the increase in white-collar crime). The run of people—and the immigrants especially—take the functioning of government and society for granted, unaware of the intricacy and intensity of the human effort needed to maintain law, order, and prosperity. At a time when moral responsibility should grow proportionally with the advancing complexity of society, it noticeably contracts.

What, then, is the public mood? At its deepest, it would seem, it is a novel existentialist vulnerability caused by many factors, by the decline of spiritual resilience to adversity (or by spiritual lassitude generally), by the rising complexity of life and the unmanageable overload, by the disorientation resulting from the subversive inroads of otherness from the outside, and by the tide of violence at home and abroad. Americans are beginning to be caught in the disorientation familiar in other parts of the world. Amidst an uncommonly good-natured people the growing unease has produced a penchant for violence and violent solutions, as a defense against insecurity that further increases insecurity. More weapons are in circulation; paramilitary "survival training"—even for mercenary service abroad—is gaining popularity; more people than before favor the death penalty (let alone reliance on nuclear superiority in the arms race).

Out of that embattled insecurity has come a divisive drift toward old verities, toward a splintering fundamentalism in moral or political creed, sometimes to the extremes of fanatical irrationality Nazi-fashion. On the opposite side too—not all of American society runs in the mainstream—attitudes harden, to the point of civil disobedience among peace groups. As long as prosperity lasts, the polarization of opinion will be limited; violence is likely to remain in the background. But what will happen in case of a recession like that of the 1930s? Have Americans retained the resilience in the face of adversity and the capacity for consensus which they so remarkably demonstrated in those years?

These changes, inevitably, have left their mark on the character of American politics. Any shrinkage of public awareness signifies an unwillingness to depart from established ways; it implies a shift in favor of conservatism. The ascendancy in public affairs of the Europe-oriented "eastern establishment," of the liberal elite trained for innovation and openness in world affairs—the intellectual guardians of the Western tradition—has come to an end. It has been replaced by a more powerful isolationist-minded elite from parts of the country geographically and intellectually farthest removed from Europe but close to ordinary folk immersed in their local and personal affairs.

In the change, American liberalism generally has lost its appeal. In the past, liberalism as a state of mind in politics and society has represented a capacity for taking a larger and more inclusive view. That readiness for adaptation to enlarged perspectives presupposed a spiritual, intellectual, and material sense of security, individually and collectively. If that vital basis is reduced, the liberal capacity for outreach suffers. Especially in regard to foreign affairs, to the non-American world, liberalism faces a crisis. Traditionally handicapped for understanding the dynamics of non-Western societies by its unconscious adherence to Western values, it is of little help in coping with the problems of the contemporary world. Liberals have no answers to the ominous tensions of global politics except to propagate the principles of American democracy and, should these be challenged, to fall back on the affirmation of national security—to the delight of authoritarian conservatives.

Liberal democracy is premised on an informed public opinion. Yet the complexity of political issues rising out of the infinite details of technology, industry, finance, and international relations—of life in all its aspects—surpasses individual grasp. Inevitably, the specialists take over. Ideologically oriented think tanks and political action committees financed by special interest groups, corporations, and industries assume the function of a defunct informed public opinion. More than before, money buys votes in Congress and in elections at large, generally in favor of vested interests, of the representative rich allied with the military-industrial complex—which lends support to a long-standing Soviet (and pre-Soviet Russian) stereotype about Western democracy.

What, then, of the overall national perspectives which in the past were embedded in public consensus? In the absence of an active consensus-building interaction, that supreme responsibility falls more than ever on the symbol and agent of national unity, on the president, the person in charge of the executive branch of the government. As the executive dominates the legislature and judiciary, government becomes authoritarian American-style. Presidential authority is geared to the mass audience in the manner of the media which stoop to the lowest common denominator as the bedrock of consensus. In the absence of any effective public capacity to judge political issues on their merits, politics becomes mood politics, psycho-politics. A skillful politician, with the help of public opinion polls, can quickly take the pulse of public feeling and manipulate it to advantage in the name of national security (parallels from other countries discussed in earlier chapters may readily come to mind).

In times of widespread unease, national security has a double face, one turned inward, toward domestic affairs. It radiates assurance of secure employment in an ever more complex economy that is baffling especially to the new generation entering the job market. The affluence-oriented new mood politics matches and offsets that deep sense of insecurity by a strong infusion of escapist cheerfulness. America's traditional optimism has been raised into an article of political faith, into a patriotic ideology. Yes, the American way of life is right; it is the best the world around. If the world needs basic adjustments, not Americans but the others have to change.

In this manner the domestic sources of insecurity add to international tension. The run of Americans approach international affairs with a pervasive double standard. It is the others whose attitudes and behavior in any conflict with American policy are wrong; American action is always morally correct. If, for instance, too many Americans catch the drug habit, it is the fault of the alien drug producers or drug smugglers, not of their American customers, nor of the public and the government, which pay all too little attention to the factors producing drug addiction. It is other governments, not their own, which engage in espionage, subversive activity, or terror. Casting doubt on American righteousness causes resentment; it adds to the widespread sense of insecurity and disorientation. If that smugness should fall apart, are Americans prepared for the intellectual and moral effort required to regain a sense of balance? Or will that smugness harden into outright ideology?

In any case, realistic political leaders have little choice but to conduct mood politics geared to the base instincts of public opinion, to self-righteous wishful thinking combined with ignorance of the nature of social harmony generally, of conditions within the country, and, most abysmally, of the world at large. Mood politics is here to stay; if skillfully conducted, it endows ignorance with a sense of patriotic righteousness and buttresses it, in the face of rising fears, with assurances of military superiority.

The current American penchant for authority manifested by force is most obvious in the militarization of the economy, of foreign policy, and American life generally. The new militarism is going hand in hand with a new consensus-padded authoritarianism, with a tightening of secrecy around government operations, especially those concerned with national security, and with a hardening of national ideology. Transformed into a veritable party line, the new American ideology is spreading from national-security personnel into the media, into the classrooms of schools and universities, and into the public at large; it affects the news coverage, especially from abroad. That invisible but effective party line makes anti-communism into a central tenet of Americanism; it joins all domestic sources of insecurity to the fear of the Soviet challenge, thereby falling into a pattern all too familiar from Soviet practice. Enemies become alike, the more so the wider they drift apart in their guiding ideology.

Escalation of hostility with the Soviet Union, it would seem then, is built into the dynamics of contemporary American life. What will happen should domestic discord grow, as is likely in the case of an economic recession? Will the domestic tensions spill over into US-Soviet relations and aggravate them further?

V

On the Soviet side we observe similar trends, although more restrained by a growing awareness among Soviet leaders of their country's weakness. The confidence displayed by Brezhnev in 1976 did not endure. In his last years the momentum of his earlier career eroded; the Brezhnev era ended in stagnation and corruption (Brezhnev died in 1982, succeeded by two short-lived party leaders). In February 1985 Mikhail Gorbachev became general secretary, opening a new phase in Soviet rule, with yet more evidence of convergence toward the enemy through invidious comparison, from the opposite end of the spectrum.

As before, the Communist Party is determined to improve the Soviet standard of living to a point where it can be safely measured against the advanced industrial countries. It is also bent on introducing more democracy, democracy Soviet-style, through popular participation in decision-making within the bounds set by the party, especially at the place of work and in the local community. That democratization also calls for an improved openness in the relations between the party and the people, for greater freedom of discussion, for greater mutual trust leading to greater social cohesion and even a sense of freedom in the far-flung and fragile body politic. In this respect too the Soviet leaders need to have their country face up to comparison with the leading industrial countries.

Yet, as before, conditions are stacked against them. Their population (slowly approaching 300 million) is culturally conservative, sometimes openly yearning for the simpler, preindustrial past; even more than Americans the Soviet people are resentful of the demands made by contemporary life. They also lack the easy socialized flexibility in private initiative which would allow the government to reduce its overstaffed and counterproductive bureaucracy. The party calls for increased initiative as a necessary spur to higher productivity (Soviet per capita industrial productivity has fallen behind that of Japan, Taiwan, and even South Korea). But, at the same time, it is in doubt how much latitude to permit. Aware of its subjects' lack of the requisite skill for socially constructive cooperation, it fears, rightly it would seem, that economic freedom will lead to corruption, to channeling profits into socially and politically undesirable activities.

An effective private enterprise system requires a highly integrated society within a cohesive body politic—conditions which the Soviet Union cannot artificially create. By party judgment, the country's internal diversity will not even permit the limited extension of the private market economy possible in a culturally much more cohesive China, although many people call for just that. Given the country's internal and external insecurity, it can hardly afford the extensive and costly sociocultural experimentation needed for the creation of an effective market economy. Relaxing the entrenched controls might indeed mobilize the country's pent-up centrifugal forces. The Soviet Union is engaged in a global power contest with the United States; it has cause to worry about its neighbors; it rightly distrusts its Eastern European satellites; it is at war in Afghanistan. Any sign of internal unrest constitutes proof of weakness.

There are other obstacles to increasing productivity and raising the standard of living. Soviet society, like Western society, shows signs of weakening in its moral fibre. It suffers from high divorce rates, drunkenness, juvenile delinquency, escapism, and disorientation. In the years ahead the Soviet

Union also faces a marked labor shortage, a factor which threatens the slim preponderance of the Russian ethnic element in that multinational empire. Lacking free access to the high technology of advanced industrial countries, it is in no position to substitute robots and computers for the missing workers. It cannot at the same time modernize its economy and improve the standard of living, let alone keep up with the American nuclear weapons programs. Its industrial efficiency remains low, both in human attitudes and mechanical equipment, except in its defense and space industries. Like the United States, the Soviet Union is committed to an ambitious space program which, considering the country's ground-floor condition, is even more escapist than its American counterpart. The exalted Soviet faith in science and technology cannot remedy the shortcomings in human motivation and social relations.

The Soviet system as a whole, furthermore, suffers from the growing inadequacy of its guiding ideology. As an artificial Russian-oriented summary of the West's organically grown cultural heritage, Marxism-Leninism could claim a practical validity in the early phases of the Soviet regime. At the end of the 20th century it is dated (as is recognized in China). It needs fresh inspiration taking it beyond its narrow materialism and class analysis, which stands in the way of realistic assessments of Soviet conditions and world problems. Yet relaxing the ideological guidelines that give official direction and meaning to Soviet life is an unsettling prospect. What will take their place? Does there exist a reliable spontaneous Soviet patriotism that could serve as a more flexible and open-minded framework? Marxism-Leninism is a global ideology; how can the patriotism of people so long cut off from the global circulation of goods, services, and human beings provide an effective substitute in an age of intense global interdependence? The peoples of the Soviet Union are far more isolationist than Americans.

Can, indeed, the Soviet regime trust its people sufficiently to allow open participation in the exchanges of the global mainstream? Invidious comparison with the advanced industrialized countries (especially the United States) still prompts disloyalty. Against that danger the country must be held together by compulsion and a fear of the outside world stiffened by a carefully cultivated yet always dubious sense of moral superiority. Hostility to the outside world, and especially to the United States as the chief source of defeat in invidious comparison, is as much a necessity of governance now as, in different forms, it has been in the Russian past. At the heart of the global power struggle lies the competition over an ultimate superiority. If the United States claims that distinction, so on the rebound does the Soviet Union. In this manner the Soviet Union too spills the dynamics of its domestic politics into international relations. It contributes its share to the upward spiral of hostility.

VI

The mechanism driving that spiral is obvious: it is challenge and response in intensifying repetition. Any gain, real or imagined, on one side has to be matched by the other; pride and self-interest demand that at least parity be preserved if superiority is out of reach. The contests cover a wide front. They include peaceful competition in space exploration, scientific discovery, artistic and literary creativity, in standards of living and the quality of life. They cover the earth: which country has more friends and allies, more bases from which to wield power? Which can contribute more to human welfare? The contestants elicit partisan support from the world's hot spots or from opposition in the enemy's camp. Some of the staunchest anti-Soviet voices in the United States are refugees or émigrés from the Soviet satellites in Eastern Europe or the Soviet Union itself. Treated as experts on Soviet affairs, they infuse the hard intolerance characteristic of their societies into the American mentality (another aspect of enemies becoming more alike).

The enemies are very much alike in dramatizing the weaknesses of each others' social and political systems, predicting their eventual collapse. Each side judges the other from its own ideals, willfully—or even quite unconsciously—downplaying its own shortcomings. In both camps criticism of domestic conditions or pessimism regarding the future is considered unpatriotic, patently so in the more fragile Soviet polity; but even in the United States one observes a hardening of attitudes in favor of the status quo. More embittered than the propaganda battle of invidious comparison is the daily covert war of espionage, intelligence operations, and surveillance, generally ignored despite its enormous costs in money and human resources. It contributes to the intensity of the arms race, which tops all other contests in its deadly significance. And, finally, there is the incessant moral mobilization, transforming the political rivalry into the ultimate contest between right and wrong, good and evil. That apocalyptic fanaticism stops all rational analysis and promotes a murderous indifference to human survival.

In this fierce competition, the United States outpaces by far the Soviet Union. By comparison its internal problems are small; it still can count on the loyalty of its population as well as on their capacity for voluntary cooperation in politics as in economic productivity. It enjoys a worldwide presence politically, economically, and culturally, even where it is not wanted (as in the Soviet Union). Viewed realistically, the Soviet Union can claim parity only in military power and nuclear weapons, its sole guarantee of external security and political survival, at least by Soviet judgment. According to hard-line American judgment, however, the fate of the Soviet system is a matter that can perhaps be decided in Washington.

Observing the profound problems faced by the Soviet leaders and the difficulties likely to prevent their resolution, anti-communist American policymakers, following the expectations expressed at the start of the Cold War, are tempted to escalate those difficulties to the utmost. They can do so by encouraging the centrifugal forces within the Soviet Union and obstructing the government's drive for a more effective economy, by increasing the Soviet leaders' sense of insecurity and forcing them to return to repressive measures that incite further opposition in their country as well as anti-Soviet revulsion in the United States, Western Europe, and Japan. This policy may perhaps lead to the collapse of the communist regime and install democracy even in the expanses of Eurasia (whatever the human consequences).

Given the worldwide resources of the United States, Washington indeed enjoys a large measure of control even over the Soviet future. It can, on the one hand, reduce the Soviet sense of insecurity and thereby relax the pace of political competition, and therefore also the harshness of Soviet rule, both among its own peoples and its small Eastern European satellites so tragically trapped between powerful rivals. On the other hand, more likely under the given conditions, it can intensify that insecurity to the point of destroying perhaps the Soviet system altogether—assuming, of course, that when driven to desperation, the Soviet leaders will not use their nuclear weapons in retaliation. Thus the contest revolves ever more centrally around the buildup of nuclear armaments.

The crux of the arms race is progressive deterrence, the capacity of both sides to prevent a nuclear war by ever more refined preparations for it. The rational core of this utterly irrational calculation lies in the assumption that neither side can gain by a full-scale nuclear war (has not deterrence prevented a war which surely would have started long ago in an age of conventional weapons?). Yet pride and fear combined still press on for nuclear superiority, certainly on the part of Americans; the Soviets, as the weaker partner, are content with parity (though how, in the complexity of nuclear weapons technology, does one measure parity?). Thus the refinement of weapons system leads to ever greater insecurity in deterrence; decision time in moments of supreme crisis is becoming shorter, threatening to entrust computers with the fatal throw of the dice.

Irrationality dogs the arms race in all its aspects. Its technicalities have long grown beyond public comprehension and judgment; its dynamics on the American side are part of the new psychopolitics, geared to the elemental instincts of an overburdened public. The relentless momentum of scientific research and technological innovation adds further acceleration: if a new weapon can be invented, it will be built and tested (or else the other side will have it first). Thus the arms race has moved into outer space, propelled by extravagantly expensive technological fantasies and justified by an ever more intense moral mobilization. The vested interests of the military-industrial-intelligence complex with its ardently anti-Soviet party line and grip on public opinion are dedicated to forcing the pace. By compelling the Soviet leaders to match the American military buildup, they wish to hamper Soviet efforts to catch up to the United States in civilian prosperity (that buildup hurts the well-being of Americans too, especially the poor).

It seems hardly an exaggeration, therefore, to argue that the arms race has acquired an independent momentum. Although its speed may vary, it is now unstoppable, even by government decision. No elected statesman possesses the power to resist the pressure of its beneficiaries and their hold over the public mind. The abdication of human control lays the outcome open to chance. The fatal dice may be thrown by many haphazard occurrences: by human error or wild fanaticism—given the rise of irrationalism in politics, the latter seems not impossible; by mechanical malfunction; by public outrage over an act of terrorism in a crisis of political confrontation; or by an impulse of desperation on the part of Soviet leaders.

Only a massive popular revulsion against the certainty of a nuclear holocaust could raise human eyes from ground-floor preoccupation to a commanding overview and stop the abdication of human control over human survival. Such a reversal of opinion is unlikely except in the wake of a massive catastrophe costing millions of lives—if there be any survivors capable of making such decision. Even the explosion of a small part of the available nuclear warheads could produce catastrophic consequences in atmospheric conditions affecting the weather around the world. A full-scale nuclear war would mean the end of human civilization as evolved over the past ten thousand years, not counting the destruction of many other forms of life.

And that is the awesome, well nigh incomprehensible truth. There hangs a Damocles sword of utter desolation over all humanity, hung there foremost by Americans, the pacesetters of the 20th century. There is no escape for Americans from that supreme responsibility until they learn to put themselves, as Leibniz advocated, into the position of their adversary. Viewing the world for once through Soviet eyes means admitting the legitimacy of the Soviet system, acknowledging its achievements and shortcomings, respecting its troubles (not forgetting the American contribution to them), and helping to minimize its problems rather than aggravate them; it means diluting the customary cognitive imperialism with a dash of cultural relativism. In that manner Americans will also promote human rights in lands where thus far adversity has prevented the refinement of humaneness. And they will give new hope to all humanity. But here analysis turns into utopianism—as an antidote to an all-too-justified pessimism.

25

The Liberating Discipline
of Globalism

In conclusion, the argument shifts to a different way of looking at the present, escaping from the assessment of "what is expected" to an exploration of how to achieve "what is hoped for."

I

What is expected by the analysis here set forth bodes ill for the future. The world revolution of Westernization has run its course. It has created an interdependent world supporting five and more billions of human lives, a large percentage existing in comparative misery. Interdependence has been built on universalized Western terms, on Western accomplishments in institutions and command over nature. It forms an external and still superficial framework of human existence, within which incompatible cultures, religions, and political ambitions clash, often geared for war (even nuclear war). There exist enormous differences in standards of living and resources for human improvement; the gap between rich and poor is still widening. The poor are taunted by the presence, either physically or by image, of the rich in their midst, aspiring to live like them. Yet the human habitat on earth does not possess enough clean air or water, or land and mineral resources, to support everybody in the style of the wealthy. Can it sustain the world's ever growing population at present levels of consumption and material security? How will the scarce resources be allocated, and under what regulation? Not surprisingly, the anarchy in the global community keeps rising.

The prospects are further dimmed by the fact that the West has spent its spiritual capital, the socially oriented individual self-discipline, the deliberate structuring of the individual psyche for the purpose of communal perfection and survival. The Judeo-Christian spiritual restraints in individual and collective life, though still at work, have ceased to be a public force within the West itself. They lack justification expressed in rational and scientific language; they lack symbols drawn from contemporary life. They are further undermined by affluence, the corruption of power, and self-indulgent ways of life or attitudes drawn from less socialized cultures.

As a largely invisible ingredient the West's spiritual discipline has never effectively spread around the world as part of the outpouring of Western material culture. Where imitation of Western power is a matter of political survival (as, most prominently, in the Soviet Union), the built-in social restraints of Western culture have to be externally—and often counterproductively—imposed. Elsewhere, in the run of non-Western societies (Japan excepted), society continues in violence and crisis, while heads of states typically call for more discipline. Ignorant of the causes and consequences of its expansion, the West has left the interdependent world of its creation without guidance, without a rational explanation on how to cement the fragile life-supporting unity into a true community. Thus, both in the West and in the non-Western world, the forces of anarchy are on the march.

It is not surprising therefore that pessimists reared in the Western tradition have compared this age with the Roman Empire at its decline, undermined by decadence within and threatened by the barbarians without. Now in the anarchic global community, it would seem, ignorance and indiscipline threaten from all sides, and especially from those countries most responsible for creating the global confluence.

Considering their actual and potential power, let alone the unprecedented destructiveness of their weapons, the leading countries—and the United States foremost—should be the most alert and farsighted. Yet the world they have created and which they still dominate has grown over the heads not only of those peoples whom they have dragged into global interdependence but, what is more important, over their own. Trapped in the prevailing—and politically encouraged—shallow optimism, the run of people persist in their self-indulgent and petty ground-floor routines, without a sense of reality, without command over their destiny, and sometimes indulging in the most irrational Armageddon fantasies. Such barbaric attitudes influence government policies as well.

The parallel with the fate of the Roman Empire brings to mind the prescient Roman saying: *Quos deus perdere vult, prius dementat*—commonly translated as, "Those whom the gods want to destroy they first make mad." Surely, if there is no massive change of mind among the peoples in the leading industrial countries, there will be unprecedented global doom.

II

For escaping from such gloomy prophecy and for counteracting the shrinkage of awareness discussed in the previous pages, there is need to set out some hopeful and constructive perspectives, breaking the ground for recasting the prevailing state of mind.

Always keeping in mind our unconditional dependence on the society in which we live, let us in this spirit examine the individual and collective discipline required for the inescapable interdependence that at present upholds human life and creativity. With a better understanding of the webbing of dependency in contemporary society we can gain some insight into the processes leading to the more inclusive perception and tighter social discipline needed for reducing the life-threatening anarchy of the global community.

What, to start with, is the price we pay (or used to pay) for the benefits of citizenship? Commonly as much taken for granted and unexamined as the material services of social interdependence, that price deserves close attention. Citizenship means socialization, submission to a strict social discipline, all the more strict the larger the community.

Socialization begins at the start of individual life, with toilet training, with forming sounds into words and words into speech, with learning to control body and soul in interaction with parents, siblings, and neighbors, always resisting submission and always compelled or persuaded to comply. Compliance may be obtained by force or elicited by love and affection, by rewards that open up more opportunity for self-affirmation—self-affirmation that perpetuates the paradox of socialization: further submission for the sake of enlarged self-affirmation.

The process continues through childhood and adolescence, in relations with other human beings in work or play or in relation to mechanical tools. Similarly, learning to play an instrument requires relentless practice in submitting to the peculiarities of the instrument as well as to the musical score—for the liberating effect of producing life-inspiring beautiful sounds. Schooling, of course, is a crucial socializing experience, a hard, solitary struggle for the most part; yet under loving instruction it pays off well. In book-learning as in sports, or in life generally: if we like the results, we submit without awareness of submission. If we don't, we resist. If we resist, we may be compelled, though hoping for liberation. Liberation, however, comes in various forms, either by overthrowing the external compulsion or dissolving the inward objection through a change of will, through sublimation. Rebellion closes many opportunities for self-affirmation; overthrowing external compulsion courts chaos, while compliance opens new doors. Constructive compliance offers new vistas for self-affirmation, through developing new skills of self-control.

Social discipline is anchored in the core of our psyche; it presumes a skill of psychic self-manipulation, a challenge traditionally faced most effectively in religious practices that offer their own rewards. Fortunate are the individuals who learn these socializing skills early in life. Fortunate are those societies which can rely on the routine (and nonpolitical) practice of these spiritual skills. These skills do not imply submission to an unjust social order, as Marx contended (though they can be abused for that purpose); they do not limit artistic freedom, as romantics have argued. Cultivated generally, they promote a liberating submission to the necessities of peaceful cooperation, the guarantee of civilized life and its creative potential.

Adult life with its many responsibilities, of course, enforces continuous socialization, in the lengthy struggles of advanced education, of apprenticeship to employment, of adjustment to work in large organizations, of marriage and raising a family, and of civic responsibility generally. Well-socialized citizens are active members of the community, willing to apply part of their resources wherever there is need (and the needs are endless). Society benefits when a large volume of public service is supported by voluntary private effort, whether in social welfare, education, religion, or the arts; when, in other words, private awareness extends into the community at large and molds individual conduct for citizenship.

Even daily routine enjoins a minimum of social discipline, for self-interest alone, as in driving a car through heavy traffic, in the necessity of holding a job, or merely through the desire to be liked by one's neighbors. One can test the quality of socialization in a given society negatively by the frequency of breakdowns in ordinary life, whether in the individual psyche, in marriage, or in social relations through crime in its various forms.

Socialization is also accomplished more subtly through respect for collective myths and symbols. The national anthem and the national flag implant a subliminal habit of loyalty to established authority that keeps civil strife to a minimum. The intensity of subconscious social discipline is revealed in the restricted diversity of party organizations emerging from political agitation; the more effective that discipline, the fewer the number of competing political parties and the greater the mutual toleration. That civic discipline goes deep; it softens individual wills for submission to majority opinion, for ready fusion with other wills, for the political deference renowned among the run of English-speaking peoples. That civic discipline links the psychic structure of the will with the community at large, with the nation as a whole.

That discipline, obviously, has to be the more intense the larger the number of fellow citizens. One self-willed individual in a small group has to adjust to relatively few other self-willed individuals. Adjusting to millions and hundreds of millions of fellow citizens calls for infinitely reduced and flexible wills—wills liberated in turn by the hugely enlarged and subtly refined opportunities for individual self-assertion that come with large-scale social cooperation. Seen from without, that tight social discipline appears as a form of submission or even enslavement, a source of discontent. Viewed from within, it is a guarantee of civilizing liberation.

Kindly external circumstances, like those enjoyed historically by Americans, certainly contribute to making wills flexible. And good habits embedded in the collective unconscious perpetuate themselves in individual conduct— up to a point. What is needed against the ever present lapses into constrictive selfishness is the daily effort in promoting the malleability of the individual will, the conscious and deliberate harnessing of the individual psyche for service to the community in its largest dimensions. That was the contribution of religion at its best, enshrined in sacred symbols because it touched the core of peaceful human cooperation, the guarantee of collective survival.

Let readers judge on their own how effectively the wellsprings of social discipline operate in their society.

III

Here the argument advances to an exploration of the transition from the nation-state to the global community at large. In order to curb the potentially fatal anarchy in the global community, the nation-bound citizens will have to rise to a yet higher perspective, to a yet wider civic awareness, to a yet tighter self-discipline. In this search for transnational citizenship we possess little practical guidance. We can merely extend the guidelines evolved for judging existing society into the yet unexplored realm of civic globalism.

Of our urgent need for such an advance beyond national perspectives there can be no question; it will have become obvious throughout this book. What is needed to start with is a clearer sense of the obstacles to the growth of an enlightened self-interest required for survival. Consider the burden of added obligations and duties, of individual and collective self-restraint. What of freedom and self-determination if we have to bow to the will of strangers or even enemies? The cultivation of a new globalism makes heavy demands on all human faculties. It calls for changes in material lifestyles, for institutional rearrangements, and most centrally for recasting the core of human motivation and awareness, the precondition for all lesser adjustments. Adding a top story to the high rise of civilization requires no less than a restructuring of basic human identity, an arduous process of reculturation even for the people most skilled in social cooperation.

In the first place—to survey the necessary changes in an ascending order of importance—promoting more effective transnational cooperation obviously costs money, money spent for international agencies, for aid and assistance to people needing support in distant lands, and even for education at home in languages and cultures far removed from daily experience. As a result individuals, corporations, and the government will be deprived of hitherto available funds and resources—or they will have to work harder to make up for the loss. The widespread resistance to taxation certainly bodes ill for spending money inside and outside the country for purposes so distant from customary routines.

Enlarged international cooperation, furthermore, imposes new restrictions and obligations on individuals and government. Both have to adjust to additional pressures and comply with the needs of strangers hitherto inconsequential, disregarded, or considered hostile. Decision-making, already overcomplicated, will be still more laborious and protracted. Uncertainty is bound to grow in proportion with the number and diversity of the participants. Their dependence on others will be increased, their freedom of action curtailed, their role diminished, their ego correspondingly humbled. In addition, forward-looking participants in inter-governmental cooperation are weighed down by the necessity of coping with new information and work conditions, all taxing their patience, flexibility, and brainpower. They must work in the absence of supporting symbols comparable to those sustaining national identity, pioneers in international cooperation always in doubt whether public opinion at home will back them up.

There lies the rub. The chief obstacle to enlarged international cooperation comes from public ignorance of foreign lands and conditions and from the natural unwillingness or incapacity of the run of people to change accustomed ways for purposes lying beyond their comprehension. Authoritarian regimes perhaps lend themselves more readily to innovation in internationalism, their governments acting as better-informed intermediaries between domestic opinion and foreign realities. In the case of democratic countries, progress will be exceedingly slow, limited to minor accommodation among allies afraid of a common enemy. The results of summit meetings among the leaders of the industrial democracies have been insubstantial and are likely to remain so. The European Economic Community certainly has made only limited progress toward economic, let alone political, integration. Nations and national economies are unwieldy bodies moving in deep grooves of habit and cultural conditioning.

Consider, then, the changes of individual lifestyles needed for rising above these grooves. In essence, the run of people will have to rearrange their routines in line with a greatly expanded community of common interests, looking at themselves from a yet higher elevation. Rising to that new perspective means dropping much useless baggage, reordering priorities, shifting energies to new pursuits supporting the enlarged vision. The human carrying capacity is limited; it is impossible to lift the full load of ground-floor life to a higher elevation and take on the additional responsibilities that go with global cooperation. Life has to be simplified in order to make room for the added responsibility. All along, less emphasis has to be given to bodily indulgence, and more to cerebral activity. It is the latter that enlarges horizons.

These changes add up to a new intensity of civic discipline. In the expansion of cooperation around the world everybody has been reduced in stature, forced into accommodation with numerous newcomers even less well known than the unseen millions of fellow citizens. Community has become still more impersonal, still lacking the symbols and myths that have given human content to the abstract unity of nations. In order to transform the lifeless concept of humanity into a living reality and to find a meaningful place in the vastly enlarged human family, individuals need a metaphysical guarantee of psychological security tying them constructively into the vast, unfamiliar human framework. The ethnocentric stereotypes and preconceptions that have contributed to national unity—and even to religious belief—in the past must be dismantled, to be replaced by more capacious symbols. Immense indeed are the revisions of established convictions that will be needed for reducing the hostility between the superpowers—and for gains scarcely visible from the ground floor of life!

How is it possible to establish common ground with strangers across the barriers of cultural incomprehension and ingrained political hostility? How can one get to the other side? These questions point to the most difficult problems of all in the next phase of human evolution. The answer, however, is obvious: we must make every possible effort to know the peoples beyond our boundaries, always comparing and contrasting their collective fortunes with our own. In this manner we can make some progress toward the other side. Yet even with the best of our knowledge we still lack—and always will lack—experiential understanding; our most expert knowledge will always remain outsiders' knowledge. Our judgments in regard to others therefore have to be provisional, implying that we have no final answers—and that, above all, we have no moral justification for hostility, let alone for war. The expansion of sociopolitical awareness and cultural perception requires a profound refinement of moral sensibilities.

These limitations indicate that, even with the best intentions all around, we—and the people on the other side too—can never reach more than a halfway point. But that halfway point is the vital middle ground on which a transcendent perspective can be achieved through experiments in cooperation. Only open-minded cooperation that recognizes the legitimacy of cultural otherness can build universal values that reach into the unconscious depths of human awareness, thereby preparing the groundwork for a truly common human nature. There is no end of urgent common projects: keeping the peace, preserving the earth's perishable resources, assisting the developing countries in their problems, strengthening civic morale everywhere; in short, reducing the anarchy in the global community by stretching the awareness of every individual from the ground floor of life to the distant global roof under which the chancy interaction of all with all decides the future of humanity.

IV

Peaceful changes of this magnitude, however desirable considering the growing anarchy in the global community, would seem to lie beyond the realm of the possible in the foreseeable future. Yet can we think of a viable alternative? The necessities of the ever growing global interdependence press for more inclusive association in line with a well-established progression toward expanded social cooperation. Human communities have grown from small lineage-bound associations to states, to empires, to nations, from blood relations to ever more impersonal and abstract communal ties. Advancing toward closer global cooperation beyond the nation-state would continue, possibly to its culmination, a well-established trend. But how can this be done under present conditions?

In the prenuclear past the formation of more inclusive associations was accomplished by war and conquest, combined, when successful, with a widespread recognition of common interests, which after conquest held the enlarged association together and transformed it into a community. The ultimate common interest patently was survival as the only alternative to destruction; submission to the conqueror and abandonment of the earlier parochial ways, however dear, was preferable to annihilation. In this manner, by a rough analysis, rose the Greek polis, the Roman Empire, the Western nation-state. And what magnificent human creativity was subsequently released by the combined resources and the enlarged scale of cooperation!

In the 20th century the world has made some progress toward worldwide cooperation through an expansion of global awareness after two world wars, the first leading to the creation of the League of Nations, a pioneer institution for the peaceful solution of international conflict, the second creating the United Nations, which strengthened the tradition started by the League without, however, living up to its promise.

Can the brutal process of awareness-raising through mass murder be continued in the age of nuclear weapons? Or will public revulsion against the mere threat of the nuclear holocaust take the place of the catastrophes which previously have jolted hesitant people from narrow into wider conceptions of self-interest? The immediate need for the latter course is not a world government, but a greater willingness to prevent a major war through peaceful negotiation, possibly within the framework of the United Nations, diverting available human resources from preparation for a suicidal war into strengthening the universals of peaceful cooperation. The United Nations certainly offers the most practicable available framework for mutual accommodation. Would that it could also assume the function of a universal church preaching the spiritual discipline required for the sense of global citizenship without which there can be no peaceful resolution of international conflict. What are the prospects?

Again, let readers cope with this question in the light of contemporary opinion within their own country and around the world.

V

History, as Hegel said, should ultimately be shaped by what is hoped for. Let this historical essay conclude openly on the visionary note which has been heard faintly throughout. What is needed, obviously, is a drastic simplification and slowing down of life's overrapid pace for the sake of a radical reconsideration of daily routines according to the priorities of global cooperation and survival. We need to strengthen our energies and clear our minds for gaining a more inclusive awareness, concentrating as much attention on the precariousness of human relations as on the perfection of technology. We should refine human sensibilities rather than machines, cross-cultural alertness rather than instruments for space exploration (which should be a joint human enterprise rather than a race between competing superpowers). In the age of the global confluence our responsibilites rest foremost with setting the human habitat in order on Planet Earth, which means expanding our sense of selfhood, down to the depths of our psyche, to global dimensions. This can be done only with the help of the culture-enhancing flywheel of an ascetic spiritual practice incorporated into daily routine, as in the Judeo-Christian religion of old.

Submission to the exacting necessities of peaceful cooperation among five and more billion human beings deeply divided by cultural experience will be gloriously rewarded in what counts most, in humaneness, in the capacity to manage the individual human psyche in harmony with all other members of the human family and with our physical environment as well—or, by old-fashioned language, in harmony with God's will. Thus may be realized the ultimate promise of the world revolution of Westernization.

Appendix: A Note on Method and on the Assumptions Made in This Book Regarding Culture and Inter-cultural Comparison

I

The approach followed throughout these pages differs from current practice in the study of history and the social sciences. Conventional emphasis rests on the careful, the "scientific," exploration of documentary or documentable data. The evidence is studied from a generally unquestioned set of assumptions, both conscious and unconscious. These assumptions in turn are based on a predetermined range of human sensibilities, of insights into human relations and social processes representative of the prevailing state of mind. In this essay more emphasis has been placed on sensibility than on data, on "soft evidence" rather than the hard evidence of the documents.

In an age of profound change like the present, it seemed more important to try expanding the range of sensibilities, to achieve a heightened awareness of the complexity of human relations in all their aspects, before getting absorbed in the evidence. As the quality of a work of art is determined by the artist's character, so the evidence in all studies of history and society is shaped by the structure and sensitivity of the observer's mind; there exists no objective evidence. The more alert and perceptive the observer's mind—or rather mind and soul, conscious and subconscious combined—the better the grasp of what is essential. If reason, according to Luther, is a harlot, so is scholarship. With an elegant finery of data chosen from the infinitude of facts, it can prove virtually anything it wants. The distinction—the truth—lies in the underlying values and in the moral purpose (on which there exists no consensus).

By traditional methods, this approach may seem unscholarly or "unscientific," but it has the advantage of perceptual flexibility and openness, offering a heightened responsiveness to the elusive reality "out there." What is "real" is not a matter of the data, but of the sensitivity that captures them and organizes them into "reality." In any creative work, insight and sensibility come first. Especially in the boundlessness of the present age it is necessary to realize how routinized and restricted our professional sensibilities have become. Now more than ever we need to engage *all* human capacities and develop to the utmost *all* dimensions of human consciousness. After we have reached a more inclusive awareness, we can proceed to search for the evidence—with forever imperfect results. The merit of the search lies in the insights offered; and the insights depend not on the abundance of the evidence but on the sensibility with which it is chosen. In the case of global history in the present century, as in any assessment

370

of the contemporary world, documenting novel insights with irrefutable fact obviously surpasses an individual scholar's capacity. The choice before historians in this age lies between pursuing a limiting and increasingly pointless specialization (writing "So What?" histories) and searching for meaningful perspectives which, though bound to remain controversial, call for moral sensibility.

How has it happened that scholars of human affairs, in history or the social sciences, have proceeded with so little awareness of the moral dimension? Is it because they wanted to be "scientific"? The natural sciences, like technology, are not concerned with refining the human alertness toward other humans on which peaceful cooperation in society depends. What social responsiveness is cultivated in quantum mechanics and Einstein's theory of relativity, or in aerodynamics? Yet how much more moral sensibility is exercised in historical studies or social science inquiries? The run of them proceed, it would seem, within a fixed groove of conventional morality, putting their creative edge into the examination of facts.

Yet, considering that any statement about human affairs constitutes a human relationship between the observer and the observed, should the observer not also exercise the moral responsibility that goes with human relationships? Do we try to ease tension and promote harmony? Do we respect the human beings under scrutiny? Do we understand them before we judge (and, all too often, condemn) them? In studies of our own society we generally follow the prevailing mores of toleration and good sense, tempering our conclusions with the prevailing moral restraint. Regarding international relations concerned with power, scholarship has shown but scant readiness for exploring the human implications of its conclusions. We are tied to our countries, at least implicitly and subtly, never endangering our civic loyalty, even in our criticism.

In the preglobal past such parochialism had its justifications; in an interdependent world mortally threatened, however, these limitations should be openly discussed. In all our statements about the world we ought to ask: what qualifications do we possess for moral judgments? Does our method of inquiry lend itself to better understanding? Do our conclusions reduce the existing tensions? Or are we engaged unawares in the morally corrupting power politics of moral righteousness?

As regards the answers, a further moral ingredient should be mentioned in the method employed throughout these pages (and most noticeable toward the end): its prophylactic pessimism. Moralists have always pleaded that human nature is corrupt, that human reason is fallible, and that pride goeth before a fall. In this spirit, morally alert observers are obliged to turn their sociocultural analysis towards the shortcomings of the existing order and stress what still needs to be done—or else people grow careless and self-indulgent, bringing about the very calamities foretold by the pessimists. In order to counteract the shallow optimism that shrinks human awareness, the analysis here has emphasized the dark side in contemporary affairs and the yet darker prospects before us.

Such prophylactic pessimism also justifies the determinism underlying the analysis here presented. Whatever the philosophical justifications for determinism, it is a useful heuristic device. We stand to gain from the humble assumption that we do not know enough yet about the factors shaping the course of human events. We must still perfect our analysis of the multitude of forces at work in our world (and improve our understanding of the past as well) in order to gain a more liberating command over our destiny. The widespread ideological optimism about free choice and an open-ended future engenders complacence and ignorance; it courts disaster.

The issues of prophylactic pessimism and determinism, although crucial for any realistic assessment of our times, are merely raised here for discussion rather than systematic investigation. The approach followed in the appendix, as in the main part of the body, is that of an exploratory essay.

II

As for culture, let it here be defined simply as the tool of collective human survival, as a group instrument handled by a politically organized sovereign body of people determined to affirm their collective identity under a specific set of external conditions. Different groups within roughly similar settings may share common cultural skills, as was the case in Western culture; but the essence of culture, the core of cultural creativity, lies within a polity, a politically constituted group interacting in competition or cooperation (or both) with other, similarly constituted groups. Viewed in this manner, culture covers the totality of human skills for cooperation on which group survival depends.

For the purpose of analysis, that totality has been traditionally divided into three categories. The first deals with the social dimension: human beings relating to other human beings. The second focuses on the relations between human beings and nature, on science and technology, on material production, food, clothing, and shelter, on command over the given physical environment (including climate, natural resources, and geographical location—factors often disregarded in inter-cultural comparison). The third category, the most important but least understood, has to do with human beings in relation to their inward selves and to the human and physical world around them; it is concerned with the religious dimension of life.

All three categories of cultural skills are inseparably fused with each other. None can function without the others; none can be understood without reference to the others—culture enforces a holistic approach to the study of human affairs. Social existence is inseparable from the volume and manner of material productivity; the larger and more secure the cultural community, the ampler must be the volume of material production. Both require creative individual participation, which is determined by what in these pages has been called "spiritual discipline," that is, the organization of the raw libidinous individual psyche for the tasks of social organization and material production. The more complex the cultural community, the more demanding that spiritual discipline.

Since the first two categories have been much studied and are reasonably well understood, we will briefly comment on the third, starting with the Cartesian assumption of the centrality of the thinking individual. Culturally integrated groups are composed, in the last analysis, of thinking individuals. Their thoughts are socially and culturally determined, yet they always have an independent edge sharpened by an existentialist individual insecurity. Deep down, all individuals are alone and helpless in the face of isolation and death, at the mercy of other human beings or of nature. Escape from that helplessness is quintessentially an individual effort. Society can help, but not in the inner workings of the individual psyche.

III

The most basic defense against that helplessness has been religion. Religion stands at the core of culture, central to social relations and material productivity. Traditionally, it has offered a psychological defense system projecting human helplessness outward toward that ever present elusive superior power—a divine power—at whose mercy life ran its course. That power could be propitiated by proper human action, by observance of collective rituals, by moral conduct, and by attuning thought and feeling in obedience to its divine commands. Whatever the nature of that divinity, at its best it socialized individual conduct and structured the individual psyche; it thereby became the sacred guardian of collective survival.

Religion, of course, changes in rhythm with other factors at work in culture. The radical monotheism of the ancient Hebrews, evolved for self-protection in the incessantly threatening movement of peoples in the eastern Mediterranean, was a major advance in the human capacity for enhancing collective life. One God, absorbing the power of all local deities and standing incomprehensibly and even unmentionably far above human beings, laid down a drastic code of disciplined conduct, a code enormously strengthening the cohesion of society by transforming, at the very core of their being, self-willed individuals into community-oriented, God-fearing citizens. Individual or collective misfortune, however accidental, was considered a form of divine punishment. Suffering had to be atoned for by still more diligent observance of God's law. The response, pragmatically considered, resulted in yet greater individual and collective strength—not of material force but, more centrally, of will. No gods made such intense demands on the individual and the community as Yahweh.

The discipline enjoined by that all-powerful God and supported by Greek philosophy—both serving a civic polity—was carried to its climax in the ascetic monasticism of the Christian church. What could be more arduous than forcing libidinous bodies into celibacy and—even more difficult—prurient minds into chasteness? Obviously, such discipline could not be enforced on the bulk of the population. But there was no lack of saintly examples to show that such ascetic and selfless heroism was socially constructive and therefore desirable. And above the saints towered Christ Crucified. Seeing the tormented body of Christ on the Cross—and nobody could escape seeing it—who dared complain of hardship and suffering or turn personal misfortune against the community? Christ's fate offered a socially and individually constructive—a metaphysical—answer to the deep sense of human insecurity. That symbol imprinted its mark on Western society for centuries. It was the key contributor to Western superiority, giving rise to all other boons (as well as to a counterproductive pride—the pride of the ascetic over the sensuous man, never more obvious than in the early contacts of Europeans with Africans).

Western culture and the Western state were built around Judeo-Christian religion, around its capacity to structure the believers' ever rebellious psyche for social cooperation and control over the forces of nature. Under that command, raw bodily energy was transformed into cerebration, corporal sensation into abstract thought. Without such transformation literacy was impossible; without literacy there could be no science and technology, no high culture. Without individuals elaborately disciplined at the core of their libidinous selves there would be neither an effective society nor a powerful state, neither social peace nor individual happiness. The external security provided by society had to be forever matched—and undergirded—by the individual's psychic capacity to take life's adversities in stride and to maximize the individual's capacity for service to the community, come what may. How could that be done?

The process may be briefly explained in nonreligious terms as a form of deliberate control over the stream of consciousness linking body and mind. In the natural interaction of both, the right thoughts, the socially constructive thoughts, are encouraged to prevail and allowed to sink into the subconscious, where they become part of the body's subtle chemistry. From the body and the subconscious they are fed back into spontaneous behavior and the tenor of thinking, in an endless circular process controlled by an alert mental censorship sorting out socially constructive and destructive—good and evil—thoughts. Such censorship was part of spiritual life and self-control, in charge of creating an ordered inward universe as complex as the outer one. Thinking individuals created their civic identity and through that identity structured their entire existence.

In recent times, obviously, the practice of these techniques for the proper structuring of the individual psyche has suffered. It was too intimately associated with symbols and rituals hardly consonant with enlightened ways of thinking; yet it is not entirely forgotten. It is also partially perpetuated by spontaneous conformity to standard behavior; the Judeo-Christian discipline enforced over centuries in the Western tradition has become deeply embedded in the discipline of urban-industrial life. Yet that self-enforced discipline needs constant revitalization, especially when social relations are daily becoming more complex and adjustment to cultural otherness is called for. One wishes that it be restated in contemporary concepts more socially aware than psychoanalysis.

What is lacking in the present precariousness of human existence, one might argue, is an updated analysis of the crucial contributions made in the past by religion, to rescue that contribution from its increasingly outdated terminology—or even from an outdated conception of God. As a cultural artifact, the definition of God too has to change with the times; we need to search for a universal God suitable to the global age, beyond all previous revelations. In any case, whatever the creedal formulations, the capacity for structuring the human psyche for service to the community—now the global community—has to be considered sacred. The inward order of the psychic universe and the outward order have to be harnessed together; their unity is the central guarantee of human survival.

It is also the key element in any analysis of culture geared to contemporary need, especially to the problems of "development." What are the existing practices for structuring the individual psyche in the Third World, especially in Africa? How far have people advanced in the skills of transforming bodily—or sensual—energy into cerebration? How ascetic is their lifestyle, how adjusted their ego to the needs of close-knit interdependence? Corrupted by the example of material indulgence set by the socially much more integrated wealthy metropolitan centers, the poor and underdeveloped peoples unfortunately are not likely to turn to what they need most: practice in bodily self-denial for the sake of intensified social cooperation; in the face of famine indeed physical vitality deserves priority. The implicit asceticism of advanced industrial societies is based on solid material security; while a properly structured psyche is essential, it still depends on all other factors in the instrumentarium of culture. How, then, can the people in developing countries learn the skills of asceticism when material security is provided from the outside (as in famine relief)? Will they not continue in their old ways—and remain dependent?

In any case, the world revolution of Westernization has imposed a universal obligation for strengthening that spiritual discipline. The anarchy in the global community can only be reduced by a common awareness that the causes of violence, of suffering, and of catastrophe lie ultimately within the hidden structure of the individual will, an independently manageable factor in human existence. The technical resources needed for global cooperation pose no problems; they grow with the spiritual refinement of human wills. What lies ahead—to take a very long and hopeful view in the manner of Teilhard de Chardin or Saint Augustine—is a further spiritualization of human existence, a further refinement of the subliminal structures of the human psyche for the sake of more intensive and subtle cooperation among more people and their physical environment. Viewed in this light, human evolution is a process of liberation through an ever more socially oriented ascetic self-discipline.

Unfortunately, the analysis of the inward universe of the individual subconscious as an independent variable in human affairs has yet barely begun, even in regard to our own society. How real, how distributed in individual awareness and conduct, is the asceticism here ascribed to a society which even in 1909 repelled Gandhi as hedonistic? How effective was it in past middle-class lifestyles; how effective is it now? As for the rest of the world, how much socialization flows from Islamic monotheism, from Buddhist meditation, from Confucianism or Shintoism? What other factors contribute to making religion—the structuring of the inward universe—a politically unifying, power-oriented force? As for the perfection of personal asceticism, Indian mystics perhaps are unrivalled, yet with so little social and political effect—why? The dynamics of culture at its innermost mechanisms remain still very obscure.

IV

But—to continue the argument—let it be contended here that culture, like the individual mind, is a complex universe of which the major part is hidden in the vast recesses of the subconscious. Culture resembles an iceberg; only a very minor part of its substance is visible to the conscious mind; the bulk lurks below the surface of awareness, although an intrinsic part of it. The collective subconscious is the reservoir of social discipline, the repository of all past actions that have become part of the skills of survival under a given set of unique circumstances. It is embedded in language and folkways, in music, art, literature, in the organizations and institutions that hold society together, in the form of government. It is communicated without awareness and passed on from one generation to the next through informal unconscious assimilation. The dimensions of those transvisible and transaudible messages, it would seem, are still unexplored.

The collective subconscious also is the invisible ingredient that makes the concepts and techniques of the conscious mind functionally effective. Rational thought or rational conduct, for instance, are made rational only in the contexts of the hidden assumptions embedded in the collective subconscious. Freedom creates the psychological sensation of being free only because of the hidden structure of the socially conditioned individual psyche; it inspires social cooperation only because of the unconscious collective discipline trained by a long historic evolution. Without those hidden promptings from below the surface of awareness, freedom degenerates into anarchy. No visible accomplishment of any culture can be lifted out of its invisible contexts and successfully implanted in another cultural universe. In the case of industrialism or statehood, the alien universe has to be recultured down to its depths, a formidable, painful, and perhaps even impossible task. Even more subtle aspects of culture, such as respect for human rights, cannot be transplanted unless the conditions that make them practicable are also transplanted.

V

In the light of these assumptions about the nature of culture, the plea for cultural relativism as a necessary tool for the study of the new globalism in human affairs seems justified. Considered as self-contained universes built up in response to unique circumstances, cultures are mutually incompatible. They are therefore also mutually incomprehensible. All individuals are products of their culture; their cognitive equipment is shaped by their culture. As the anthropologist E. T. Hall has written: "no matter how hard man tries, it is impossible for him to divest himself of his own culture, for it has penetrated the roots of his nervous system and determines how he perceives the world."[1] Put differently, when dealing with other cultures we—and everybody else around the world—practice "cognitive imperialism." Our perceptions are shaped for use within our own culture; going outside it, we are culture-blind.

There exist, in short, no cultural universals providing a common language for transcultural understanding; like poetry, cultures are not translatable. We have no choice but to interpret the others by our own lights, in our own cultural vernacular, never able to see the insiders in other cultures as they see themselves. Given the inescapability of cognitive imperialism, we have to ask in all questions of cross-cultural understanding: who understands whom on whose terms? In the last analysis, cross-cultural understanding is a matter of raw power: who has the power to make his own understanding prevail? Similarly, in all cross-cultural comparison, the question is: who compares himself with whom on whose terms? Who has the power to impose their own terms in the comparison? Who provides the premises of comparison? How much, in short, do we really understand?

Let us admit, then, that our Western knowledge of the non-Western world, so impressively compiled by ethnographers, anthropologists, historians, and political scientists, is surface knowledge; it is a Western facade tacked onto a non-Western world. We have Westernized the world also in the images of our minds, in the structures of our thoughts, essentially unchallenged. No other people has been able to impose its way of looking at the world on so many others outside its culture. The only effective challenge in recent times has come perhaps from the Soviet Union; the superpower rivalry is at its core concerned with the question: who imposes their view of reality, of the world, upon whom? Wherever we look, the perception of reality both inside and outside one's own culture is based on power. Power protects culture from unsettling external interference and allows it to be projected outward. Or else it is in disarray, its focus blurred, its purpose defeated. Culture cannot exist without a firm power base or a firm ethnic-spiritual commitment, as in the case of the Jews (before the modern state of Israel).

VI

Are we, then, trapped into ignorance by our cultural conditioning and the resulting cultural relativism? Are all foreign relations reduced to the culture-free universal language of threats of force or force itself? Are our intellectual efforts to understand others limited by the prevailing power relations? So it would seem. To repeat: our mental picture of the world is based on the power relationships established by Western political and cultural expansion. In the case of powerful enemies past or present, say, fascism or communism, our understanding is shaped by our political instincts, by our will to win. The bitterness engendered by World War II, for instance, still prevents understanding the sociopolitical dynamics that shaped Hitler's ambition and power. And in the studies of the Soviet system the political contexts are writ large over every page, even in the most conciliatory approach (like this author's). Obviously, we are trapped, far more than is generally realized. This contention, no doubt, will seem extreme, but it lends itself to a more realistic assessment of inter-cultural relations than results from the common disregard of cultural differences.

And with a realistic understanding of the obstacles the trap can be sprung, at least partially. First, there has to be a moral decision. As the challenge of otherness lures us into the defensive affirmation of our cultural blinders, we can by an act of will loosen that defensiveness; to some extent the prison is of our own making. We can deal separately and consciously with the moral ingredient in the act of cognition; we can guide our observations with a determination to keep our culturally conditioned gut reactions under control, dismantling all temptations of hostility as they arise. As peacemaking in personal relations requires the selfless downgrading of personal preference, so transcendence in inter-cultural relations calls for a progressive loosening of our cultural bondage and a corresponding expansion of our human outreach. These mental-spiritual techniques are implied in the moral command of the Sermon of the Mount: "love thine enemy," and in Goethe's observation that we cannot understand what we do not love. Breaking out of the prison of cultural incomprehension is a difficult but not impossible moral decision. It goes together with the assessment of the obstacles that are faced in the advance to the other side, with understanding the facts of cultural relativism and cultural determinism. When we have recognized the nature of these obstacles, we can humbly begin to overcome them, aware of our limitations and therefore properly cautious about the validity of our judgments (which assists the process of emancipation from our cultural conditioning).

Our minds prepared, we then proceed with the obvious: learning the language, reading the literature, appreciating the arts, and gradually expanding our explorations of the alien culture, always bent on probing into its totality through careful study of *all* relevant factors: geography, climate, physical conditions, natural wealth (or poverty), the problems of government, the attitudes of the people, the nature of external influences and the country's position in the world, as well as the cumulative experience of rulers and ruled through the centuries. Having once passed through the laborious process of exploring an alien culture in this fashion, we are prepared to approach other cultures more sensibly, always aware of our limitations.

Thus equipped, we approach the halfway mark to the other side, forced to admit that the rest of the way is closed to us. As outsiders we are barred from the conditioning-in-depth rising out of the collective subconscious. At the halfway mark we are stopped, like first-generation immigrants in the United States who even after a lifetime in that country remain outsiders, conforming on the surface yet structured differently in their depths. The halfway mark, however, makes an excellent meeting place, if from the other side there comes a similar effort at transcendence. The main challenge lies in setting an example through demonstrable progress toward that halfway point.

VII

Progress toward that halfway mark, toward understanding the world in which we live, depends on our capacity to comprehend cultures in their entirety. We must deliberately practice a holistic approach, even when concerned only with limited aspects. No social, political, intellectual, or artistic phenomenon can be understood except in its total—its cultural—contexts. That is why in these pages the term "capitalism" has always been put into quotation marks. However impressive the scholarship devoted to the development and role of "capitalism," it does not explain the causes of the extraordinary progress in the economic productivity of the West. For a more helpful approach, these pages have offered an experimental "culturalist" method of analysis, downplaying the economic factors whose importance has been overestimated. Would that all the other relevant factors responsible for the hothouse pace of Western evolution had been studied with similar zeal!

The "culturalist" approach tying all other approaches together, and given the central place in the present analysis, concentrates on power in all its aspects, power rooted in the psychic depths of individuals. The existential need for security and self-affirmation is socialized for greater effect through all human activities (including economics and religion) within the outer framework of culture. What counts most in the affirmation of collective power is the capacity for disciplined cooperation among all members of a given culture, in politics, economics, and in their relations to the physical environment. That capacity is promoted by the peaceful arts of service and selflessness rather than by compulsion. For its fullest development it requires external security, a prerequisite as well as a product of an effective culture.

Only a power-oriented holistic approach to culture can produce useful insights in comparative studies, especially when we ask large questions. Why did Europe rise to global preeminence? Why did Japan match Western technological progress and not China? What can be done to help developing countries overcome their agonies? In comparative studies, comparing small features in different countries without reference to their complete cultural envelopes is not intellectually respectable; the results offer no constructive conclusions.

In the age of the Global Confluence we are morally obliged to help resolve the mounting conflicts; we must lessen the tensions arising out of uncomprehended and unanalyzed inter-cultural contact by taking into account the full range of factors determining the dynamics of domestic politics in each state as well as of the interaction of states in their global competition. We furthermore have to consider each state as a self-contained cultural system. Culture has a political base in statehood, now more than ever; international relations therefore are inter-cultural relations as well. And foreign policy should be considered culturally, set into the broadest applicable contexts and treated as inter-cultural policy.

How the proposed approach can be applied to the daily operations of government, or to scholarship designed to prepare the way by illuminating the vital interconnections, cannot here be spelled out. The intention of this book is merely to show a way of escaping from sterile and outdated routines in scholarly analyses and to open up, tentatively and experimentally, perspectives more appropriate to the new age of globalism. In historical research too we must accept the challenge of the times, taking into account the fullness of global reality without being mentally crushed by the overload of detail. There is no intellectual security or moral reward in petty specialization. Critically selected details must be presented in meaningful and constructive perspectives. We should concentrate on existential thinking rather than on mindless research; we have to grope for the essentials and forget the rest, travelling light into the uncertain future.

We live in an entirely novel and unprecedentedly perilous age, and intellectuals who do not think in scale with that age, whatever the risks, betray their moral as well as their intellectual responsibility.

Notes

Preface

1. Ralf Dahrendorf, *Gesellschaft und Demokratie in Deutschland* (Munich, 1965), p. 11.
2. Johann Wolfgang von Goethe, *Sämtliche Werke* (Jubiläumsausgabe), Vol. 39 (Stuttgart, 1902), p. 72.
3. Francis Bacon, *Novum Organum,* Aph. cxxix.
4. James Boswell, *Life of Johnson,* 19 September, 1777.

Chapter 1

1. Betty Balfour, ed., *The History of Lord Lytton's Indian Administration, 1876–1880, compiled from letters and official papers* (London, 1899), pp. 510–512.
2. John Flint, *Cecil Rhodes* (Boston, 1974), pp. 32–33.
3. Raymond J. Sontag, *Germany and England, Background to Conflict, 1849–1894* (New York, 1938), p. 309.
4. Benjamin Disraeli, *Selected Speeches of the Late Right Honorable the Earl of Beaconsfield,* ed. T. E. Kebbel (London, 1889), Vol. 2, p. 534.

Chapter 2

1. W. Theodore de Bary, ed., *Sources of Indian Civilization* (New York, 1958), p. 601.
2. Ryusaku Tsunoda, W. Theodore de Bary, and Dennis Keene, eds., *Sources of Japanese Civilization* (New York, 1958), p. 628.

Chapter 3

1. Ibid., pp. 630–631.
2. Thomas Hobbes, *Leviathan* (London, 1914), p. 88.
3. de Bary, op. cit., p. 800.
4. Ibid., p. 803.
5. Ibid., p. 652.
6. Wilfred Cartey and Martin Kilson, eds., *The Africa Reader: Independent Africa* (New York, 1970), p. 29ff.
7. James Africanus Horton, *West African Countries and Peoples, 1868* (Edinburgh, 1969), pp. 59, 61.
8. Cartey and Kilson, op. cit., pp. 48–56.
9. Richard Meinertzhagen, *Kenya Diary, 1902–1906* (Edinburgh, 1957), entry for September 13, 1905.

Chapter 4

1. Johann Gottlieb Fichte, *Addresses to the German Nation* (New York, 1968), p. 228.
2. Max Weber, *Gesammelte Politische Schriften* (Tübingen, 1958), p. 23.
3. Karl Marx, *Early Writings* (New York, 1963), p. iii.
4. Robert C. Tucker, ed., *The Marx-Engels Reader* (New York, 1972), p. 335ff.

5. Ibid., pp. 583–587.

6. Fyodor Dostoyevsky, *Notes from Underground* (New York, 1961), pp. 93–115.

7. Alexander Herzen, *The Russian People and Socialism: An Open Letter to Jules Michelet* (London, 1956), pp. 190–191.

8. Josiah Strong, *Our Country* (Cambridge, Mass., 1963), p. 214.

9. Karl Pearson, *National Life from the Standpoint of Science* (London, 1901), p. 21.

Chapter 5

1. Dwight D. Lee, *Europe's Crucial Years: The Diplomatic Background of World War I, 1902–1914* (Hanover, N.H., 1974), p. 440.

2. Fritz Fischer, *Griff nach der Weltmacht* (Düsseldorf, 1962), p. 80.

3. Ibid., p. 140.

Chapter 6

1. Vladimir I. Lenin, *The Lenin Anthology*, ed. Robert C. Tucker (New York, 1975), p. 297.

2. Ibid., p. 435.

3. Ibid., p. 83.

4. Vladimir I. Lenin, *Lenin on Politics and Revolution*, ed. James E. Connor (New York, 1968), p. 183.

5. *The Lenin Anthology*, op. cit., p. 619ff.

Chapter 7

1. Benito Mussolini, *Fascism, Doctrine and Institutions* (Rome, 1935), p. 31.

2. Ibid., pp. 8, 19.

3. Ibid., p. 11.

4. Ibid., p. 13.

5. Ibid., p. 14.

6. Herman Finer, *Mussolini's Italy* (New York, 1965), p. 545.

7. Adolf Hitler, *Mein Kampf* (Boston, 1943), p. 468.

8. Ibid., p. 353; see also p. 331.

9. Ibid., p. 336.

10. William R. Louis, *Imperialism at Bay* (London, 1978), p. 5.

11. Hitler, op. cit., p. 645.

12. Ibid., p. 654.

13. Ibid., p. 664.

Chapter 8

1. Imanuel Geiss, *The Pan-African Movement* (London, 1974), pp. 229–230.

2. Ssu-Yu Teng and John K. Fairbank, eds., *China's Response to the West* (Cambridge, Mass., 1954), p. 242.

3. Jonathan D. Spence, *The Gate of Heavenly Peace* (New York, 1982), pp. 159–160.

4. Sun Yatsen, *The Three Principles of the People, San Min Chu I* (Taipei, [1963]), pp. 10, 19.

5. Ibid., p. 5.

6. Ibid., p. 14.

7. Ibid., p. 49.

8. Ibid., p. 67.

9. Ibid., p. 107.

10. Ibid., p. 212.

11. Maurice Meisner, *Li Ta-chao and the Origins of Chinese Marxism* (Cambridge, Mass., 1967), p. 68.

12. Ibid., p. 65.

13. Ibid., p. 68.

14. Conrad Brandt, Benjamin Schwartz, and John K. Fairbank, eds., *A Documentary History of Chinese Communism* (New York, 1966), p. 80.

15. Geiss, op. cit., p. 230.

16. Colin Legum, *Pan-Africanism: A Short Political Guide* (New York, 1965), p. 28.

17. Ibid., p. 29.

18. Geiss, op. cit., p. 260.

19. Ibid., p. 265.

20. Ibid., p. 269.

21. Ibid., p. 320.

22. Ibid., p. 329.

Chapter 10

1. George R. Urban, "Stalin Closely Observed: A Conversation with Boris Bazhanov," *Survey* 25 (Summer 1980), p. 100.

2. Anton Antonov-Ovseyenko, *The Time of Stalin: Portrait of a Tyranny* (New York, 1981), p. 41.

3. Leo Tolstoy, *Resurrection* (London, 1936), pp. 427–428.

4. Robert V. Daniels, ed., *A Documentary History of Communism* (New York, 1960), Vol. 2, p. 23.

5. Alexander Solzhenitsyn, *The Gulag Archipelago*, Vol. 1 (New York, 1973), p. 160.

Chapter 11

1. Franz Schurmann and Orville Schell, eds., *The China Reader: Republican China* (New York, 1967), p. 184.

Chapter 12

1. Ibid., p. 152.

2. Chiang Kai-shek, *Collected Wartime Messages of Generalissimo Chiang Kai-shek* (New York, 1946), p. 180.

3. Ibid., p. 106.

4. Arthur F. Wright, ed., *Studies in Chinese Thought* (Chicago, 1953), p. 286ff.

5. As described in Kuo-heng Shih, *China Enters the Machine Age* (Cambridge, Mass., 1944), passim. (See especially Chapter 10, p. 151ff.)

6. Stuart R. Schram, *The Political Thought of Mao Tse-tung* (New York, 1974), p. 290.

7. Ibid., p. 277.

8. Ibid., p. 264.

9. Conrad Brandt, Benjamin Schwartz, and John K. Fairbank, eds., *A Documentary History of Chinese Communism* (New York, 1966), pp. 379, 409.

10. Schram, op. cit., p. 179.

11. Brandt, op. cit., pp. 386, 392.

12. Ibid., p. 326.

13. Ibid., p. 388.

14. Ibid., p. 400.

15. Ibid., p. 273.

16. Schram, op. cit., p. 191.

17. Mark Selden, *The Yenan Way in Revolutionary China* (Cambridge, Mass., 1971), p. 194.

18. Brandt, op. cit., p. 278.

19. Selden, op. cit., p. 242ff.

20. Schram, op. cit., p. 316.

Chapter 13

1. Albert Speer, *Inside the Third Reich* (New York, 1970), p. 109.
2. Ibid., p. 217.
3. Joachim C. Fest, *Hitler* (New York, 1975), p. 540.
4. Speer, op. cit., pp. 221–222.
5. Ibid., p. 222.
6. Ibid., p. 219.

Chapter 14

1. *Hitler's Secret Conversations, 1941–1944* (New York, 1961), pp. 91–92.
2. Henry Luce, Editorial, *Life*, February 11, 1941.
3. J. Samuel Walker, *Henry A. Wallace and American Foreign Policy* (Westport, Conn., 1976), pp. 84–85.
4. Robert Jay Lifton and Nicholas Humphrey, *In a Dark Time: Images for Survival* (Cambridge, Mass., 1984), p. 46.
5. Speer, op. cit., p. 469.
6. *Charter of the United Nations*, Preamble.

Chapter 15

1. Arthur Mann, *The One and the Many* (Chicago, 1979), p. 59.
2. Ibid., p. 60.
3. Fyodor Dostoyevsky, *The Possessed* (London, 1931), p. 231.
4. Mann, op. cit., p. 69.
5. Ibid., p. 83
6. Ibid., p. 69.
7. James Bryce, *The American Commonwealth*, Vol. 3 (London, 1888), p. 50.
8. Ibid., p. 55.
9. Ibid., p. 48.
10. Ibid., p. 24.
11. Ibid., p. 352.
12. Ibid., p. 607.
13. Francis Jennings, *The Invasion of America: Indians, Colonization and the Cant of Conquest* (New York, 1975), p. 183.
14. Frederick Merk and L. B. Merk, *Manifest Destiny and Mission in American History* (New York, 1963), p. 29.
15. M. C. Perry, *Paper by Commodore M. C. Perry, U.S.N., March 6, 1856* (New York, 1856), p. 28.
16. Samuel Eliot Morison and Henry Steele Commager, *The Growth of the American Republic*, Vol. 2 (New York, 1937), p. 322.
17. Ibid., p. 323.
18. Ibid., p. 324.
19. Akira Iriye, *From Nationalism to Internationalism* (Chicago, 1977), p. 214.
20. William Appleman Williams, *American-Russian Relations* (New York, 1971), p. 106.

Chapter 16

1. Geoffrey Hodgson, *America in Our Time* (New York, 1976), p. 18.
2. George F. Kennan, *Memoirs, 1925–1950* (Boston, 1967), p. 351.
3. Ibid., p. 547ff.
4. George F. Kennan, "The Sources of Soviet Conduct," *Foreign Affairs*, July 1947, passim.
5. Council on Foreign Relations, *The United States in World Affairs, 1947* (New York, 1948), p. 33.

6. Hodgson, op. cit., p. 13.

7. Kennan, *Memoirs*, p. 559.

8. *NSC-68: A Report to the National Security Council by the Executive Secretary on United States Objectives and Programs for National Security, April 14, 1950* (Reprinted in the *Naval War College Review*, May–June 1975), passim.

9. Barry Goldwater's acceptance speech for the Republican presidential nomination, 1964.

10. Warner R. Schilling et al., *Strategy, Politics and Defence Budgets* (New York, 1962), p. 309.

11. Lawrence S. Wittner, *Cold War America* (New York, 1974), p. 177.

12. George C. Herring, *America's Longest War* (New York, 1979), p. 74.

13. Charles L. Sanford, "The Intellectual Origins and New-Worldliness of American Industry," *Journal of Economic History* 18 (1958), p. 16.

14. Richard J. Barnet and Ronald E. Muller, *Global Reach: The Rise of Multi-National Corporations* (New York, 1974), pp. 14–15.

15. Ibid., p. 19.

16. *UN Yearbook 1973* (New York, 1973), p. 603ff.

17. William E. Leuchtenburg, *A Troubled Feast* (Boston, 1979), p. 70.

18. Wittner, op. cit., pp. 360–361.

19. Thomas M. Franck and Edward Weisband, *Word Politics: Verbal Strategy among the Super Powers* (New York, 1972), p. 79.

20. Wittner, op. cit., p. 265.

21. Ibid., p. 168.

22. *Alleged Assassination Plots Involving Foreign Leaders: An Interim Report of the Select Committee to Study Governmental Operations with Respect to Intelligence Activities* (New York, 1976), p. xxiii.

23. Wittner, op. cit, p. 223.

24. Hodgson, op. cit., p. 310.

25. Wittner, op. cit., p. 344.

26. *Alleged Assassination Plots*, p. xxiff.

27. Ibid., p. xiii.

Chapter 17

1. Klaus Mehnert, *Soviet Man and His World* (New York, 1961), p. 30.

2. Thomas Riha, ed., *Readings in Russian History*, Vol. 2 (Chicago, 1964), p. 299.

Chapter 18

1. *Current Soviet Policies, Documentary Record of the 19th Party Congress* (New York, 1953), p. 119.

2. Seweryn Bialer, *Stalin's Successors* (New York, 1980), p. 238.

3. Stephen F. Cohen, ed., *An End to Silence: Uncensored Opinion in the Soviet Union from Roy Medvedev's Underground Political Diary* (New York, 1982), p. 141.

4. *Current Soviet Policies II* (New York, 1957), p. 29ff.

5. Ibid., p. 31.

6. *Current Soviet Policies III* (New York, 1960), p. 43ff.

7. Ibid., p. 65.

8. *Current Soviet Policies IV* (New York, 1962), p. 64.

9. *Current Soviet Policies III*, p. 56.

10. *Current Soviet Policies IV*, p. 55.

11. Ibid., p. 15.

12. Ibid., p. 67.

13. Ibid., p. 67.

14. Ibid., pp. 27–28.

15. Leonid I. Brezhnev, *Vospominaniia* (Moscow, 1983), p. 108.

16. Bialer, op. cit., p. 149.

17. *Current Soviet Policies VII* (Columbus, Ohio, 1976), p. 31.
18. Howard L. Parsons, *Ethics in the Soviet Union Today* (New York, 1965) p. 11.

Chapter 19

1. George McT. Kahin, *The Asian-African Conference, Bandung, Indonesia* (Ithaca, N.Y., 1956), p. 42.
2. Ibid., p. 42.
3. Frantz Fanon, *The Wretched of the Earth* (New York, 1963), p. 32.
4. Michael Brecher, *Nehru, a Political Biography* (London, 1959), p. 6.
5. Hiranyappa Venkatasubbiah, *India's Economy Since Independence* (Bombay, 1958), p. 283.
6. Ibid., p. 283.
7. Ibid., p. 283.
8. Ibid., p. 287.
9. Jean Filliozat, *India, the Country and its Traditions* (London, 1962), p. 1.
10. Brecher, op. cit., p. 341.
11. Bernard Dahm, *Sukarno and the Struggle for Indonesian Independence* (Ithaca, N.Y., 1969), p. 338.
12. Ibid., p. 337ff.
13. Ibid., p. 345.
14. Ibid., p. 345.
15. Ibid., p. 346.
16. Ibid., p. 348.
17. Ibid., p. 299.
18. Gamal Abdel Nasser, *Egypt's Liberation; The Philosophy of the Revolution* (Washington, 1955), p. 61.
19. R. Hrair Dekmajian, *Egypt under Nasir* (Albany, N.Y., 1971), p. 51.
20. Nasser, op. cit., p. 61.
21. Ibid., p. 11.
22. Ibid., pp. 32–34.
23. Ibid., pp. 35–36.
24. Ibid., pp. 68–70.
25. Ibid., p. 41.
26. Ibid., pp. 71–74.
27. Ibid., pp. 84, 88.
28. Ibid., pp. 110, 111, 112, 113, 114.
29. Ibid., p. 114.
30. Kwame Nkrumah, *The Autobiography of Kwame Nkrumah* (Edinburgh, 1957), pp. 197–198.
31. Ibid., p. 196.
32. Ibid., p. 26.
33. Ibid., p. vii.
34. Ibid., p. 27.
35. Ibid., p. 48.
36. Ibid., p. 303.
37. Ibid., p. 56.
38. Ibid., p. 292.
39. Ibid., p. 291.
40. Dennis Austin, *Politics in Ghana, 1946–1960* (London, 1964), p. 17.
41. Trevor Jones, *Ghana's First Republic, 1960–1966* (London, 1976), p. 62.
42. Nkrumah, op. cit., p. viii.
43. Ibid., p. x.
44. Ibid.
45. Ibid., pp. 205–206.
46. Kahin, op. cit., p. 12.

47. Ibid., p. 76ff.

48. Gunnar Myrdal, *Asian Drama: An Inquiry into the Poverty of Nations*, Vol. 1 (New York, 1968), p. 277.

49. Anour Abdel-Malek, *Egypt's Military Society* (New York, 1968), p. 335.

50. Raymond W. Baker, *Egypt's Uncertain Revolution under Nasser and Sadat* (Cambridge, Mass., 1978), p. 114.

51. Ibid., p. 47.

52. *Kwame Nkrumah*, Panaf Great Lives Series (London, 1974), p. 155.

53. Ibid., p. 92.

54. Jones, op. cit., p. 160.

55. T. Peter Omari, *Kwame Nkrumah: The Anatomy of an African Dictatorship* (London, 1970), p. 149.

56. Ibid., p. 216.

57. Paul Johnson, *Modern Times: The World from the Twenties to the Eighties* (New York, 1985), p. 513.

58. Kwame Nkrumah, *I Speak of Freedom* (New York, 1961); *Africa Must Unite* (New York, 1964); *Consciencism* (New York, 1964); *Neo-Colonialism* (New York, 1965).

59. Omari, op. cit., p. 195 (italics added).

60. W. Scott Thompson, *Ghana's Foreign Policy, 1957–1966* (Princeton, 1969), pp. 12–13.

61. Johnson, op. cit., pp. 493–495, 515 (Hammarskjöld seems needlessly maligned).

Chapter 20

1. Stuart Schram, *The Political Thought of Mao Tse-tung* (New York, 1974), pp. 167–168.

2. Jerome Ch'en, *Mao and the Chinese Revolution* (New York, 1965), p. 312.

3. Conrad Brandt, Benjamin Schwartz, and John K. Fairbank, eds., *Documentary History of Chinese Communism* (New York, 1966), pp. 449–461.

4. Roger Garside, *Coming Alive: China After Mao* (New York, 1981), p. 336.

5. Brandt, op. cit., p. 460.

6. Dick Wilson, ed., *Mao Tse-tung in the Scales of History* (Cambridge, U.K., 1973), p. 181.

7. Schram, op. cit., pp. 343–344.

8. Wilson, op. cit., p. 51.

9. Schram, op. cit., p. 346.

10. Ibid., p. 301.

11. Mao Tse-tung, *Chairman Mao Talks to the People: Talks and Letters, 1956–1971* (New York, 1974), p. 61ff.

12. Ibid., p. 83.

13. Schram, op. cit., p. 308.

14. Maurice Meisner, *Mao's China: A History of the People's Republic* (New York, 1977), p. 190.

15. *Chairman Mao Talks to the People*, p. 91.

16. Ibid., p. 91.

17. Ibid., pp. 92–95.

18. Ibid., p. 106.

19. Ibid., p. 115.

20. Ibid., p. 119.

21. Ibid., p. 122.

22. Wilson, op. cit., p. 212.

23. *Chairman Mao Talks to the People*, p. 132ff.

24. Ibid., p. 146.

25. Ibid., p. 175.

26. Ibid., p. 175.

27. Mao Tse Tung, *Quotations from Chairman Mao* (Peking, 1966).

28. *Chairman Mao Talks to the People*, pp. 203, 205, 207.

29. Ibid., p. 231.

30. David Milton, Nancy Milton, and Franz Schurmann, eds. *The China Reader: People's China* (New York, 1974), p. 285.

31. Ibid., pp. 269–270.

32. *Chairman Mao Talks to the People,* pp. 253, 255.

33. Milton, op. cit., p. 271.

34. *Decision of the Central Committee of the Chinese Communist Party Concerning the Great Proletarian Cultural Revolution. Adopted on August 8, 1966* (Peking, 1966), passim.

35. *Quotations from Chairman Mao,* pp. 108, 111–112, 178–179, 204–205.

36. *Chairman Mao Talks to the People,* p. 268.

37. *Ibid.,* pp. 271, 274.

38. Stanley Karnow, *Mao and China: Inside China's Cultural Revolution* (New York, 1972), p. 407.

39. Ibid., p. 441.

40. Meisner, op. cit., p. 361.

41. *Chairman Mao Talks to the People,* p. 283.

42. Wilson, op. cit., pp. 80–81.

43. Meisner, op. cit., pp. 376–377.

44. Fox Butterfield, *China: Alive in the Bitter Sea* (New York, 1982), p. 16.

45. *Chairman Mao Talks to the People,* p. 180.

46. Chie Nakane, *Japanese Society* (Berkeley, Calif., 1970), p. 150.

Chapter 21

1. *North-South: A Program for Survival. The Report of the Independent Commission on International Development Issues under the Chairmanship of Willy Brandt* (Cambridge, Mass., 1980), p. 32.

2. William E. Smith, *We Must Run while They Walk* (New York, 1971), p. 5.

3. Hélène Castel, *World Development: An Introductory Reader* (New York, 1971), p. 18.

4. Frantz Fanon, *The Wretched of the Earth* (New York, 1963), p. 30.

5. Ibid., p. 35.

6. Ibid., p. 76.

7. Ibid., p. 133.

8. Ibid., p. 157.

9. Ibid., p. 116.

10. Ibid., p. 196.

11. Ibid., p. 199.

12. Ibid., p. 79.

13. Ibid., p. 83.

14. Paul Johnson, *Modern Times: The World from the Twenties to the Eighties* (New York, 1985), p. 537.

Chapter 22

1. *Charter of the United Nations,* Article 11, (2).

2. Ibid., Preamble.

3. Ibid., Article 55.

4. "Universal Declaration of Human Rights," *UN Yearbook 1948/49* (New York, 1949), p. 535ff.

5. UN Charter Chapter I, Article 2, (1).

6. *UN Yearbook 1960* (New York, 1960), p. 49.

7. Herbert G. Nicholas, *The United Nations as a Political Institution* (London, 1975), p. 64.

8. *UN Yearbook 1966* (New York, 1966), p. 419ff.

9. Ibid., p. 423ff.

10. *UN Yearbook 1974* (New York, 1974), p. 324ff.

11. Ibid., p. 403.

12. *UN Yearbook 1981* (New York, 1981), p. 15.

13. Ibid., p. 3.

Chapter 23

1. Javier Perez de Cuéllar, "A Foundation to Build On . . . " *UN Chronicle* 22 (Sept. 1985), p. 2.

Chapter 24

1. Amos Oz, *Im Lande Israel* (Frankfurt, 1984), p. 80.

Appendix

1. Edward T. Hall, *The Hidden Dimension* (Garden City, N.Y., 1969), p. 188.

Index

In an interpretive essay on twentieth century history it is impossible to index every detail in the text. Only items of substance are indexed.